# Middleware Solutions for Wireless Internet of Things

# Middleware Solutions for Wireless Internet of Things

Special Issue Editors

**Paolo Bellavista**
**Carlo Giannelli**
**Sajal K. Das**
**Jiannong Cao**

MDPI • Basel • Beijing • Wuhan • Barcelona • Belgrade

*Special Issue Editors*
Paolo Bellavista
University of Bologna
Italy

Carlo Giannelli
University of Ferrara
Italy

Sajal K. Das
Missouri University of Science and Technology
USA

Jiannong Cao
The Hong Kong Polytechnic University
Hong Kong

*Editorial Office*
MDPI
St. Alban-Anlage 66
4052 Basel, Switzerland

This is a reprint of articles from the Special Issue published online in the open access journal *Sensors* (ISSN 1424-8220) from 2018 to 2019 (available at: https://www.mdpi.com/journal/sensors/special_issues/Middleware_Solutions_for_Wireless_IoTs)

For citation purposes, cite each article independently as indicated on the article page online and as indicated below:

LastName, A.A.; LastName, B.B.; LastName, C.C. Article Title. *Journal Name* **Year**, *Article Number*, Page Range.

**ISBN 978-3-03921-036-7 (Pbk)**
**ISBN 978-3-03921-037-4 (PDF)**

# Contents

# About the Special Issue Editors

**Paolo Bellavista** is a full Professor of Distributed and Mobile Systems at the Dept. of Computer Science and Engineering (DISI), Alma Mater Studiorum—Università di Bologna. He is co-author of around 80 articles in international journals/magazines (in publication venues that are considered excellent in his research fields, such as *ACM Computing Surveys, IEEE Communication Surveys&Tutorials, IEEE T. Software Engineering, IEEE J. Selected Areas in Communications, ACM T. Internet Technology, IEEE T. Vehicular Networks, Elsevier Pervasive and Mobile Computing J., Elsevier J. Network and Computer Applications, Elsevier Future Generation Computer Systems, IEEE Computer, IEEE Internet Computing, IEEE Pervasive Computing, IEEE Access, IEEE Communications, IEEE Wireless Communications).* Moreover, he is co-author of around 15 chapters in international peer-reviewed books and of around 110 additional papers presented in international conferences. About organization and reviewing activities, he is Editor-in-Chief of the MDPI *Computers* Journal (2017-) and he is (or has been) member of the Editorial Boards of several international journals/magazines, among which *IEEE T. Computers (2011–2015), IEEE T. Network and Service Management (2011-), IEEE T. Services Computing (2008–2017), IEEE Communications Magazine (2003–2011), IEEE Comm. Surveys&Tutorials (2019-), Elsevier Pervasive and Mobile Computing J. (2010-), and Elsevier J. Network and Computing Applications (2015-).* About international projects he has led and has participated to several H2020/FP7 projects, such as IoTwins, SimDome, Arrowhead, Mobile Cloud Networking, and COLOMBO.

**Carlo Giannelli** received the Ph.D. degree in computer engineering from the University of Bologna, Italy, in 2008. He is currently an Assistant Professor in computer science with the University of Ferrara, Italy. His primary research activities focus on Industrial Internet of Things, Software Defined Networking, heterogeneous wireless interface integration, and hybrid infrastructure/ad hoc and spontaneous multi-hop networking environments based on social relationships.

**Sajal K. Das** is the Daniel St. Clair Endowed Chair Professor at the Computer Science department at Missouri University of Science and Technology (S&T). During 2008–2011 he served the US National Science Foundation as a Program Director in the division of Computer Networks and Systems. His research interests include wireless and sensor networks, mobile and pervasive computing, smart environments and smart health care, pervasive security, biological networking, applied graph theory and game theory. His research on wireless sensor networks and pervasive and mobile computing is widely recognized as pioneering. He is a Fellow of the IEEE.

**Jiannong Cao** is currently a Chair Professor of Department of Computing at The Hong Kong Polytechnic University, Hong Kong. He is also the director of the Internet and Mobile Computing Lab in the department and the director of University's Research Facility in Big Data Analytics. His research interests include parallel and distributed computing, wireless sensing and networks, pervasive and mobile computing, and big data and cloud computing. He has co-authored 5 books, co-edited 9 books, and published over 600 papers in major international journals and conference proceedings. He received Best Paper Awards from conferences including, IEEE Trans. Industrial Informatics 2018, IEEE DSAA 2017, IEEE SMARTCOMP 2016, IEEE/IFIP EUC 2016, IEEE ISPA 2013, IEEE WCNC 2011, etc.

# Preface to "Middleware Solutions for Wireless Internet of Things"

The proliferation of powerful but cheap devices, together with the availability of a plethora of wireless technologies, has pushed for the spread of the Wireless Internet of Things (WIoT), which is typically much more heterogeneous, dynamic, and general-purpose if compared with the traditional IoT. The WIoT is characterized by the dynamic interaction of traditional infrastructure-side devices, e.g., sensors and actuators, provided by municipalities in Smart City infrastructures, and other portable and more opportunistic ones, such as mobile smartphones, opportunistically integrated to dynamically extend and enhance the WIoT environment.

A key enabler of this vision is the advancement of software and middleware technologies in various mobile-related sectors, ranging from the effective synergic management of wireless communications to mobility/adaptivity support in operating systems and differentiated integration and management of devices with heterogeneous capabilities in middleware, from horizontal support to crowdsourcing in different application domains to dynamic offloading to cloud resources, only to mention a few.

The book presents state-of-the-art contributions in the articulated WIoT area by providing novel insights about the development and adoption of middleware solutions to enable the WIoT vision in a wide spectrum of heterogeneous scenarios, ranging from industrial environments to educational devices. The presented solutions provide readers with differentiated point of views, by demonstrating how the WIoT vision can be applied to several aspects of our daily life in a pervasive manner.

**Paolo Bellavista, Carlo Giannelli, Sajal K. Das, Jiannong Cao**
*Special Issue Editors*

Article

# Managing Devices of a One-to-One Computing Educational Program Using an IoT Infrastructure

**Felipe Osimani [1,†], Bruno Stecanella [1,†], Germán Capdehourat [2,†], Lorena Etcheverry [1,†] and Eduardo Grampín [1,*,†]**

1 Instituto de Computación (INCO), Universidad de la República (UdelaR), Montevideo 11300, Uruguay; felipe.osimani@fing.edu.uy (F.O.); bruno.stecanella@fing.edu.uy (B.S.); lorenae@fing.edu.uy (L.E.)
2 Centro Ceibal para el Apoyo a la Educación de la Niñez y la Adolescencia (Plan Ceibal), Montevideo 11500, Uruguay; gcapdehourat@ceibal.edu.uy
* Correspondence: grampin@fing.edu.uy
† These authors contributed equally to this work.

Received: 13 November 2018; Accepted: 21 December 2018; Published: 25 December 2018

**Abstract:** *Plan Ceibal* is the name coined in Uruguay for the local implementation of the One Laptop Per Child (OLPC) initiative. Plan Ceibal distributes laptops and tablets to students and teachers, and also deploys a nationwide wireless network to provide Internet access to these devices, provides video conference facilities, and develops educational applications. Given the scale of the program, management in general, and specifically device management, is a very challenging task. Device maintenance and replacement is a particularly important process; users trigger such kind of replacement processes and usually imply several days without the device. Early detection of fault conditions in the most stressed hardware parts (e.g., batteries) would permit to prompt defensive replacement, contributing to reduce downtime, and improving the user experience. Seeking for better, preventive and scalable device management, in this paper we present a prototype of a Mobile Device Management (MDM) module for Plan Ceibal, developed over an IoT infrastructure, showing the results of a controlled experiment over a sample of the devices. The prototype is deployed over a public IoT infrastructure to speed up the development process, avoiding, in this phase, the need for local infrastructure and maintenance, while enforcing scalability and security requirements. The presented data analysis was implemented off-line and represents a sample of possible metrics which could be used to implement preventive management in a real deployment.

**Keywords:** one-to-one computing educational program; Mobile Device Management; Internet of Things

---

## 1. Introduction

One Laptop Per Child (OLPC) projects involve distributing low-cost laptop computers in less developed countries with the intent to increase opportunities for students. Nicolas Negroponte, the primary advocate for this project, announced his idea of a low-cost laptop at the World Economic Forum in Davos, in 2005, suggesting that laptops and the Internet can compensate for shortcomings in the educational system, and therefore, children should have access to a computer on a daily basis. Following the first XO laptop prototype presented in 2005, there are currently OLPC implementations in 40 countries including Uruguay, Ethiopia, Afghanistan, Argentina, and the United States [1].

Uruguayan government launched Plan Ceibal in 2007 on one single school and rapidly spread over the country reaching full primary schools coverage in the first couple of years. In subsequent years the coverage was extended to high schools, the devices were updated and diversified, comprising several hardware platforms and operating systems. The wireless infrastructure was partially deployed in-house but mainly outsourced to the state-owned telecom operator ANTEL. Managing about one million devices, together with other components of the ICT infrastructure of Plan Ceibal has proven to be a hard task, involving software updates, hardware maintenance and repair, inventory tracking, and many other tasks.

The problem of Information and Communication Technologies (ICT) management is not new. Back in the eighties, as a consequence of the growing dependency of ICT, several guidelines were proposed by governments and the industry, eventually leading to the appearance of the IT Infrastructure Library (ITIL), originated as a collection of books, each covering a specific practice within IT service management [2]. The ICT operations management sub-process enables technical supervision of the ICT infrastructure, being responsible for (i) a stable, secure ICT infrastructure, (ii) an up to date operational documentation library, (iii) a log of all operational events, (iv) the maintenance of operational monitoring and management tools, and (v) operational scripts and procedures. ICT operations management involves many specific sub-processes, such as output management, job scheduling, backup and restore, network monitoring/management, system monitoring/management, database monitoring/management, storage monitoring/management. ITIL processes implementation has been aided by several software suites, which have evolved following the standard frequent updates (https://www.iso.org/committee/5013818/x/catalogue/p/1/u/0/w/0/d/0). The arrival of mobile devices and specifically the Bring Your Own Device (BYOD) movement has presented new challenges to ICT operations management, leading to the appearance of the Mobile Device Management (MDM) concept, and the need of integration with legacy management processes. We will explore these concepts further in Section 2.

The Internet-of-Things (IoT) is a ubiquitous concept that evolves from sensor networks, and includes distributed devices, communications, and cloud platforms for data storage and processing, comprising both analytics and decision-making activities, which may involve actuation over the devices in response to certain conditions [3], fulfilling an *observe-analyze-act* cycle. These concepts are hardly standardized, and the deployment of IoT applications is heavily dependent on verticals, i.e., health, agriculture, smart cities, industries, among others. Mobile devices such as notebooks, tablets, and smart-phones can be considered as sensors with networking and processing capabilities, and therefore may easily accommodate in the IoT architecture; in fact, smart-phones frequently operate as sensor devices in crowd-sourcing applications. Nevertheless, devices usually sense external variables such as temperature, humidity, location, among many others. In the present case, the devices are used to sense internal variables such as CPU usage, battery status, power-on time, and also external variables such as signal strength and WiFi SSIDs in range; push actions, configuration changes, and software updates can also be implemented. Therefore, we focus on a device management module, which can also be integrated into the network management modules, i.e., measuring network traffic. We will further develop these ideas in Sections 3 and 4.

Our contribution is twofold: on the one hand, we provide an analysis of the applicability of IoT application development to the MDM problem, integrated with a one-to-one computing educational program management. On the other hand, we implement and deploy a prototype solution to the MDM problem using a standard IoT infrastructure. Furthermore, we perform simple, preliminary analytics over the gathered data, which allows envisioning the potential of the proposed solution.

## 2. Management Systems in a One-to-One Program

Managing a one-to-one educational program requires some particular tools and functionalities, besides typical enterprise management systems such as Enterprise Resource Planning (ERP) and Customer Relationship Management (CRM) systems. The current solution at Plan Ceibal considers four key elements: ERP and CRM, devices management, network management, and learning platforms management and integration. Figure 1 summarizes this scenario.

In this section, we first describe the different elements that are currently considered by Plan Ceibal management systems. To give a general understanding of the particularities of this domain, we discuss existent solutions to manage each element, but without the intent to do an exhaustive review. Then, we focus on the new module presented in this work, which enhances the management and monitoring of the devices.

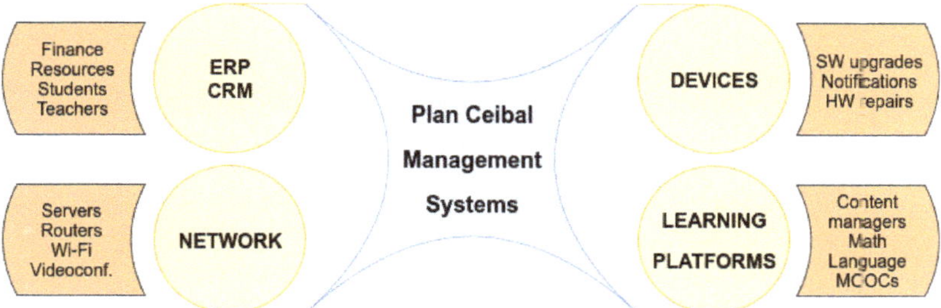


**Figure 1.** Management systems at Plan Ceibal.

### 2.1. ERP and CRM Systems

ERP systems [4,5] cover the main functional areas of any organization. They are typically composed of several modules that deal with core business processes. For example, the finance and accounting modules handle the budgeting and costing, fixed assets, cash management, billing and payments to suppliers, among others. Another standard module has to do with human resources, involving aspects such as recruiting, training, rostering, payroll, benefits, holidays, days off and retirement. A full ERP system may have many other modules, covering aspects such as work orders, quality control, order processing, supply chain management, warehousing and project management.

CRM systems [6] deal with customer interaction, including aspects such as sales and marketing, and the management of several communication channels (website, phone, email, live chat, social media), which typically involves running a call center or a contact center. Initially, these tasks were part of ERP systems responsibilities. However, given the importance and the complexity of managing customer relationships in almost any business, it is often operated by a different software solution that exchanges data with the ERP.

Plan Ceibal uses standard commercial ERP/CRM solutions. Nevertheless, in the context of this one-to-one program, a fundamental requisite for these systems is the capability to integrate data from the educational system successfully. Students, teachers, and schools act like customers in this scenario, since the main goals of a one-to-one computing initiative are to deliver a laptop or tablet to every teacher and student, and also to provide Internet access to every school. The one-to-one management systems do not own educational system data, but must use it to fulfill its goals, i.e., they must have access to school data (such as location, or prints), and teachers and students data (e.g., listings for each level in every school, and contact info). Due to the sensibility of this information, which includes personal data, data management must take into account the privacy of the program beneficiaries and adhere to existent legislation.

### 2.2. Device Management

The first component specific to a one-to-one program corresponds to the systems that manage delivered devices, which are known as Mobile Device Management (MDM) systems [7–9]. Common MDM features include device inventory and monitoring, security functions (e.g., to lock lost or stolen devices), operating system updates, application distribution, and user notifications. This module is particularly critical due to the large number of devices that are usually handled by one-to-one programs. Several commercial MDM solutions are available in the market for major commercial platforms (Android, iOS, Windows, MacOS) installing specific software agents, but do not seem easy to integrate into a general multi-platform environment, and particularly on Linux based systems, which represent the vast majority of the devices delivered by Plan Ceibal. Therefore, no commercial MDM tools are currently deployed; software updates and other simple tasks are performed using home-made tools. For these reasons, it is essential to deploy an MDM module which may help to improve device management, and this is the primary driver for our prototype.

### 2.3. Network Management

The network management module, also specific to this scenario, deals with the administration of the connectivity infrastructure, which provides Internet access, and supports the video-conference equipment and other networking services among schools. The operation and maintenance of all these services typically involve a Network Management System (NMS) [10]. NMS solutions aim to reduce the burden of managing the growing complexity of network infrastructure. They usually collect information from network devices using protocols including, but not limited to, SNMP, ICMP, and CDP. This information is processed and presented to network administrators to help them quickly identify and solve problems such as network faults, performance bottlenecks, and compliance issues. They may also help in provisioning new networks, dealing with tasks such as installing and configuring new equipment and assist in the maintenance of existing networks performing software updates and other tasks.

Typically, network operators and IT administrators deal with several systems and applications to manage their infrastructure, and one-size-fits-all NMS solutions are rare. Most proprietary products have their management systems, which are typically not possible to integrate with other solutions. Moreover, each subsystem such as routing and switching equipment, WLAN solutions and video conference infrastructure has a different management system, and, even for the same provider, all these subsystems are not easy to integrate. Since global management solutions are quite expensive, administrators usually prefer to manage each subsystem separately to avoid this costs. There exist several open-source solutions, and although some of them are very popular, they also face difficulties in integrating with proprietary systems.

### 2.4. Learning Platforms

The fourth component corresponds to learning platforms. In this case, no general solution incorporates the broad range of possibilities involved in providing educational content. On the one hand, we have learning management systems (LMS) which are general content managers for teachers, which enable to create courses and interact with students. On the other hand, intelligent tutoring systems (ITS) are typically tailored for specific topics such as math or language and represent a different platform flavor that should also be managed. In the latter, the goal is to reduce the teacher assistance, automatically guiding the students with previous exercise results. Finally, traditional websites and digital libraries can also be considered educational content providers.

### 2.5. Interactions

While usually each of the previously mentioned systems work independently, it is clear that there must be information exchange between them and data consistency among the data saved in each of the corresponding databases. For example, a router should be present in the ERP database as it corresponds to an asset of the organization, while the same equipment should be identified and monitored in the NMS for its operation. Another clear example is all the information regarding end-users, which is managed by the CRM, but it is also needed for the operation of the learning platforms, with the corresponding privileges for each kind of user. Without breaking this loosely-coupled relationship model, a more advanced integration of these systems would enable better management of the educational program. For instance, it would be nice to integrate the user experiences, typically collected from the CRM, with the network management and operation, thus enabling a Quality of Experience (QoE) [11] based service operation.

## 3. Developing Applications Over IoT Platforms

As already mentioned, the Internet-of-Things (IoT) is a ubiquitous concept that includes (i) distributed devices (a.k.a *things*) that can be identified, controlled, and monitored, and (ii) communications, data storage and processing capabilities that may comprise both analytic and decision-making activities. The IoT is inherently heterogeneous due to the vast variety of devices, communication protocols, APIs, middleware components, and storage options. However, there is also a myriad of approaches to develop and deploy IoT solutions. Ranging from very domain-specific applications (e.g., healthcare domain, traffic, and transportation), to high-level general purpose frameworks and platforms, from open-source solutions to hardware and vendor-specific approaches, the amount of options is enormous.

In this scenario, developers and practitioners must deal with the difficult task of choosing the right tools and approaches to undertake their projects. There are multiple IoT platforms, many of them open-source, which may be deployed and controlled by the application owner. For example, the Hadoop (https://hadoop.apache.org/) ecosystem provides tools that can be combined to fulfill the requirements of an IoT platform. An architecture for smart cities based on these tools is discussed in [12], while in [13] an IoT cloud-based car parking middleware implementation is presented, using Apache Kafka (https://kafka.apache.org/) and Storm (http://storm.apache.org/).

Also, in the last years, the idea of integrating IoT and Cloud Computing has gained momentum [3,14]. Most of the major providers of public Cloud Computing environments started offering IoT features, which try to ease the development and deployment of IoT solutions, exploiting the already available infrastructure. For instance, a cloud-based IoT platform for ambient assisted living using Google Cloud Platform (https://cloud.google.com/) is presented in [15]. We next present an overview of IoT cloud-based platforms, focusing on general-purpose IoT platforms.

### *3.1. An Overview of IoT Cloud Platforms*

Despite many standardization efforts, there is still a lack of a reference architecture for IoT applications and platforms. After performing a comprehensive survey on the matter, Al-Fuqaha et al. [16] collect four common architectures. Early approaches applied a simple three-layer architecture, borrowing concepts from network stacks. The authors claim that this approach hinders the complexity of IoT, while the middleware and SOA-based architectures are not suitable for all applications since they impose extra energy and communications requirements to resolve service composition and integration. Finally, they conclude that a five-layer architecture is the most applicable model for IoT applications (Figure 2). We now briefly sketch this approach.

The **Objects** or perception layer represents the physical sensors and actuators that collect data. These objects interact with the **Object Abstraction Layer**, which is responsible for transferring the data produced by the Objects Layer to the Service Management layer through secure channels. Various technologies such as RFID, 3G, GSM, UMTS, WiFi, Bluetooth Low Energy, infrared, or ZigBee are used to transfer data. The **Service Management Layer** pairs services and requesters based on addresses and names, allowing IoT application programmers to work with various objects hiding the specificities of the underlying hardware. Finally, customers and clients either interact with the **Application** or the **Business layers**. The former is the interface by which end-users interact with the devices and query for data (e.g., it may provide temperature measurements) while the latter manages the overall system activities, monitoring and managing the underlying four layers. Usually, this layer supports high-level analysis and decision-making processes. Due to all its responsibilities, the Business Layer usually demands higher computational resources than the other layers.

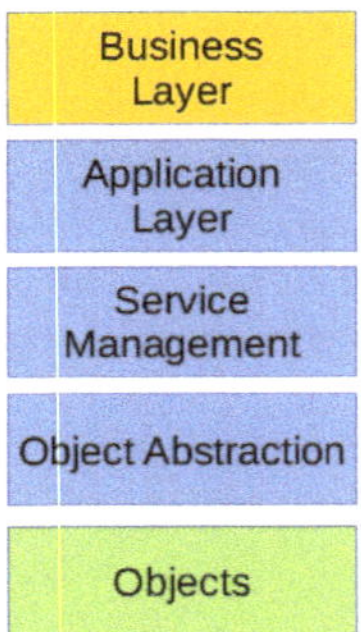


**Figure 2.** Five-layer IoT architecture (adapted from Al-Fuqaha et al., 2015 [16]).

Since cloud computing environments already provide distributed and scalable hardware and software resources, the idea of implementing some of the layers of an IoT platform using these environments seems straightforward. Cavalcante et al. [17] performed a systematic mapping study on the integration of the IoT and Cloud Computing paradigms, where they focused on two research questions: which are the strategies for integrating IoT and Cloud Computing and which are the existing architectures supporting the construction and execution of cloud-based IoT systems. The authors characterize the integration of IoT and Cloud Computing according to the distribution of responsibilities among the three traditional cloud layers: Infrastructure as a Service (IaaS), Platform as a Service (PaaS) and Software as a Service (SaaS). They identify three integration strategies: minimal, partial and full integration. In the case of the minimal integration, a cloud environment (either in the IaaS or the PaaS) is used to deploy the IoT middleware, and to use it to visualize, compute, analyze, and store the collected data in a scalable way. In the case of partial integration, not only the IoT middleware is deployed in a cloud environment, but the platform also provides new service models based on abstractions of smart objects. Therefore, service models such as Smart Object as a Service (SOaaS) and Sensing as a Service (S2aaS) are provided to hide the heterogeneity of devices and virtualize their capabilities. Finally, the full integration strategy proposes new service models that extend all the conventional Cloud Computing layers to encompass services provided by physical objects, allowing physical devices to expose their functionalities as standardized cloud services. Most of the reviewed approaches either apply the minimal or partial integration strategies. Regarding the architectures that support the construction and execution of cloud-based IoT systems, the authors report that the reviewed solutions are significantly distinct from each other, but the majority of them adopted traditional approaches such as smart gateways, Web services based on REST or SOAP (Simple Object Access Protocol) and drivers or APIs deployed in the SaaS cloud layer. Also, they found that most approaches use the PaaS layer to support the deployment of tools and services for developing applications, as well as the IaaS layer as the underlying infrastructure for hosting and executing applications.

Existing surveys on IoT platforms compare different aspects such as the communication protocols, the device management, and the analytics capabilities, among others; the interested reader may refer for example to [18,19]. In our case, we considered three public cloud IoT platforms, namely Microsoft Azure, IBM Watson, and AWS IoT, and a couple of open-source platforms that can be locally-deployed: Kaa and Fiware. While the later offer a clear advantage concerning privacy and control, they also impose a significant steep learning curve and management efforts for local IT administrators. Given that the primary purpose of this project was to test the feasibility of our approach, we decided to focus on public cloud IoT platforms that allowed us to develop and deploy our prototype quickly.

A detailed analysis of the platforms mentioned above is out of the scope of this paper; nevertheless, among the reasons to choose AWS IoT platform for our prototype, it is worth mentioning the strong device authentication services, the maturity of the documentation, and the usability of Lambda functions. Nevertheless, it is advisable to carefully analyze local deployment of the solution in the commissioning phase of the project, mainly for security and privacy concerns, given that Plan Ceibal manages students and teachers data. In the following, we briefly sketch the main components of AWS IoT platform.

### 3.2. AWS IoT Architecture

AWS IoT is an Amazon Web Services platform that allows to collect and analyze data from internet-connected devices and sensors, feeding that data into AWS cloud applications and storage services [20]. In this approach, most of the layers discussed in Section 3.1 are implemented by developers on the cloud, using the features provided by the platform. We next describe AWS IoT main components and functionalities.

#### 3.2.1. Communications

IoT cloud platforms support multiple communication protocols, usually at least MQTT[21] and HTTP, which permit to send data to the server and directives to the devices. AWS IoT supports HTTP, MQTT and WebSockets communication protocols between connected devices and cloud applications through the Device Gateway, which provides secure two-way communication while limiting latency. The Device Gateway scales automatically, removing the need for an enterprise to provision and manage servers for a pub/sub messaging system, which allows clients to publish and receive messages from one another.

#### 3.2.2. Device Registry

IoT cloud platforms provide a registry of devices, a system of permissions, and mechanisms to add new devices to that registry programmatically. In AWS IoT the Device Registry feature lets a developer register and track devices connected to the service, including metadata for each device such as model numbers and associated certificates. This scheme simplifies the task of adding a new device to the solution, where system administrators only have to provide and install a daemon in the desired devices to perform the registration. The platform will keep track of the devices and their permissions afterward. Also, developers can define a Thing Type to manage similar devices according to shared characteristics.

#### 3.2.3. Authentication/Authorization

AWS requires devices, applications, and users to adhere to authentication policies via X.509 certificates, AWS Identity and Access Management credentials or third-party authentication. AWS encrypts all communication to and from devices. The AWS IoT platform features strong authentication, incorporates fine-grained, policy-based authorization and uses secure communication channels.

#### 3.2.4. Device Shadows

IoT platforms have to deal with the fact that devices may not always be connected. In AWS IoT, Device Shadows provide a uniform interface for all devices, regardless of connectivity limitations, bandwidth, computing ability, or power. This feature enables an application to query data from devices and send commands through REST APIs.

#### 3.2.5. Event-Based Rule Engine

Most IoT cloud platforms provide a rule engine that acts based on events, allowing developers to program behavior into the platform. The capabilities of rule engines vary among platforms, but typical actions include storing in databases, sending messages to devices, and running arbitrary code. In AWS

IoT, rules specified using a syntax that's similar to SQL can be used to transform and organize data. This feature also allows developers to configure how data interacts with other AWS services, such as AWS Lambda, Amazon Kinesis, Amazon Machine Learning, Amazon DynamoDB, and Amazon Elasticsearch Service.

#### 3.2.6. Storage and Analytic Services

Other services, such as databases, can extend IoT platforms capabilities. Together with the rule engine, these services can be used to store data and error logs, process complex events and issue alerts, keep track of firmware versions and update them, and in general to implement the Business Layer features.

#### 3.2.7. Development Environment

AWS IoT developers can manage and develop their solutions with the AWS Management Console, software development kits (SDKs) or the AWS Command Line Interface. The platform also provides AWS IoT APIs to perform service configuration, device registration and logging (in the control plane) and data ingestion in the data plane. Several open-source AWS IoT Device SDKs can be used to optimize memory, power and network bandwidth consumption for devices. Amazon offers AWS IoT Device SDKs for different programming languages, including C and Python.

## 4. Description of Our Proposal

In this section we present our prototype of an MDM module for Plan Ceibal, using AWS IoT infrastructure as the back-end. This module is capable of periodically logging network quality of service metrics (QoS) as seen by the devices, as well as relevant usage metrics, for example, CPU load, RAM usage, battery health, hard drive usage, and application monitoring. Additionally, the module is capable of on-demand data collection: an administrator can order the devices to log at an arbitrary time. A monitoring agent running on the mobile devices, specifically on Linux-based laptops, gathers the data and sends them to a collecting platform. All of the back-end infrastructures reside within the AWS ecosystem, including an MQTT broker, rules over the AWS event-based rule engine, data process and formatting algorithms, logging system, the primary database, and the device discovery and registry modules. The data collecting platform stores the data which may be used to run online or on demand analysis.

The proposed solution meets two fundamental requirements: it can discover managed devices with a standard security level automatically, and it supports to scale up to a million devices. We argue that the features offered by cloud-based IoT platforms, such as Amazon IoT or Microsoft Azure IoT, are well suited to serve as the back-end of our MDM module, mainly because they permit to achieve the scalability requirement easily. Our solution also presents additional functionalities that are not frequent in traditional MDM systems. For example, we can collect network performance data from end-user devices, allowing a significant improvement in the management of the wireless network. Measuring device specific and network specific data with the same tool, potentially avoiding the deployment of additional network monitoring solutions that include specialized and out-of-band sensors like the ones presented in [22–24], is a definite advantage. The benefit is twofold. On the one hand, it reduces management budget, and on the other hand, allows us to have a better wireless monitoring solution using measurements from the end-user devices. We next describe the architecture of our prototype.

### *4.1. Prototype Architecture*

Figure 3 presents the architecture of our prototype. The **Sensor** layer can be loosely mapped to the **Objects** layer of the five-layer model by Al-Fuqaha et al. [16], while the remaining there layers, namely **Sensor data retrieval**, **Data processing**, and **Data storage** layers permit to implement a **Business** layer on top. In our case, this layer could implement proactive management processes based on online analytics. As mentioned throughout the paper, the prototype implements data retrieval and storage, while we provide some off-line data analysis tasks as a sample of what can be achieved with the system. We now review the architecture.

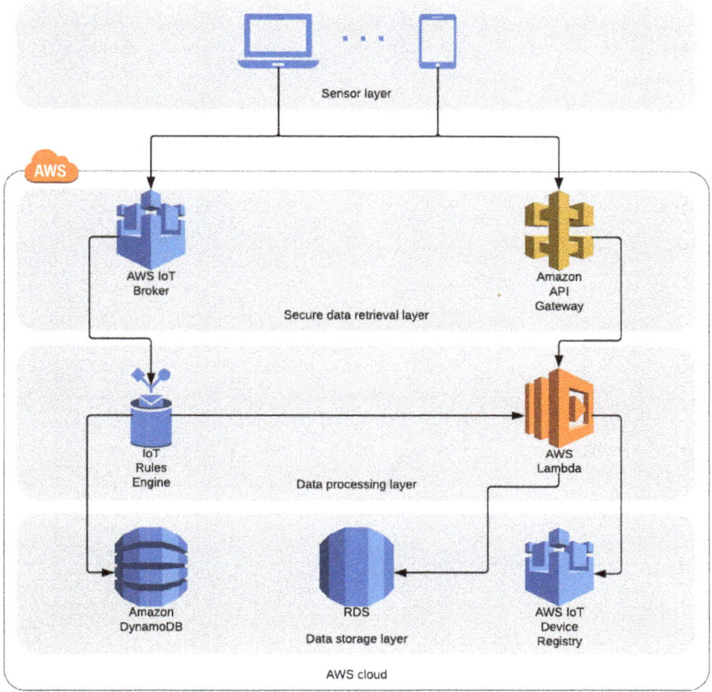


**Figure 3.** Prototype architecture.

#### 4.1.1. Sensor Layer

This layer is responsible for data collection using the utilities provided by the Linux OS. We developed a daemon module in *Python* that runs in the computers of Plan Ceibal. It executes periodically using `cron`, collects data, and sends it to the broker using MQTT messages. The collected data is shown in Table 1. We used standard Linux tools for data collection (including `date`, `ifconfig`, `iwconfig`, `acpi`, `df`, and information in `/proc`, among others) and Eclipse Paho for communication. Regarding portability, changing the back-end would only imply changing the registration and messaging URLs at the device side.

#### 4.1.2. AWS Layers

The three layers implemented over the AWS platform (Sensor data retrieval, Data processing, and Data storage) permit to implement the two primary use cases of the prototype: (i) device registration and (ii) data and error logging. Such architecture, based on platform components, is currently referred to as a serverless architecture. It is worth mentioning that the term serverless has been previously used with different meanings. In particular, within the database community, it refers to database engines that run within the same process as the application, thus avoiding to have a server process listening at a specific port and reducing the communication costs involved in sending requests and receiving responses via TCP/IP or other protocols. SQLite (https://www.sqlite.org), H2 (http://www.h2database.com/), and Realm (https://realm.io/) are examples of this kind of databases. In this project, the term serverless means that the solution does not require the provisioning or managing of any servers by the developers. The servers still exist, but they are abstracted away, and their issues are taken care of by the cloud services provider [25]. This term also encompasses the way server-side logic is executed. In the Function-as-a-Service (FaaS) paradigm, the code is run in event-triggered stateless containers that are managed by cloud service providers. AWS Lambda is Amazon implementation of FaaS. In the following, we describe how we use AWS components to implement the use cases.

**Table 1.** Logged data from devices.

| Attribute | Description |
|---|---|
| `mac_addr` | Device's MAC address. |
| `serial_number` | Device's serial number. |
| `ip_addr` | Device's public IP address. |
| `timestamp` | Added timestamp. |
| `ap_mac_addr` | MAC address of the access point the device is connected to. |
| `frequency` | Frequency of the network the device is connected to. |
| `rssi` | Relative received signal strength of the network the device is connected to. |
| `tx_packets_quantity`<br>`tx_packets_overruns`<br>`tx_packets_carrier`<br>`tx_packets_errors`<br>`tx_packets_dropped`<br>`tx_excessive_retries` | Information about the packages transmitted by the interface. |
| `rx_packets_quantity`<br>`rx_packets_overruns`<br>`rx_packets_frame`<br>`rx_packets_errors`<br>`rx_packets_dropped`<br>`rx_bytes` | Information about the packages received by the interface. |
| `charging` | Indicates if the device is charging or not. |
| `battery_temp` | Battery temperature. |
| `battery_power` | Battery charge level. |
| `uptime` | Time the device has been powered-on. |
| `boot_time` | Last boot time. |
| `load_avg_5_min` | Average load of the CPU in the last five minutes. |
| `total_memory_kb` | Total RAM capacity. |
| `free_memory_kb` | Free RAM. |
| `total_swap_memory_kb` | *Swap* memory size. |
| `free_swap_memory_kb` | Free *swap* memory. |
| `cached_memory_kb` | Page cache size. |
| `buffers_memory_kb` | I/O *buffers* size. |
| `root_dir_total_disk_space_kb` | Total space on the device's /root directory. |
| `root_dir_free_disk_space_kb` | Free space on the device's /root directory. |
| `home_dir_total_disk_space_kb` | Total space on the device's /home directory. |
| `home_dir_free_disk_space_kb` | Free space on the device's /home directory. |

Devices Registration

The prototype extends AWS Device Registry functionality, developing a *device discovery feature*. Devices register automatically; on install, the daemon registers the device in the platform using its serial number, obtaining credentials and a topic to publish new data. Administrators only need to install the daemon on the devices, and the platform will keep track of them, scaling automatically with the demand. Besides, the devices subscribe to topics, which allows the system to send them messages. Plan Ceibal staff were particularly interested in these features for on-demand data collection and software updating.

A daemon can be implemented, using this approach, for any device with an Internet connection. The only requirements are the proper registration and to send data to the right endpoint. Any device can register, but policies that restrict the access are set on the server side. AWS API Gateway provides a simple and effective way to create and manage *RESTFul* APIs that scale automatically. We developed an API capable of managing HTTP POST requests, registering the devices and creating unique certificates for the devices to connect to the broker. A POST request to the API triggers a *Lambda function* that creates an *X.509 certificate*, used to establish MQTT connections between devices and the broker. It also attaches a policy—a JSON document specifying a device's permissions in AWS—to that certificate, so the device will only be allowed to publish to a specific MQTT topic created for it. Figure 4a shows a flow diagram for this process.

Data and Errors Logging

Our implementation uses AWS IoT MQTT broker and the event-based AWS IoT Rules Engine to process collected data and log errors. First, the Rules Engine sorts incoming messages from already registered and authenticated devices by their type, according to which MQTT topic they are sent. Then, data messages trigger a Lambda function that processes the data, formats it and stores it in a PostgreSQL database running inside AWS RDS, while error messages are forwarded and stored in DynamoDB.

To log errors, messages are sent to device-specific MQTT broker topics that adhere to the following pattern: devices/<CERTIFICATE-ID>/errors. Then, the AWS IoT Rules Engine triggers a custom-made rule that stores the error message in a DynamoDB table. Listing 1 shows this rule. Line 2 contains a pseudo-SQL query that describes the messages that this rule should process; in this case all the messages from any errors topic. Lines 5–14 describe the actions to perform. The roleArn parameter identifies a role entity inside AWS that has write access on the errors-table, while lines 9 to 12 specify the key-value pair to store (the device -id and the error message) and the timestamp.

**Listing 1.** A rule to process and store error messages into DynamoDB.

```
1  {
2    "sql": "SELECT * FROM 'devices/*/errors'",
3    "ruleDisabled": false,
4    "awsIotSqlVersion": "2016-03-23",
5    "actions": [{
6        "dynamoDB": {
7            "tableName": "errors-table",
8            "roleArn": "arn:aws:iam::123456789012:role/my-iot-role",
9            "hashKeyField": "device-id",
10           "hashKeyValue": "${topic(2)}",
11           "rangeKeyField": "timestamp",
12           "rangeKeyValue": "${timestamp()}"
13       }
14   }]
15 }
```

This error logging system complements the default error logging on AWS (CloudWatch) and allows us to view errors affecting the daemon without physical access to the devices. Of course, this approach only works it the errors are not related to the connection itself; in that case, it is necessary to check the error logs stored on the device. Figure 4 shows the complete flow of the data and error logging functionalities of the system.

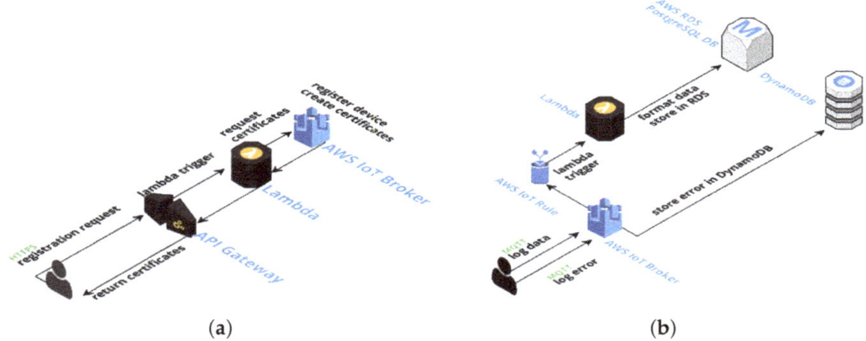


(a) (b)

**Figure 4.** Register new device and error logging flow and architecture of the prototype. (**a**) Register new device flow and architecture. (**b**) Data and error logging flow and architecture

### 4.2. Proof of Concept

A pilot test was carried out with end-user devices delivered by Plan Ceibal to validate the prototype. The data collection daemon was installed in more than 2000 devices using Plan Ceibal software update tools. This test lasted one month in October–November 2017, and it was useful not only to verify the platform operation but also to collect parameters from the devices which had never been monitored before by Plan Ceibal. In effect, parameters are measured using standard Linux OS utilities running on the device, and therefore the possibilities for parameter monitoring is only limited by OS capabilities, in the case of the prototype, Ubuntu 14.04. Plan Ceibal uses different hardware platforms and operating systems; the prototype comprised two models with Intel Celeron N3160 4-core CPU @1.60Ghz and IEEE 802.11b/g wireless connectivity: (i) Clamshell laptops with 2 GB RAM and 32 GB of eMMC memory, and (ii) Positivo laptops with 2 GB RAM and 16GB SSD mSATA memory.

It is important to consider the cost of running the system over a public IoT platform, in this case, AWS. In particular, let's consider Lambda cost analysis. Lambda charges the user based on the number of function requests and the duration, i.e., the time it takes for the code to finish executing rounded up to the nearest 100ms. This last charge depends on the amount of RAM the user allocates to their functions. A basic free tier is offered for the first one million requests and 400,000 GB-Seconds of computing time per month. At the time the test was performed, the charge was 0.2 US Dollars per one million requests, and 0.00001667 US Dollars for every GB-Second used after that [26]. The deployed prototype did not surpass the free-tier limits while running.

Currently, the system stores collected data for later analysis, but it may be used to feed on-line analytics. To illustrate the kind of analysis that can be performed over collected data, we present two examples. The first one uses the information corresponding to the battery charge of the devices, while the second one analyzes the locations where the devices are used, distinguishing *at school* use from *at home* use.

#### 4.2.1. Battery Charge Analysis

Battery charge is an important parameter that allows analyzing relevant aspects such as battery performance, the typical users' habits for charging the devices, and the equipment autonomy drift as time passes. Moreover, this type of analysis may be used to trigger preventive maintenance or replacement actions, in order to shorten downtimes caused by hardware failures. Figure 5a shows the battery charge empiric distribution obtained from the collected data, considering every measurement individually, and without any aggregation per device. It is worth noting that fully charged devices connected to an external power supply explain the peak in the last bin. We guess so because the reported battery charge parameter value is 100% in those cases. Collected data approximates quite well to a uniform distribution leaving aside the last bin. Next, the data is aggregated taking the median value by device. Figure 5b corresponds to the resulting distribution, which as expected is well approximated by a Gaussian distribution since the median and the average are quite similar for the collected data. Finally, Figure 5c shows a time series example for a specific device. The missing values correspond to periods when the device was off or had no connection to the network; therefore no data is reported. In this case, the discharge/charge cycles of the device can be observed looking at the temporal evolution. This information would be beneficial for device management (e.g., for selecting better battery suppliers, detecting a malfunction, among many others).

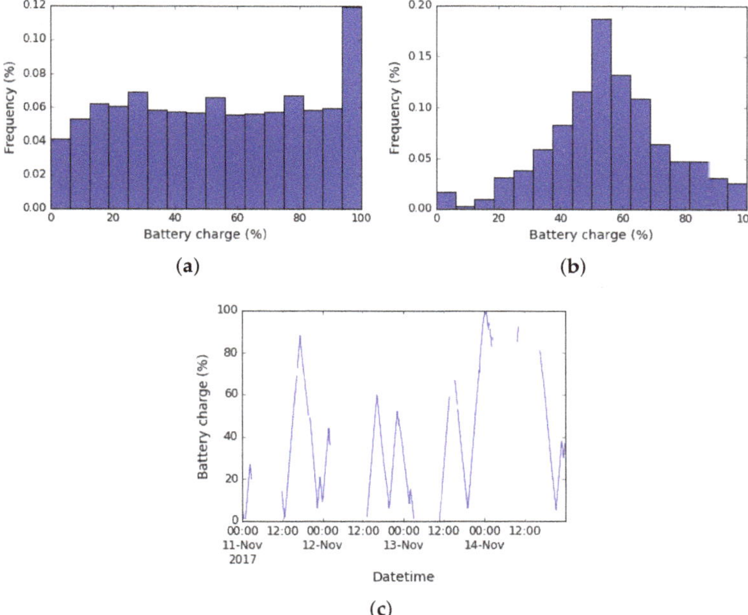


**Figure 5.** Time series example: battery charge. (**a**) Battery charge histogram. (**b**) Battery charge histogram, considering the median value per device. (**c**) Example of battery charge time series, corresponding to one particular device during four days.

We delve into collected data to analyze the battery performance, considering only the two laptop models with the most significant amount of devices involved in the pilot (called A and B from now on). First, we processed the time series of the battery charge measurements and calculated the discharge coefficient for each device. For this purpose, a linear regression of the discharge curves was performed, which seemed an appropriate model for the observed data (cf. Figure 5c). Finally, in order to have only one discharge rate per device, we took the median value from all the estimates.

Figure 6a shows the empiric distribution for the estimated discharge rates for each laptop, considering the values for devices of type A and B. The discharging coefficient is expressed in percentage of charge per unit of time, thus it indicates the battery discharge per second. Using these values, it is possible to calculate the device estimated autonomy (i.e., the time a fully charged laptop takes to discharge its battery completely).

Considering battery performance, it is more useful to examine the devices autonomy as a metric instead of the discharging coefficient. Therefore, for the two laptop models analyzed, we compared the devices' estimated autonomy with measurements taken in lab conditions by Plan Ceibal using their standard equipment evaluation processes. For each device, two autonomy tests are usually carried out by Plan Ceibal, one in low power usage conditions (screen with low brightness and the device running no activity) and another in high battery consumption conditions (maximum screen brightness and device playing HD video). Figure 6b,c, present the estimated autonomy empiric distribution for each device. The vertical red lines correspond to the lab measurements in low and high consumption conditions respectively. As we can see, for both laptop models, most of the devices have estimated values (calculated from the field measurements) within the range given by the minimum and maximum autonomy values measured in the lab environment.

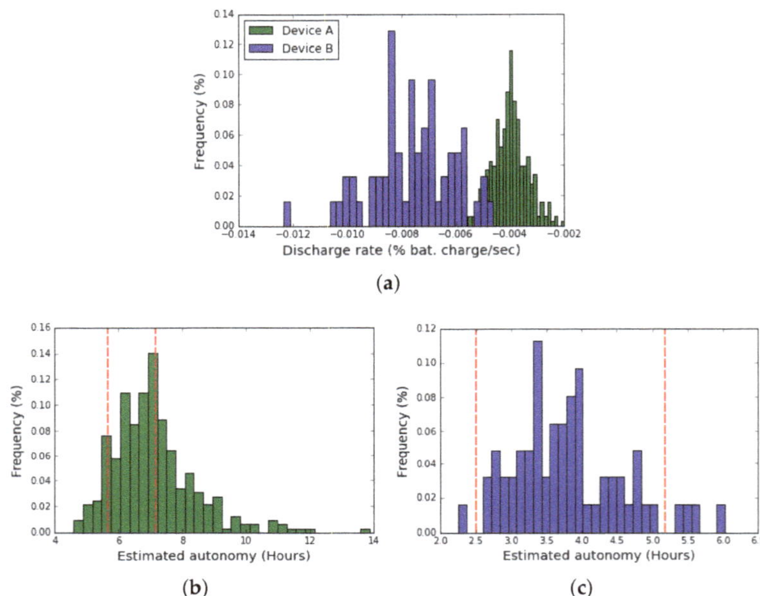


**Figure 6.** Discharge rate distribution and estimated autonomy analysis. (**a**) Discharge rate distribution for devices of type A and B. (**b**) Estimated autonomy for devices of type A. (**c**) Estimated autonomy for devices of type B.

For type A devices, 51% of them fall within the range from lab measurements, while for type B devices the percentage rises to 92%. Devices with an estimated autonomy above the maximum measured in the lab may correspond to low power usage conditions, which consume less battery than the extreme conditions used in the lab test. Furthermore, with this comparison, it is possible to identify those batteries with performance worse than expected (i.e., those below the minimum autonomy measured in the lab environment).

#### 4.2.2. Use by Location

As already mentioned, every student that assists to primary and secondary public schools obtains in property a device with wireless capabilities. They carry their laptops from home to school and back every day. It is therefore interesting to compare the use of the device along the day and to distinguish at school from at home use. This analysis is relevant, because making decisions based only on at school use may lead to the wrong conclusions. Although in many cases devices are hardly used at school, they are widely used at home. This fact has to be considered when assessing the impact of Plan Ceibal and the fulfillment of its goal of fighting the digital divide. We can indirectly measure if the device is used, and where it is used, using our tool, by considering that the device is in use if we receive data from it. It is important to notice that this may underestimate at home use since we are not able to measure if the device is off-line, while at school off-line use is not frequent since internet access is always available. We first analyze the average amount of days that each device reported data during our pilot study, and using the IP addresses and WiFi SSID information we distinguish at school use from outside school use. Figure 7a presents these data, showing that devices are used more outside the school than at school. We then refine this analysis. Figure 7b shows the average connected hours per day, while Figure 7c shows the percentage of time connected to the internet that corresponds to at school connections. This last figure clearly shows that beneficiaries use their devices at home more than at school. Finally, we analyze the evolution of this indicator. Figure 7d presents the average connection time per device and per day, showing that at school use is always below 50% and that during the weekend at school use drops to least than 10%.

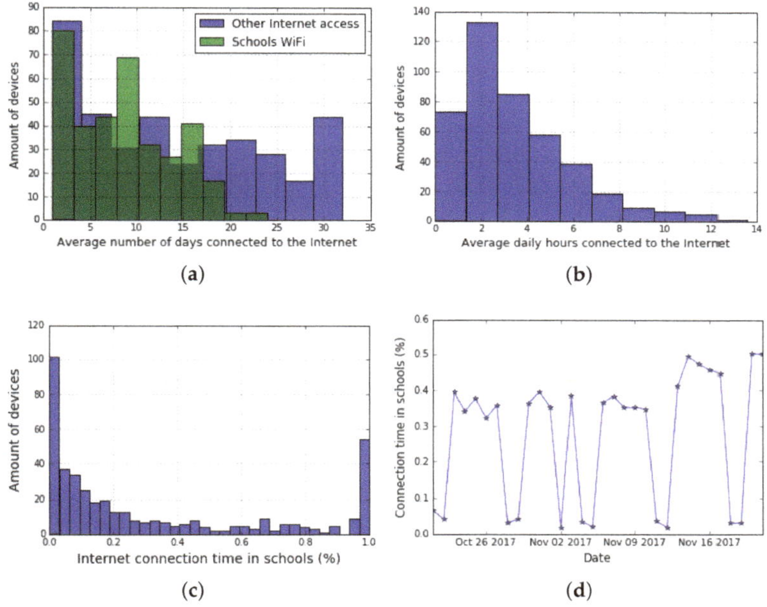


**Figure 7.** Use by location example. (**a**) Average number of connected days. (**b**) Average connected hours per day. (**c**) Percentage of at school connection time. (**d**) Percentage of at school connection time by day.

These simple examples show the potential offered by the developed platform. The first one shows that it can be used for preventive maintenance, in this case, applied to batteries. It is possible to identify batteries in lousy condition, enabling, for example, automatic user alerts for battery replacement, contributing to improve user experience, as mentioned before. The second one shows that it can also be used to analyze user behavior, and with the associated analytics it can provide useful insight to decision making processes. The system administrators may develop particular data analysis based on the collected data, and, depending on the envisioned use cases, they can tune the parameters under monitoring, by merely developing scripts running on the devices, which automatically will collect data after a software update.

As mentioned in Section 4.1, the prototype implements data retrieval and storage, offering capabilities to build a business layer on top. In particular, we envision that proactive management processes based on online analytics should be the core of such a business layer, driven by Plan Ceibal needs; the aforementioned preventive battery replacement is a clear example. Other possible scenarios include (i) proactive capacity provisioning for school networks which exhibit QoS degradation signs, and (ii) laptop upgrades based on CPU, memory and storage usage, among many other possibilities.

## 5. Conclusions

One-to-one computing educational programs comprise a growing management complexity, including the interaction of ERP/CRM, Network, Device, and Learning Platforms management systems, in order to build advanced services, such as QoS guarantees to students and teachers depending on their behavior. Mobile Device Management of different terminals such as laptops, tablets, smart-phones with diverse hardware and operating systems is particularly challenging, and commercial solutions, apart from being costly, frequently do not cover every possibility. In this paper, we propose to apply IoT concepts to the MDM problem, where each mobile device acts as a sensor/actuator and the primary target is the device itself, i.e., gathering internal variables such as CPU and battery usage and temperature, on-off periods, application usage among many other possibilities. Devices can also help network management, for example collecting network traffic statistics, WiFi signal strength and coverage combined with location.

The development of IoT applications is hardly standardized, and therefore finding design patterns is of supreme importance. After surveying state of the art, we focused on a particular platform, AWS IoT. Developing an MDM module over an IoT cloud platform provides many advantages over a traditional approach. First, it is simple to develop and deploy daemons for new devices. As long as they implement the protocol, there is no need for changes in the back-end modules. Second, the new devices will be automatically added to the registry and will be able to send messages afterward. Third, since the module is in the cloud, scaling is simple and happens automatically.

The major downside of our approach is its dependency on a proprietary platform. We identify platform-independent aspects of the development, stating a clear migration path to other cloud-based (or owned) platforms, preserving application logic and data. If migration were necessary, every service should be re-implemented using the equivalent service of the target platform. However, our solution does not use any platform specific feature, especially we avoid MQTT broker advanced features such as *Thing Shadows*; the Lambda functions can be run in a dedicated machine if necessary, while the primary database could be run on any machine, and a standard SQL database or non-proprietary No-SQL database can implement the error logs database. The daemon running in the devices does not use Amazon's MQTT library, making it easier to migrate hosts. The most difficult parts of the system to migrate would be the rules system and permissions since even though every platform has one, the implementations vary greatly.

We deployed our prototype over 2000 devices for over a month, which enabled us to test the intended functionality and gather some sample data to perform preliminary analytics. A simple analysis of battery performance found a strong correlation between the battery usage measured with the prototype, compared to the battery tests performed in laboratory conditions. Measuring real battery performance would permit to perform preventive maintenance, improving the user experience. We also analyzed the usage of the device along the day, seeking to distinguish at school from at home usage. We found that at home usage is longer on average than at school usage, and this result is relevant considering that one of the main goals of Plan Ceibal is bridging the digital divide.

This kind of analytics over the gathered data may help decision-making and planning processes, seeking for improved user experience, and consequently a higher impact of the solution. Indeed, giving the platform better proactive properties is being considered as future work, together with the careful analysis of the architecture (including the cloud platform used) needed for the commissioning phase.

Overall, the proposal delivers a fair, scalable and extensible solution to MDM in the context of one-to-one programs.

**Author Contributions:** Conceptualization, F.O., B.S., L.E. and E.G.; Data curation, G.C.; Investigation, F.O., B.S., L.E. and E.G.; Resources, G.C.; Software, F.O. and B.S.; Supervision, L.E. and E.G.; Validation, G.C.; Writing—original draft, G.C., L.E. and E.G.; Writing—review & editing, G.C., L.E. and E.G.

**Acknowledgments:** This work was undertaken as part of the regular work by G. Capdehourat at Plan Ceibal and L. Etcheverry and E. Grampín at UdelaR. F. Osimani and B. Stecanella implemented the prototype as part of their undergraduate thesis, and therefore, no specific funding was needed. In case of acceptance, UdelaR will cover the costs to publish in open access.

**Conflicts of Interest:** The authors declare no conflicts of interest.

## References

1. Moscatelli, E. The Success of Plan Ceibal. Master's Thesis, University of Texas at Austin, Austin, TX, USA, 2016.
2. Clifford, D.; Van Bon, J. *Implementing ISO/IEC 20000 Certification: The Roadmap*; Van Haren Publishing: Zaltbommel, The Netherlands, 2008.
3. Botta, A.; de Donato, W.; Persico, V.; Pescapé, A. Integration of Cloud computing and Internet-of-Things: A survey. *Future Gener. Comput. Syst.* **2016**, *56*, 684–700. [CrossRef]
4. van Everdingen, Y.; van Hillegersberg, J.; Waarts, E. Enterprise Resource Planning: ERP Adoption by European Midsize Companies. *Commun. ACM* **2000**, *43*, 27–31. [CrossRef]
5. Themistocleous, M.; Irani, Z.; O'Keefe, R.M.; Paul, R. ERP problems and application integration issues: An empirical survey. In Proceedings of the 34th Annual Hawaii International Conference on System Sciences, Washington, DC, USA, 3–6 January 2001. [CrossRef]
6. Goodhue, D.L.; Wixom, B.H.; Watson, H.J. Realizing business benefits through CRM: Hitting the right target in the right way. *MIS Q. Exec.* **2002**, *1*, 79–94.
7. Ortbach, K.; Brockmann, T.; Stieglitz, S. Drivers for the Adoption of Mobile Device Management in Organizations. In Proceedings of the 22nd European Conference on Information Systems, Tel Aviv, Israel, 9–11 June 2014.
8. Patten, K.P.; Harris, M.A. The need to address mobile device security in the higher education IT curriculum. *J. Inf. Syst. Educ.* **2013**, *24*, 41.
9. Harris, M.A.; Patten, K.P. Mobile device security considerations for small-and medium-sized enterprise business mobility. *Inf. Manag. Comput. Secur.* **2014**, *22*, 97–114. [CrossRef]
10. van der Hooft, J.; Claeys, M.; Bouten, N.; Wauters, T.; Schönwälder, J.; Pras, A.; Stiller, B.; Charalambides, M.; Badonnel, R.; Serrat, J.; et al. Updated Taxonomy for the Network and Service Management Research Field. *J. Netw. Syst. Manag.* **2018**, *26*, 790–808. [CrossRef]
11. Schatz, R.; Hoßfeld, T.; Janowski, L.; Egger, S., From Packets to People: Quality of Experience as a New Measurement Challenge. In *Data Traffic Monitoring and Analysis: From Measurement, Classification, and Anomaly Detection to Quality of Experience*; Springer: Berlin/Heidelberg, Germany, 2013; pp. 219–263. [CrossRef]
12. Diaconita, V.; Bologa, A.R.; Bologa, R. Hadoop Oriented Smart Cities Architecture. *Sensors* **2018**, *18*, 1181. [CrossRef] [PubMed]
13. Ji, Z.; Ganchev, I.; O'Droma, M.; Zhao, L.; Zhang, X. A Cloud-Based Car Parking Middleware for IoT-Based Smart Cities: Design and Implementation. *Sensors* **2014**, *14*, 22372–22393. [CrossRef] [PubMed]
14. Mineraud, J.; Mazhelis, O.; Su, X.; Tarkoma, S. A gap analysis of Internet-of-Things platforms. *Comput. Commun.* **2016**, *89*, 5–16. [CrossRef]
15. Cubo, J.; Nieto, A.; Pimentel, E. A Cloud-Based Internet-of-Things Platform for Ambient Assisted Living. *Sensors* **2014**, *14*, 14070–14105. [CrossRef] [PubMed]
16. Al-Fuqaha, A.; Guizani, M.; Mohammadi, M.; Aledhari, M.; Ayyash, M. Internet-of-Things: A Survey on Enabling Technologies, Protocols, and Applications. *IEEE Commun. Surv. Tutor.* **2015**, *17*, 2347–2376. [CrossRef]
17. Cavalcante, E.; Pereira, J.; Alves, M.P.; Maia, P.; Moura, R.; Batista, T.; Delicato, F.C.; Pires, P.F. On the interplay of Internet-of-Things and Cloud Computing: A systematic mapping study. *Comput. Commun.* **2016**, *89*, 17–33. [CrossRef]
18. Derhamy, H.; Eliasson, J.; Delsing, J.; Priller, P. A survey of commercial frameworks for the Internet-of-Things. In Proceedings of the 2015 IEEE 20th Conference on Emerging Technologies Factory Automation (ETFA), Luxembourg, 8–11 September 2015; pp. 1–8. [CrossRef]
19. Hejazi, H.; Rajab, H.; Cinkler, T.; Lengyel, L. Survey of platforms for massive IoT. In Proceedings of the 2018 IEEE International Conference on Future IoT Technologies (Future IoT), Eger, Hungary, 18–19 January 2018; pp. 1–8. [CrossRef]

20. AWS IoT Developer Guide. Available online: http://docs.aws.amazon.com/iot/latest/developerguide/aws-iot-how-it-works.html (accessed on 29 December 2017).
21. MQTT Protocol. Available online: http://mqtt.org/ (accessed on 30 December 2017).
22. 7Signal Sapphire Eye Wi-Fi Sensors. Available online: http://7signal.com/products/sapphire-eye-wlan-sensors/ (accessed on 26 December 2017).
23. Cape Networks Testing Sensor. Available online: https://capenetworks.com/#wireless (accessed on 26 December 2017).
24. Ekahau Sidekick. Available online: https://www.ekahau.com/products/sidekick/overview/ (accessed on 26 December 2017).
25. Serverless Architectures with AWS Lambda. Available online: https://d1.awsstatic.com/whitepapers/serverless-architectures-with-aws-lambda.pdf (accessed on 13 December 2018).
26. AWS Lambda Pricing. Available online: https://aws.amazon.com/lambda/pricing/ (accessed on 13 December 2018).

 sensors

Article

# Performance Analysis of Latency-Aware Data Management in Industrial IoT Networks †

**Theofanis P. Raptis *, Andrea Passarella and Marco Conti**

Institute of Informatics and Telematics, National Research Council, 56124 Pisa, Italy; andrea.passarella@iit.cnr.it (A.P.); marco.conti@iit.cnr.it (M.C.)

* Correspondence: theofanis.raptis@iit.cnr.it

† This paper is an extended version of our paper published in The 4th IEEE International Workshop on Cooperative Wireless Networks (CWN) 2017, which was co-located with the IEEE 13th International Conference on Wireless and Mobile Computing, Networking and Communications (WiMob).

Received: 6 July 2018; Accepted: 9 August 2018; Published: 9 August 2018

**Abstract:** Maintaining critical data access latency requirements is an important challenge of Industry 4.0. The traditional, centralized industrial networks, which transfer the data to a central network controller prior to delivery, might be incapable of meeting such strict requirements. In this paper, we exploit distributed data management to overcome this issue. Given a set of data, the set of consumer nodes and the maximum access latency that consumers can tolerate, we consider a method for identifying and selecting a limited set of proxies in the network where data needed by the consumer nodes can be cached. The method targets at balancing two requirements; data access latency within the given constraints and low numbers of selected proxies. We implement the method and evaluate its performance using a network of WSN430 IEEE 802.15.4-enabled open nodes. Additionally, we validate a simulation model and use it for performance evaluation in larger scales and more general topologies. We demonstrate that the proposed method (i) guarantees average access latency below the given threshold and (ii) outperforms traditional centralized and even distributed approaches.

**Keywords:** Industry 4.0; data management; Internet of Things; performance analysis; experimental evaluation

---

## 1. Introduction

Industry 4.0 refers to the fourth industrial revolution that transforms industrial manufacturing systems into Cyber-Physical Production Systems (CPPS) by introducing emerging information and communication paradigms, such as the Internet of Things (IoT) [1]. Two technological enablers of the Industry 4.0 are (i) the communication infrastructure that will support the ubiquitous connectivity of CPPS [2] and (ii) the data management schemes built upon the communication infrastructure that will enable efficient data distribution within the factories of the future [3]. Industrial IoT networks (Figure 1) are typically used, among others, for condition monitoring, manufacturing processes and predictive maintenance [4]. To maintain the stability and to control the performance, those industrial applications impose stringent end-to-end latency requirements on data communication between hundreds or thousands of network nodes [5], e.g., end-to-end latencies of 1–100 ms [6]. Missing or delaying important data may severely degrade the quality of control [7].

Edge computing, also referred to as fog computing, implements technical features that are typically associated with advanced networking and can satisfy those requirements [8]. Fog computing differs from cloud computing with respect to the actual software and hardware realizations, as well as in being located in spatial proximity to the data consumer (for example, the user could be a device in the industrial IoT case). In particular, components used to realize the fog computing architecture can be characterized by their non-functional properties. Such non-functional properties are, for example, real-time behavior, reliability and availability. Furthermore, fog nodes can follow industry-specific standards (e.g., IEEE 802.15.4e [9] or WirelessHART [10]) that demand the implementation, as well as verification and validation of software and/or hardware to follow formal rules.

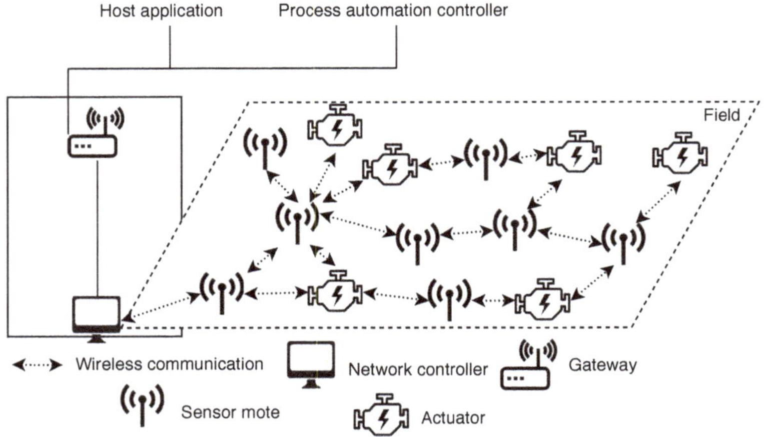


**Figure 1.** A typical industrial IoT network for condition monitoring.

Distributed data management, a key component of fog computing, can be a very suitable approach to cope with these issues [11]. In the context of industrial networks, one could leverage the set of nodes present at the edge of the network to distribute functions that are currently being implemented by a central controller [12]. Many flavors of distributed data management exist in the networking literature, depending on which edge devices are used. In this paper, we consider a rather extreme definition of distributed data management and use the multitude of sensor nodes present in an industrial physical environment (e.g., a specific factory) to implement a decentralized data distribution, whereby sensor nodes cache data they produce and provide these data to each other upon request. In this case, the choice of the sensor nodes where data are cached must be done to guarantee a maximum delivery latency to nodes requesting those data.

In this paper, we exploit the Data Management Layer (DML), which operates independently of and complements the routing process of industrial IoT networks. Assuming that applications in such networks require that there is (i) a set of producers generating data (e.g., IoT sensors), (ii) a set of consumers requiring those data in order to implement the application logic (e.g., IoT actuators) and (iii) a maximum latency $L_{\max}$ that consumers can tolerate in receiving data after they have requested them, the DML offers an efficient method for regulating the data distribution among producers and consumers. The DML selectively assigns a special role to some of the network nodes, that of the proxy. Each node that can become a proxy potentially serves as an intermediary between producers and consumers, even though the node might be neither a producer, nor a consumer. If properly selected, proxy nodes can significantly reduce the access latency; however, when a node is selected as a proxy, it has to increase its storing, computational and communication activities. Thus, the DML minimizes the number of proxies, to reduce as much as possible the overall system resource consumption (the coherency of data that reside on proxies can be achieved in a variety of ways [13] and is beyond the scope of this paper). More specifically, our contributions are the following:

- We propose a distributed data management approach to store data in a number of locations in an industrial environment, as opposed to the current industrial state-of-the-art approaches where all data are centrally stored and served from a unique location. We exploit the DML for minimizing

the number of proxies in an industrial IoT network and to reduce as much as possible the overall system resource consumption.
- We provide a multi-faceted performance evaluation, both through experiments and through simulations, for achieving scales much larger than what available experimental conditions allow. At first, we implement the DML with 95 real devices and evaluate its performance on the FIT IoT-LAB testbed [14]. Then, we use the simulation model, validate it against the experimental results and evaluate the DML performance in larger network sizes and more general topologies.
- We demonstrate that the proposed method (i) guarantees that the access latency stays below the given threshold and (ii) significantly outperforms traditional centralized and even distributed approaches, both in terms of average data access latency and in terms of maximum latency guarantees.
- We also demonstrate an additional flexibility of the proposed approach, by showing that it can be tuned both to guarantee that the average of the mean latency stays below $L_{\max}$ or that the average of the worst-case latency stays below $L_{\max}$.

Roadmap of the paper: In Section 2, we provide a brief summary of concepts related to this paper. In Section 3, we provide the model of the settings we consider, as well as the necessary notation. In Section 4, we introduce the DML and the problem that it addresses. In Section 5, we evaluate the performance of the DML in comparison with two other methods used in industrial environments. We also validate the simulation model used afterwards. In Section 6, we present simulation results in scenarios that are not possible to evaluate with the available experimental testbeds. Finally, we conclude and provide insights for future work in Section 7.

## 2. Literature Review

A note on distributed data management: Traditionally, industrial application systems tend to be entirely centralized. For this reason, distributed data management has not been studied extensively in the past, and the emphasis has been put on the efficient computer communication within the industrial environment. The reader can find state-of-the-art approaches on relevant typical networks in [5,15,16]. In [17], we defined the concept of a data management layer separate from routing functions for industrial wireless networks. The DML considered in the current paper is similar to the one defined in [17], although here, we make it more practical. More specifically, in [17], we focused on a graph-theoretic approach, which might sometimes be unrealistic when real industrial implementations with technological constraints are at hand. Furthermore, in [17], we did not provide an extensive experimental evaluation of our methods, as is done in the current paper, but we verified the methods solely via simulations. Even more importantly, in this paper, we provide (i) a real implementation of the DML, (ii) experimental results on its performance, (iii) a validation of a DML simulation model and (iv) a large-scale performance evaluation of the DML using the validated simulation model.

Related works: We now provide some additional interesting related works. In [18], although the authors considered delay and real-time aspects, the main optimization objectives were the energy efficiency and reliability. They presented a centralized routing method, and consequently, they did not use proxies. Furthermore, the model of this paper assumes that the network is operating under different protocols (e.g., 802.11). In [19], the authors addressed a different optimization objective, focusing on minimizing the maximum hop distance, rather than guaranteeing it as a hard constraint. Furthermore, they assumed a bounded number of proxies, and they examined only on the worst-case number of hops. Finally, the presented approach was somewhat graph-theoretic, which made it hard to apply to real industrial IoT networks. In [20], the authors, given the operational parameters required by the industrial applications, provided several algorithmic functions that locally reconfigured the data distribution paths, when a communication link or a network node failed. They avoided continuously recomputing the configuration centrally, by designing an energy-efficient local and distributed path reconfiguration method. However, due to the locality of the computations and the complete absence of a central coordination, this method may result in violating the latency requirements. In [21], the authors presented a cross-layer approach, which combined MAC-layer and cache management techniques for adaptive cache invalidation, cache replacement and cache prefetching. Again, the model is different, as we assume a completely industrially-oriented MAC layer, based on IEEE802.15.4e, and a different problem, focusing on the delay aspects, instead of cache management. In [22], the authors considered a different problem than ours: replacement of locally-cached data items with new ones. As the authors claimed, the significance of this functionality stemmed from the fact that data queried in real applications were not random, but instead exhibited locality characteristics. Therefore, the design of efficient replacement policies, given an underlying caching mechanism, was addressed. In [23], although the authors considered delay aspects and a realistic industrial IoT model (based on WirelessHART), their main objective was to bound the worst-case delay in the network. Furthermore, they did not exploit the potential presence of proxy nodes, and consequently, they stuck to the traditional, centralized industrial IoT setting. In [24], the authors considered a multi-hop network organized into clusters and provided a routing algorithm and cluster partitioning. Our DML concepts and algorithm can work on top of this approach (and of any clustering approach), for example by allocating the role of proxies to cluster heads. In fact, clustering and our solution address two different problems. In [25], the authors considered industrial networks with a fixed number of already deployed edge devices that acted as proxies and focused on the maximization of the network lifetime. They considered as lifetime the time point until the first node in the network died. As a result, in this paper, they assumed a different model, different optimization problem and different performance targets. Finally, a relevant application domain for distributed data management was also the workshop networks in smart factories [26,27], in which a large amounts of data was transmitted, bringing big challenges to data transfer capability and energy usage efficiency.

Motivating examples: We also present two indicative application areas where the DML and the relevant algorithm can provide additional value. In [28], the authors presented a typical situation in an oil refinery where miles of piping were equipped with hundreds and thousands of temperature, pressure, level and corrosion sensors, which were deployed in a large geographical area. Those sensor motes not only performed industrial condition monitoring tasks, but they also used the industrially-oriented communication technology TSCH. In [29], the authors presented two large-scale deployments of hundreds of nodes; one in a semiconductor fabrication plant and another on-board an oil tanker in the North Sea. They used Mica2 and Intel Mote nodes, very similar to our sensor motes of choice. In both those applications, due to the exact fact that the sensor motes were not able to communicate directly with the controller (transmission range restrictions), the system designers naturally considered a multi-hop propagation model. The targeted decentralization of the data distribution process in those large-scale condition monitoring and predictive maintenance application areas could lead to economic benefits for the industrial operator and maintenance of some important metrics in the network at good levels, while ensuring that the end-to-end latency is acceptable, without introducing overwhelming costs in the system for the purchase of expensive equipment.

## 3. System Modeling

The network: We consider networks of industrial IoT devices that usually consist of sensor motes, actuators and controller devices. We model those devices as a set of $S = \{s_1, s_2, ..., s_n\}$ nodes, with a total number of $|S| = n$ nodes, deployed in an area of interest $\mathcal{A}$. The central network controller $C$ is set as the first device in the network, $C = s_1$. The communication range $r_u$ of a node $u$ varies according to the requirements of the underlying routing protocol and the constraints of the technological implementation. Nodes $u, v \in S$ are able to communicate with each other if $r_u, r_v \geq \epsilon(u, v)$, where $\epsilon(u, v)$ is the Euclidean distance between $u$ and $v$.

We assume that the controller $C$ is able to maintain centralized network knowledge. This is usual in industrial applications, in which the locations of the nodes are known, traffic flows are deterministic and communication patterns are established a priori. We assume that $C$ knows all the shortest paths in the network and comes with an $n \times n$ matrix $\mathbf{D}$, where $\mathbf{D}_{u,v}$ is the length of the shortest path between nodes $u$ and $v$ (the offline shortest path computation between two nodes is a classic problem in graph theory and can be solved polynomially, using Dijkstra's algorithm [30]). Note that only the control of the data management plane is centralized, while, when using the DML, the data plane itself can be distributed and cooperative, by storing the data in proxies. A proxy $p$ is a node that is able to store data that can be accessed in a timely manner from the consumer nodes of the network. The set of all proxies is denoted as $P$, with $p \in P \subset S$. The DML ensures the effective proxy cooperation so as to achieve the optimal performance of the network according to the objective function defined next. The network controller $C$ is also serving as a proxy, and thus, we have that $|P| \geq 1$ in all cases.

Data production and consumption: In typical industrial applications, like condition monitoring, sensor nodes perform monitoring tasks (producers), and in some cases, their sensor data are needed either by other sensor nodes, which could need additional data to complement their own local measurement, or by actuator nodes, which use the sensor data so as to perform an actuation (consumers). When needed, a consumer $u$ can ask for data of interest using the primitives defined by the underlying routing protocol from a sensor node $v$ (ideally from a proxy $p$, when using the DML). We define the set of consumers as $S_c \subset S$, with $|S_c| = m < n$. When a consumer $u$ needs data, it requests the data via a multi-hop routing path, from the corresponding proxy $p$. When $p$ receives the data request from $u$, it sends the requested data along the same routing path, starting at $p$ and finishing at the consumer $u$. Note that the length of this individual data delivery path is twice the length of the path between $u$ and $p$. We assume that the data generation and access processes are not synchronized. Specifically, we assume that data consumers request data at an unspecified point in time after data have been generated by data producers and transferred to the proxies

The latency constraint: Let $l_{u,v}$ be the single-hop data transmission latency from a node $u \in S$ to another node $v \in S$. We define as access latency $L_{u,p}$ the amount of time required for the data to reach consumer $u$, after $u$'s request, when the data follow a multi-hop propagation between $u$ and $p$. We denote access latency as $L_{u,p} = l_{u,v_1} + ... + l_{v_i,p} + l_{p,v_i} + ... + l_{v_1,u}$. An example of the access latency composition is depicted in Figure 2. We denote as $\overline{L}$ the average access latency across all consumers as the mean value of all the latencies $L_{u,p}$. More specifically, $\overline{L}$ can be denoted as the quantity:

$$\overline{L} = \frac{\sum_{\forall u \in S_c} L_{u,p}}{m}, \quad (1)$$

where $m$ is the number of consumers.

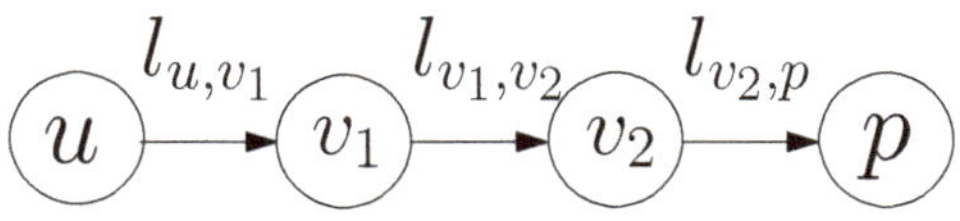

(**a**) Data request and related latencies.

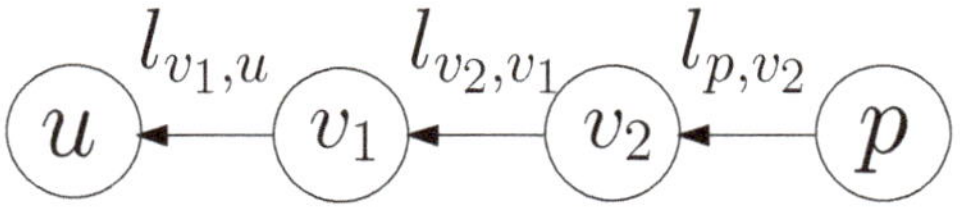

(**b**) Data delivery and related latencies.

**Figure 2.** Example of data access latency. In this case, the data access latency is $L_{u,p} = l_{u,v_1} + l_{v_1,v_2} + l_{v_2,p} + l_{p,v_2} + l_{v_2,v_1} + l_{v_1,u}$.

Industrial applications are typically time-critical, and consequently, the industrial operator requires a maximum data access latency threshold $L_{\max}$. This is an important constraint in the network, and the implementation of a data delivery strategy should ensure that the average multi-hop access latency does not exceed the threshold value. In other words, the following inequality should hold: $\overline{L} \leq L_{\max}$. Note that this formulation is amenable to different purposes. If $L_{u,p}$ is the mean latency between $u$ and $p$, the above inequality guarantees that the average of the mean latencies is below $L_{\max}$. If it is the worst-case latency between $u$ and $v$, the inequality provides a guarantee on the average worst-case latency. In the following, we show that the DML can be used in both cases.

## 4. The Data Management Layer

In order to manage the data distribution process and decrease the average access latency in the network, we consider a DML similar to the one defined in [17], which we recap here for the reader's convenience. The basic function of the DML is the decoupling of the data management plane from the network plane, as shown in Figure 3. The DML provides solutions for the selection of some nodes that will act as proxies and the establishment of an efficient method for data distribution and delivery, using the proxies. More specifically, the role of the DML is to define a set $P \subset S$, the elements of which are the selected proxies. The number of the proxies can range from 1–$n-1$. The case of one proxy is equivalent to having only the controller $C$ operating as a single point of data distribution. In this case, the data distribution is functioning as in traditional industrial IoT environments.

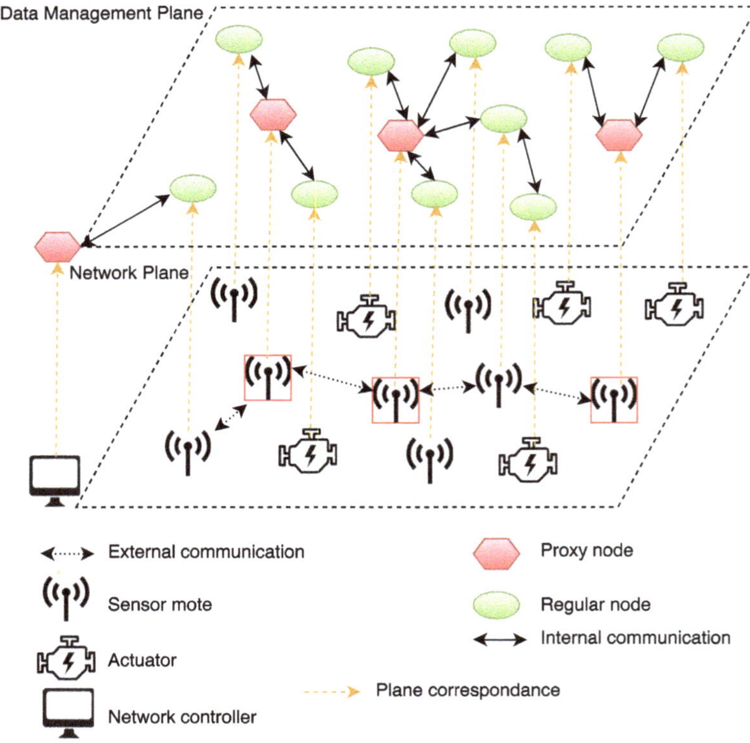


**Figure 3.** The Data Management Layer (DML): decoupling the data management plane from the network plane.

This demarcated model of data exchanges can be formulated as a publish/subscribe (pub/sub) model [31]. In a pub/sub model, a consumer subscribes to data, i.e., denotes interest for it to the corresponding proxy, and the relevant producer publishes advertisements to the proxy. The DML assumes that the pub/sub process is regulated at the central controller $C$, which maintains knowledge on the sets of producers, consumers and requests. Based on this, $C$ can find an appropriate set of proxies based on the algorithm we present next. Inside the network, the proxies are responsible for matching subscriptions with publications, i.e., they provide a rendezvous function for storing the available data according to the corresponding subscriptions. The producers do not hold references to the consumers, neither do they know how many consumers are receiving their generated data.

The selection of the proxies should be done balancing two requirements. On the one hand, the number of proxies should be sufficient to make sure each consumer finds data "close enough" to guarantee that $\overline{L} \leq L_{\max}$. On the other hand, as the role of proxy implies a resource burden on the selected nodes, their number should be as low as possible. The proxy selection problem can thus be formulated as an integer program. More specifically, given a set $S$ of nodes, a set $S_c \subset S$ of consumers and an access latency threshold $L_{\max}$, the network designer should target the minimization of the number of proxies needed in the network so as to guarantee $\overline{L} \leq L_{\max}$. We define two sets of decision variables, (a) $x_p = 1$, if $p \in S$ is selected as proxy and zero otherwise, and (b) $y_{u,p} = 1$, if consumer $u \in S$ is assigned to proxy $p \in S$ and zero otherwise. Then, the integer program formulation is the following:

$$\text{Min.:} \quad \sum_{p \in S} x_p \tag{2}$$

$$\text{S. t.:} \quad \sum_{u \in S_c} \sum_{p \in S} \frac{L_{u,p} \cdot y_{u,p}}{m} \leq L_{\max} \tag{3}$$

$$\sum_{p \in S} y_{u,p} = 1 \qquad \forall u \in S_c \tag{4}$$

$$y_{u,p} \leq x_p \qquad \forall u \in S_c, \forall p \in S \tag{5}$$

$$x_p, y_{u,p} \in \{0,1\} \qquad \forall u \in S_c, \forall p \in S \tag{6}$$

The objective function (2) minimizes the number of proxies (note that in various industrial scenarios, some nodes might be too weak to perform any other operations than generating and propagating a minimal set of data. The problem formulation that represents those scenarios is a special case of the problem formulation that we consider in this paper, with $p \in S'$, where $S' \subset S$). Constraint (3) guarantees that $\overline{L} \leq L_{\max}$. Constraints (4) guarantee that each node has to be assigned to one and only one proxy. Constraints (5) guarantee that nodes can be assigned only to proxies. Constraints (6) guarantee that all nodes are considered for potentially being selected as proxies and that all nodes requesting data are assigned to a proxy.

As we have already shown in [17], the proxy selection problem is computationally intractable, since it can be formulated as an integer program. This means that it is impossible to calculate in polynomial time the minimum proxies needed optimally while staying below $L_{\max}$. Differently from [17], in this case, the formulation of the problem considers the latency of communication $L_{u,p}$, and not an abstract number of hops. This makes it more realistic for industrial environments, but an even more difficult problem, as it becomes also infeasible to assign the real values to $L_{u,p}$ of Constraints (3). This is due to the fact that we are not able to know the exact values of the individual transmission latencies $l_{u,v}$, before they happen. To address this issue, we introduce the `ProxySelection+` algorithm (Algorithm 1), which takes into account latencies $L_{u,p}$, instead of the number of hops. `ProxySelection+` is a myopic algorithm, which does not give the optimal solution. The use of simple heuristics like the one in `ProxySelection+` shows that the DML is able to outperform the traditional centralized methods, even when adopting simple methods.

`ProxySelection+` sets the controller $C$ as the first proxy of the network, and it gradually increases the number of proxies (counter) until it reaches a number with which the average access latency $\overline{L}$ does not violate the maximum latency threshold $L_{\max}$. In every iteration (Lines 5–9), the exact selection of the next proxy in the network is performed using a myopic greedy addition (Lines 6–8). Each candidate node is examined, and the one whose addition to the current solution reduces the average access latency the most is added to the incumbent solution. To this end, the latency between a candidate proxy ($k$ in Line 7) and a consumer ($u$ in Line 7) is estimated as the length of the shortest path $\mathbf{D}_{k,u}$ that is connecting them multiplied by the expected latency on each hop ($l^{(h)}$ in Line 7). $l^{(h)}$ needs to be initiated through preliminary measurements. This happens through an initialization phase (Line 1) during which the network designer measures different single-hop data transmission latencies within the industrial installation and gathers a sufficiently representative dataset of $l_{u,v}$ measurements from different pairs of nodes $u, v \in S$ across the network. By using the mean of measured latencies, we obtain a guarantee on the average mean latency. By using the highest measured value, we obtain a constraint on the average worst-case latency, implementing the guarantee explained in Section 3. The computational complexity of `ProxySelection+` is polynomial with a worst case time of $\mathcal{O}(V^4)$. However, this worst-case performance is very difficult to experience in practice, since in order to have $\mathcal{O}(V^4)$ time, all $n$ nodes of the network have to be chosen as proxies; something that is highly unlikely.

**Algorithm 1:** `ProxySelection+`.

**Input** : $S, S_c, r_u, L_{\max}$
1 $l^{(h)} \leftarrow$ assign value through initialization phase at the industrial installation
2 $\mathbf{D}_{u,v} = \text{Dijkstra}(S, r_u), \forall u, v \in S$ [30]
3 $P = \{C\}$
4 counter $= 1$
5 **while** counter $< n$ **and** $\overline{L} > L_{max}$ **do**
6     **for** $i = 2$ : counter **do**
7         $p = \arg\min_{v \in S} \sum_{u \in S_c} \min_{k \in P \cup \{v\}} \frac{l^{(h)} \cdot \mathbf{D}_{k,u}}{m}$
8         $P = P \cup \{p\}$
9     counter $++$
**Output:** $P$

## 5. Implementation and Experimental Evaluation

### 5.1. Experimental Strategy

Strategic purpose: The strategic purpose of the experimental evaluation with real devices is to provide a realistic demonstration of how efficient data management methods can significantly improve the data access latency in industrial IoT networks, by using a limited number of proxies. The realistic approach is of paramount importance in our implementation strategy. For this reason, we follow some important steps. For the experimental implementation and evaluation, we use the Euratech testbed from the FIT IoT-LAB platform [14]. We use a network with technical specifications representative of the industrial IoT paradigm (e.g., low-power nodes, IEEE 802.15.4 radio interface, large number of devices, etc.). We carefully choose the $L_{\max}$ threshold, according to actual industrial requirements and expert groups' recommendations. In order to have a benchmark for the performance of our method, we also implement two additional representative methods, based on routing mechanisms that are usual in current industrial IoT networks. We vary several experimental parameters so as to investigate the performance consistency of our method under different settings. Finally, we validate a simulation model based on the real-world settings, with which we can further investigate the changing parameters that are impossible or too time-consuming to investigate on the testbed.

Experiment design: We use a total number of $n = 95$ nodes in the Euratech testbed, which form a 2D horizontal grid, as shown in Figure 4a. Occasionally, during the experiments, there are some dead nodes, that is nodes that have run out of available power and are not able to function. This occasional unavailability of a subset of nodes renders the experiment even more realistic, since dead node situations frequently occur in real industrial IoT networks. The nodes that were used in the experiments are WSN430 open nodes, the design of which is displayed in Figure 4b. The WSN430 open node is a mote based on a low-power MSP430-based platform, with a set of standard sensors and an IEEE 802.15.4 radio interface at 2.4 GHz, using a CC2420 antenna [32], which can support, e.g., WirelessHART settings, typical of industrial communications. We used the TinyOS configuration for CC2420, which uses a MAC protocol that is compatible with 802.15.4 and is in principle a CSMA/CA scheme. We programmed and operated the nodes under TinyOS 2.1.1, a reference operating system for sensor nodes.

(a) Euratech testbed [14].

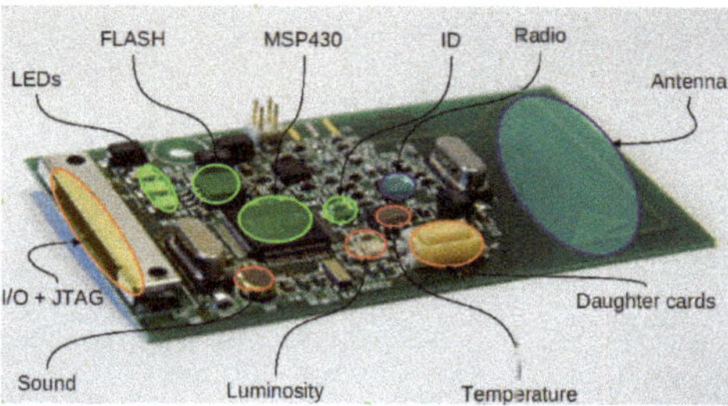


(b) WSN430 open node.

**Figure 4.** Experimental setup.

Since the testbed nodes are placed a short distance from each other, we adjust their transmission range, so as to obtain a realistic multi-hop topology. We configured the antenna TX power such that, according to the CC2420 antenna datasheet [32] and the measurements provided in [33], the transmission range is about 3 m. However, given that this value has been measured in ideal conditions, without taking into account external factors such as obstacles and interference, we program every node $u \in S$ to consider as a neighbor every other node $v \in S$ with $\epsilon(u,v) \leq 1$ m. Given this configuration, we obtain the topology that is depicted in Figure 5a. Note that the three "gaps" in the topology are a result of the dead nodes of the deployment, which are unable to communicate with other nodes. We set the percentage of requesting nodes to $m = 0.1 \cdot |S|$, selected uniformly at random from $S$, and we set $C = s_1$ as the central network controller, which corresponds to the node with `node_id` $= 1$ in the Euratech testbed (lower left node in Figure 5a).

Setting the $L_{\max}$ threshold: In order to perform the experiments in the most realistic way, it is important that the $L_{\max}$ value be aligned with the official communication requirements of future network-based communication solutions for Industry 4.0, for the targeted industrial applications. Both the WG1 of Plattform Industrie 4.0 (reference architectures, standards and norms) [34] and the Expert Committee 7.2 of ITG (radio systems) set the latency requirements for condition monitoring applications to 100 ms. However, in order to provide a complete and diverse set of results, we also measure the performance of our method for different values of $L_{\max}$.

Performance benchmarks: In order to measure the performance of the DML with respect to traditional industrial IoT alternatives, we implement two additional data delivery strategies. The first method is the most traditional data delivery strategy in current industrial IoT environments and imposes that all data requests and data deliveries are being routed through the controller $C$. More specifically, the request is routed from consumer $u$ to $C$ and then from $C$ to producer $v$. At the next step, the data are routed from $v$ again to $C$ and then from $C$ to $u$. We call this mode of operation non-storing mode, and it is obvious that it is completely centralized and not cooperative. Note that this would be the simplest data management approach that can be implemented in relevant routing mechanisms like RPL (IPv6 Routing Protocol for Low-Power and Lossy Networks [35]), where intermediate nodes are not allowed to cache data (thus, the RPL terminology non-storing mode).

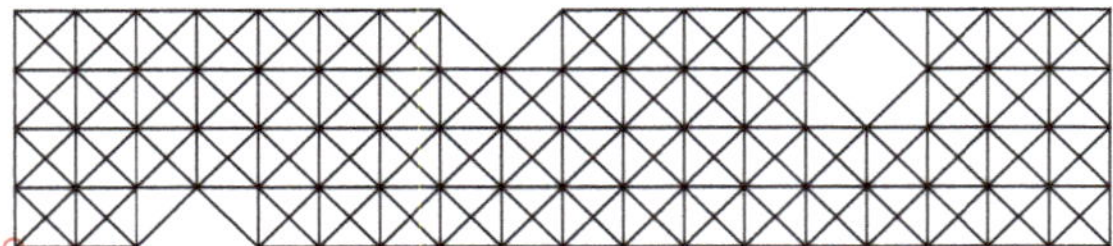

(**a**) Network topology of the Euratech testbed. In the lower left corner, the network controller C is visible as a red dot.

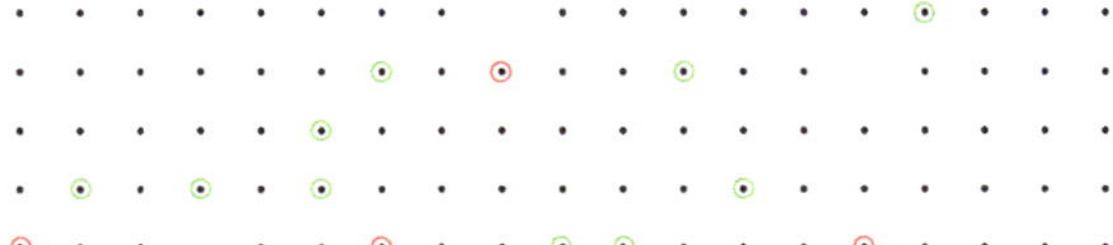

(**b**) Output of the `ProxySelection+` algorithm. The green dots represent the consumers, and the red dots represent the proxies.

**Figure 5.** Euratech topology and `ProxySelection+` output.

The second method is another, less commonly used in industrial IoT settings, but nevertheless useful alternative. It imposes that all data requests and data deliveries are being routed through the Lowest Common Ancestor (LCA) of the routing tree, routed at the controller $C$. The LCA of two nodes $u$ and $v$ in the routing tree is the lowest (i.e., deepest) node that has both $u$ and $v$ as descendants. We call this mode of operation storing mode, because the LCAs should store additional information about their descendants, and it is obvious that it is a distributed alternative. Again, this is the simplest method that one would implement with routing mechanisms like RPL in storing mode, i.e., when intermediate nodes between communication endpoints are allowed to cache content. Storing mode thus provides a distributed method.

We made those choices after careful consideration of the current realistic industrial networking status-quo. The selected protocols are standardized components of a reference and well-established communications and data management stack. In fact, they are considered the state-of-the-art, for current and future wireless industrial applications, as discussed extensively in [15,16]. More specifically, the stack is presented in detail in [15] and is considered as the de facto standard for future industrial networks. In the following, for convenience, we use the term "special nodes" when we refer to the network controller, the proxies or the LCAs.

### 5.2. Experimental Results

Running the `ProxySelection+` algorithm: We ran the initialization phase of `ProxySelection+`, so as to assign values to $l^{(h)}$, by measuring times that are needed for the data exchange of a sensor measurement from a sensor node to another. We measured the time needed for the sensor reading to be sent and received from one node to another. This latency includes time spent in the back-off phase (which cannot be predicted), time spent in sending the signal over the radio, and time spent during the propagation. We consider the propagation latency negligible, since radio waves are traveling very fast and we are not able to measure the time elapsed using the nodes' timers. In order to obtain reliable results, we repeated the propagation measurements for different pairs of transmitting and receiving nodes of the Euratech testbed, 30 times for each pair. We concluded with the measurements that are shown in the Table 1 (highest, lowest, mean value and standard deviation), after measuring the relevant times using WSN430 with CC2420 and TinyOS. We can see that the latency values of data propagation from one node to another significantly vary. While the lowest latency could be

13 ms, the highest propagation latency $l^{(\max)}$ was 23 ms. The mean latency $l^{(\text{mean})}$ of the values collected from the repetition of this experiment was 17.4 ms (other sources of latency related, e.g., to computation at the receivers have been found to be in the order of $\mu$s and, therefore, are neglected. Furthermore, the measuring methodology we used does not depend on the specific conditions under which these measures are taken). After running `ProxySelection+` with $L_{\max} = 100$ ms, $m = 0.1 \cdot |S|$ and $l^{(h)} = l^{(\text{mean})}$, we get the proxy placement that is depicted in Figure 5b. We can easily see that `ProxySelection+` is balancing the proxy allocation in the network, so as to guarantee a small data access latency to all the requesting nodes.

**Table 1.** Measured send/receive latency.

| Type of Measured Latency | Notation | Value (ms) | $\sigma$ |
|---|---|---|---|
| **Highest latency reported** | $l^{(\max)}$ | 23 | |
| **Mean latency** | $l^{(\text{mean})}$ | 17.4 | 3.2 |
| **Lowest latency reported** | - | 13 | |

Increasing and decreasing the number of proxies: Figure 6a displays the average access latency $\overline{L}$ for different numbers of proxies in the network. In order to obtain this plot, we run `ProxySelection+`, and we gradually add and remove proxies, so as to investigate the effect of changing the number of proxies on $\overline{L}$. In the case where we set $l^{(h)} = l^{(\text{mean})}$, the DML ensures that $\overline{L}$ will not surpass $L_{\max}$, by assigning four proxies in selected positions. If we further decrease the number of proxies, we have that $\overline{L} > L_{\max}$, and the latency constraint is not met. At the leftmost point of the plot, we can see the latency achieved when using only one proxy (the controller $C$, or in other words, when the DML functionalities are absent), which is much higher than when employing additional proxies. When we replace the value of $l^{(h)}$ with $l^{(h)} = l^{(\max)}$ and we re-run the algorithm, we observe similar behavior in the performance, but in this case, with eight selected proxies.

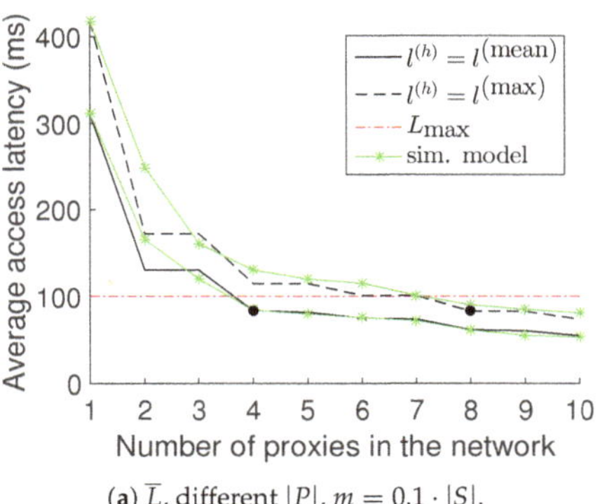


(**a**) $\overline{L}$, different $|P|$, $m = 0.1 \cdot |S|$.

**Figure 6.** *Cont.*

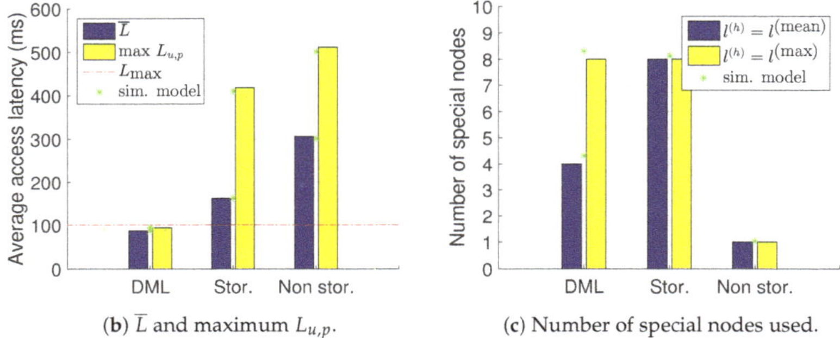


(**b**) $\overline{L}$ and maximum $L_{u,p}$. (**c**) Number of special nodes used.

**Figure 6.** Experimental results in the IoT-LAB Euratech testbed. The validation of the simulation model is displayed in green.

$\overline{L}$ achieved: Figure 6b displays the results on the average access latency for the three alternative methods. The yellow bar for the DML method is the $\overline{L}$ value when we consider the worst case of $l^{(h)} = l^{(\max)}$. This is an important point to make, as the figure shows that, by adapting the number of proxies, DML is able to always guarantee the constraint, irrespective of whether $l^{(h)}$ is formulated as an average of mean latencies or as an average of worst-case latencies. We can see that the efficient management of proxies provided by the DML results in a better performance compared to the other two alternatives. This fact is explained by the nature of `ProxySelection+`, which receives as input the $L_{\max}$.

Number of proxies used: We compare the three methods with respect to the number of special nodes that they use. The DML is using proxies; the non storing mode is using the controller $C$; and the storing mode is using LCAs. The use of special nodes is wasteful of resources. For example, the proxies store the data requested and the correspondence of producers and consumers, and the LCAs hold routing information about their descendants. In Figure 6c, we can see that the DML is performing really well compared to the storing mode and uses less special nodes. Of course, the non-storing mode is using just one special node, but this has a severe impact on the latency achieved, as shown in Figure 6b. Even when the DML uses more proxies to guarantee worst-case latencies, their number is comparable to the case of storing mode. However, the DML drastically reduces the latency in this case, thus achieving a much more efficient use of proxies.

## 6. Large-Scale Simulations

The testbed environment gives us an important ability to test the methods on real conditions and derive useful indications. However, at the same time, it does not allow us to perform larger scale, or variable experiments, easily and fast. For this reason, we developed a simulation model based on the system modeling presented in Section 3. The simulation environment we use is MATLAB. We verify that the simulations are meaningful via validation, by comparing the results obtained with the simulation model to those of the testbed experiments, and then, we extend our performance evaluation through simulations.

### 6.1. Validation of the Simulation Model and Simulation Settings

We constructed, in simulation, instances similar to the one that was tested in the Euratech testbed. The results obtained are displayed with green color in Figure 6. It is clear that the results obtained by the simulation model are very similar to the results obtained during the real experiment, and therefore, we can extract reliable conclusions from the simulation environment.

Figure 7a displays a typical network deployment of 500 nodes, with the corresponding wireless links and with the controller $C$ lying on the far right edge of the network, depicted as a red circle.

Figure 7b displays the locations of the final set $P$ of proxies depicted as red circles after running ProxySelection+. The spatial display of Figure 7b shows that the final selection results in a balanced proxy selection, ensuring that even isolated nodes, which are located near sparse areas of the network, also have access to a proxy.

In the simulations, we focus on showcasing different aspects of the data management and distribution process. We construct larger and different deployments and topologies than the ones of the Euratech testbed; we investigate different values of $L_{\max}$; we consider diverse percentages of requesting nodes; and we also measure the energy consumption. The deployment area $\mathcal{A}$ is set to be circular, and the nodes are deployed uniformly at random. We construct networks of different numbers of nodes, inserting the additional nodes in the same network area and at the same time decreasing the communication range $r_u$ appropriately, so as to maintain a single strongly-connected component at all times. An example of a generated network of 500 nodes is depicted in Figure 7a. In the following, we present results where $l^{(h)}$ is measured as the mean of latencies. Figure 7c shows the value of $\overline{L}$ obtained in the case of Figure 7a, qualitatively confirming the results shown in Figure 6a.

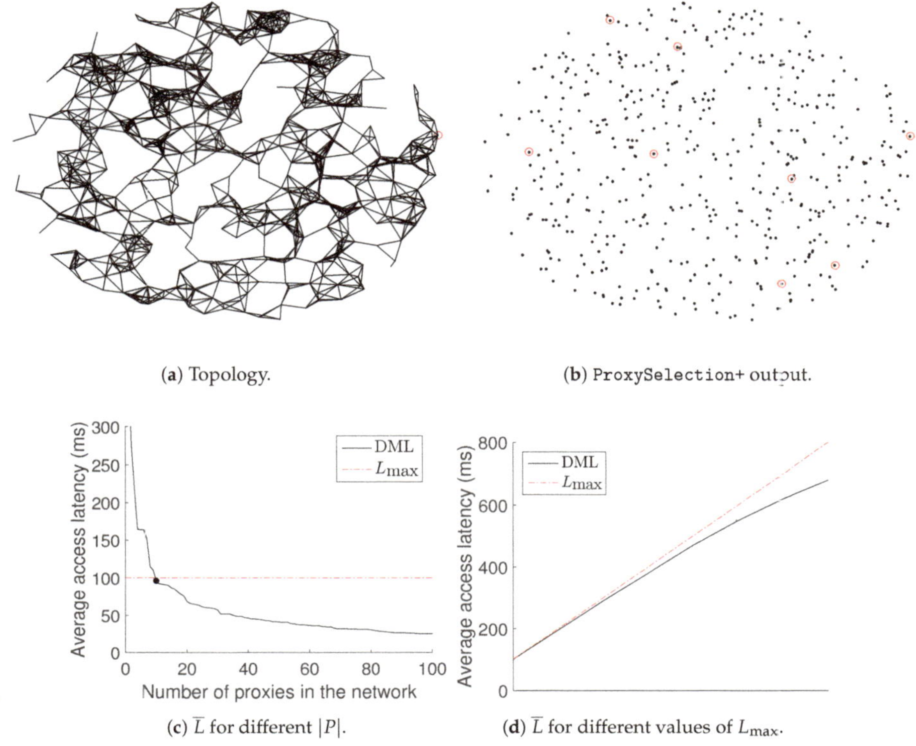


(a) Topology. (b) ProxySelection+ output.

(c) $\overline{L}$ for different $|P|$. (d) $\overline{L}$ for different values of $L_{\max}$.

**Figure 7.** Network with $n = 500$, $m = 0.4 \cdot |S|$.

## 6.2. Simulation Results

Different values of $L_{\max}$: We tested the performance of the DML for different values of $L_{\max}$, in networks of 500 nodes, with $m = 0.4 \cdot |S|$. The results are shown in Figure 7d. The red points represent the values for the maximum latency threshold $L_{\max}$ provided by the industrial operator. The average access latency achieved by the DML is always below the threshold, due to the provisioning of the ProxySelection+ algorithm. In fact, we can see that the more the value of $L_{\max}$ is increased, the larger the difference between $\overline{L}$ and $L_{\max}$ becomes. This happens because for higher $L_{\max}$ values, the latency constraint is more relaxed, and lower $\overline{L}$ can be achieved more easily.

Number of proxies used: We compare the three methods with respect to the number of special nodes that they use. As usual, the DML is using proxies; the non-storing mode is using the controller $C$; and the storing mode is using LCAs. As we mentioned earlier, the use of special nodes is wasteful of resources. In Figure 8a, we can see that the DML is performing really well compared to the storing mode and uses much less special nodes. Of course, the non-storing mode is using just one special node for any network size, but this has a severe impact on the latency achieved.

Different percentages of requesting nodes: Another possible factor that could affect the $\overline{L}$ achieved is the percentage of consumers. In Figure 8b, we can see that $\overline{L}$ remains constant for any percentage of requesting nodes, in all three alternatives. This shows that DML is able to automatically adapt the number of proxies, so as to guarantee the latency constraints irrespective of the number of consumers.

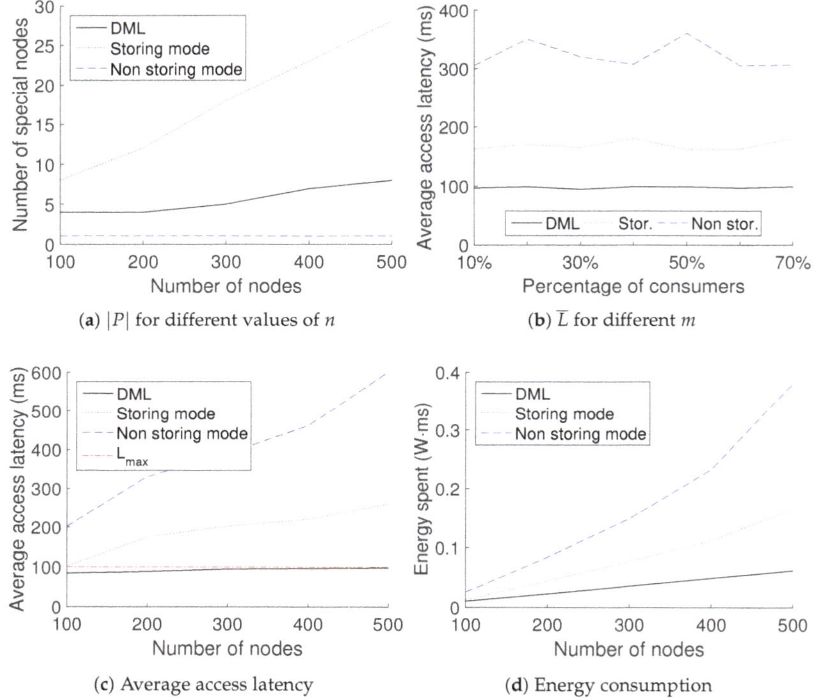


**Figure 8.** Comparison of the three methods.

Average latency achieved: Figure 8c displays the results on the average access latency for the three alternatives, for different numbers of nodes in the network. We can see that the efficient management of proxies provided by the DML results in a better performance compared to the other two alternatives. $\overline{L}$ achieved by the DML respects the latency constraint and always remains lower than $L_{\max}$ (red line).

Energy consumption: Another aspect that we can easily evaluate in the simulation is the energy cost in terms of communication, related to data access. We evaluate this as the cost of transmissions required to serve consumers' requests. In order to obtain the desired results in units of energy, we transform the dBm units provided in the CC2420 datasheet [32] to mW and we multiply with the time that each node of the network is operational. Figure 8d displays the energy consumption in the entire network for the three alternatives, for different numbers of nodes. The energy consumption for communication is lower in the case of the DML because low latency comes with less transmissions in the network, resulting in fewer energy demands.

## 7. Conclusions and Future Work

In this paper, we efficiently regulate the data distribution in industrial IoT networks using proxies. Given a set of data, the set of consumer nodes and the maximum access latency that consumers can tolerate, we consider a method for identifying a limited set of proxies in the network where data are cached. We implement the method and evaluate its performance using the IoT-LAB testbed. Additionally, we validate a simulation model and use it for performance evaluation in larger scales and more general topologies. We demonstrate that the proposed method guarantees average access latency below the given threshold and outperforms traditional centralized and even distributed approaches. The next step is to take into account limited bandwidth in the network, which can lead to congestive collapse, when incoming traffic exceeds outgoing bandwidth.

**Author Contributions:** Conceptualization, T.P.R., A.P. and M.C. Methodology, T.P.R., A.P. and M.C. Software, T.P.R. Validation, T.P.R., A.P. and M.C. Formal analysis, T.P.R. Investigation, T.P.R. Writing, original draft preparation, T.P.R. Writing, review and editing, T.P.R., A.P. and M.C. Visualization, T.P.R. Supervision, M.C. Project administration, A.P. and T.P.R. Funding acquisition, A.P. and M.C.

**Funding:** This work has been partly funded by the European Commission through the FoF-RIA Project AUTOWARE: Wireless Autonomous, Reliable and Resilient Production Operation Architecture for Cognitive Manufacturing (No. 723909).

**Conflicts of Interest:** The authors declare no conflict of interest.

## References

1. Drath, R.; Horch, A. Industrie 4.0: Hit or Hype? [Industry Forum]. *IEEE Ind. Electron. Mag.* **2014**, *8*, 56–58. doi:10.1109/MIE.2014.2312079. [CrossRef]
2. Huang, V.K.L.; Pang, Z.; Chen, C.J.A.; Tsang, K.F. New Trends in the Practical Deployment of Industrial Wireless: From Noncritical to Critical Use Cases. *IEEE Ind. Electron. Mag.* **2018**, *12*, 50–58. doi:10.1109/MIE.2018.2825480. [CrossRef]
3. Lucas-Estan, M.C.; Raptis, T.P.; Sepulcre, M.; Passarella, A.; Regueiro, C.; Lazaro, O. A software defined hierarchical communication and data management architecture for industry 4.0. In Proceedings of the 2018 14th Annual Conference on Wireless On-demand Network Systems and Services (WONS), Isola, France, 6–8 February 2018; pp. 37–44, doi:10.23919/WONS.2018.8311660. [CrossRef]
4. Molina, E.; Lazaro, O.; Sepulcre, M.; Gozalvez, J.; Passarella, A.; Raptis, T.P.; Ude, A.; Nemec, B.; Rooker, M.; Kirstein, F.; et al. The AUTOWARE Framework and Requirements for the Cognitive Digital Automation. In Proceedings of the 18th IFIP WG 5.5 Working Conference on Virtual Enterprises (PRO-VE), Vicenza, Italy, 18–20 September 2017.
5. Gaj, P.; Jasperneite, J.; Felser, M. Computer Communication Within Industrial Distributed Environment—Survey. *IEEE Trans. Ind. Inf.* **2013**, *9*, 182–189. doi:10.1109/TII.2012.2209668. [CrossRef]
6. Pang, Z.; Luvisotto, M.; Dzung, D. Wireless High-Performance Communications: The Challenges and Opportunities of a New Target. *IEEE Ind. Electron. Mag.* **2017**, *11*, 20–25. doi:10.1109/MIE.2017.2703603. [CrossRef]
7. Han, S.; Zhu, X.; Mok, A.K.; Chen, D.; Nixon, M. Reliable and Real-Time Communication in Industrial Wireless Mesh Networks. In Proceedings of the 2011 17th IEEE Real-Time and Embedded Technology and Applications Symposium, Chicago, IL, USA, 11–14 April 2011; pp. 3–12, doi:10.1109/RTAS.2011.9. [CrossRef]
8. Steiner, W.; Poledna, S. Fog computing as enabler for the Industrial Internet of Things. *Elektrotech. Inf.* **2016**, *133*, 310–314. doi:10.1007/s00502-016-0438-2. [CrossRef]
9. *IEEE Standard for Local and Metropolitan Area Networks—Part 15.4: Low-Rate Wireless Personal Area Networks (LR-WPANs) Amendment 1: MAC Sublayer*; IEEE Std 802.15.4e-2012 (Amendment to IEEE Std 802.15.4-2011); IEEE: Piscataway, NJ, USA, 2012; pp. 1–225, doi:10.1109/IEEESTD.2012.6185525. [CrossRef]
10. Chen, D.; Nixon, M.; Mok, A. *WirelessHART$^{TM}$*; Springer: Berlin, Germany, 2010.

11. Mahmud, R.; Kotagiri, R.; Buyya, R. Fog Computing: A Taxonomy, Survey and Future Directions. In *Internet of Everything: Algorithms, Methodologies, Technologies and Perspectives*; Di Martino, B., Li, K.C., Yang, L.T., Esposito, A., Eds.; Springer: Singapore, 2018; pp. 103–130, doi:10.1007/978-981-10-5861-5_5. [CrossRef]
12. Xu, L.D.; He, W.; Li, S. Internet of Things in Industries: A Survey. *IEEE Trans. Ind. Inf.* **2014**, *10*, 2233–2243. doi:10.1109/TII.2014.2300753. [CrossRef]
13. Rodriguez, P.; Sibal, S. SPREAD: Scalable platform for reliable and efficient automated distribution. *Comput. Netw.* **2000**, *33*, 33–49. doi:10.1016/S1389-1286(00)00086-4. [CrossRef]
14. Adjih, C.; Baccelli, E.; Fleury, E.; Harter, G.; Mitton, N.; Noel, T.; Pissard-Gibollet, R.; Saint-Marcel, F.; Schreiner, G.; Vandaele, J.; et al. FIT IoT-LAB: A large scale open experimental IoT testbed. In Proceedings of the 2015 IEEE 2nd World Forum on Internet of Things (WF-IoT), Milan, Italy, 14–16 December 2015; pp. 459–464, doi:10.1109/WF-IoT.2015.7389098. [CrossRef]
15. Watteyne, T.; Handziski, V.; Vilajosana, X.; Duquennoy, S.; Hahm, O.; Baccelli, E.; Wolisz, A. Industrial Wireless IP-Based Cyber-Physical Systems. *Proc. IEEE* **2016**, *104*, 1025–1038. doi:10.1109/JPROC.2015.2509186. [CrossRef]
16. Guglielmo, D.D.; Brienza, S.; Anastasi, G. IEEE 802.15.4e: A survey. *Comput. Commun.* **2016**, *88*, 1–24. doi:10.1016/j.comcom.2016.05.004. [CrossRef]
17. Raptis, T.P.; Passarella, A. A distributed data management scheme for industrial IoT environments. In Proceedings of the 2017 IEEE 13th International Conference on Wireless and Mobile Computing, Networking and Communications (WiMob), Rome, Italy, 9–11 October 2017; pp. 196–203, doi:10.1109/WiMOB.2017.8115846. [CrossRef]
18. Heo, J.; Hong, J.; Cho, Y. EARQ: Energy Aware Routing for Real-Time and Reliable Communication in Wireless Industrial Sensor Networks. *IEEE Trans. Ind. Inf.* **2009**, *5*, 3–11. doi:10.1109/TII.2008.2011052. [CrossRef]
19. Kim, D.; Wang, W.; Sohaee, N.; Ma, C.; Wu, W.; Lee, W.; Du, D.Z. Minimum Data-latency-bound K-sink Placement Problem in Wireless Sensor Networks. *IEEE/ACM Trans. Netw.* **2011**, *19*, 1344–1353. doi:10.1109/TNET.2011.2109394. [CrossRef]
20. Raptis, T.P.; Passarella, A.; Conti, M. Distributed Path Reconfiguration and Data Forwarding in Industrial IoT Networks. In Proceedings of the 16th IFIP International Conference on Wired/Wireless Internet Communications (WWIC), Shanghai, China, 7–11 May 2018.
21. Antonopoulos, C.; Panagiotou, C.; Keramidas, G.; Koubias, S. Network driven cache behavior in wireless sensor networks. In Proceedings of the 2012 IEEE International Conference on Industrial Technology, Hong Kong, China, 10–13 December 2012; pp. 567–572, doi:10.1109/ICIT.2012.6209999. [CrossRef]
22. Panagiotou, C.; Antonopoulos, C.; Koubias, S. Performance enhancement in WSN through data cache replacement policies. In Proceedings of the 2012 IEEE 17th International Conference on Emerging Technologies Factory Automation (ETFA 2012), Krakow, Poland, 17–21 September 2012; pp. 1–8, doi:10.1109/ETFA.2012.6489575. [CrossRef]
23. Saifullah, A.; Xu, Y.; Lu, C.; Chen, Y. End-to-End Communication Delay Analysis in Industrial Wireless Networks. *IEEE Trans. Comput.* **2015**, *64*, 1361–1374. doi:10.1109/TC.2014.2322609. [CrossRef]
24. Li, J.; Zhu, X.; Gao, X.; Wu, F.; Chen, G.; Du, D.Z.; Tang, S. A Novel Approximation for Multi-hop Connected Clustering Problem in Wireless Sensor Networks. In Proceedings of the 2015 IEEE 35th International Conference on Distributed Computing Systems, Columbus, OH, USA, 29 June–2 July 2015; pp. 696–705, doi:10.1109/ICDCS.2015.76. [CrossRef]
25. Raptis, T.P.; Passarella, A.; Conti, M. Maximizing industrial IoT network lifetime under latency constraints through edge data distribution. In Proceedings of the 2018 IEEE Industrial Cyber-Physical Systems (ICPS), Saint Petersburg, Russia, 15–18 May 2018; pp. 708–713, doi:10.1109/ICPHYS.2018.8390794. [CrossRef]
26. Luo, Y.; Duan, Y.; Li, W.; Pace, P.; Fortino, G. Workshop Networks Integration Using Mobile Intelligence in Smart Factories. *IEEE Commun. Mag.* **2018**, *56*, 68–75. doi:10.1109/MCOM.2018.1700618. [CrossRef]
27. Luo, Y.; Duan, Y.; Li, F.W.; Pace, P.; Fortino, G. A Novel Mobile and Hierarchical Data Transmission Architecture for Smart Factories. *IEEE Trans. Ind. Inf.* **2018**, 14, 3534–3546. doi:10.1109/TII.2018.2824324. [CrossRef]
28. Zats, S.; Su, R.; Watteyne, T.; Pister, K.S.J. Scalability of Time Synchronized wireless sensor networking. In Proceedings of the IECON 2011—37th Annual Conference of the IEEE Industrial Electronics Society, Melbourne, Australia, 7–10 November 2011; pp. 3011–3016, doi:10.1109/IECON.2011.6119789.

[CrossRef]

29. Krishnamurthy, L.; Adler, R.; Buonadonna, P.; Chhabra, J.; Flanigan, M.; Kushalnagar, N.; Nachman, L.; Yarvis, M. Design and Deployment of Industrial Sensor Networks: Experiences from a Semiconductor Plant and the North Sea. In Proceedings of the 3rd International Conference on Embedded Networked Sensor Systems, San Diego, CA, USA, 2–4 November 2005; ACM: New York, NY, USA, 2005; pp. 64–75, doi:10.1145/1098918.1098926. [CrossRef]
30. Dijkstra, E.W. A note on two problems in connexion with graphs. *Numer. Math.* **1959**, *1*, 269–271. doi:10.1007/BF01386390. [CrossRef]
31. Eugster, P.T.; Felber, P.A.; Guerraoui, R.; Kermarrec, A.M. The Many Faces of Publish/Subscribe. *ACM Comput. Surv.* **2003**, *35*, 114–131. doi:10.1145/857076.857078. [CrossRef]
32. CC2420 Datasheet. Available online: http://www.ti.com/lit/ds/symlink/cc2420.pdf (accessed on 14 November 2017).
33. Kotian, R.; Exarchakos, G.; Liotta, A. Data Driven Transmission Power Control for Wireless Sensor Networks. In Proceedings of the 8th International Conference on Internet and Distributed Computing Systems, Windsor, UK, 2–4 September 2015; Springer: New York, NY, USA, 2015; Volume 9258, pp. 75–87, doi:10.1007/978-3-319-23237-9_8. [CrossRef]
34. Network-Based Communication for Industrie 4.0. Publications of Plattform Industrie 4.0. 2016. Available online: www.plattform-i40.de (accessed on 14 November 2017).
35. Winter, T.; Thubert, P.; Brandt, A.; Hui, J.; Kelsey, R.; Levis, P.; Pister, K.; Struik, R.; Vasseur, J.; Alexander, R. *RPL: IPv6 Routing Protocol for Low-Power and Lossy Networks*; RFC 6550; RFC Editor: Fremont, CA, USA, 2012.

*Article*

# A Virtual Reality Soldier Simulator with Body Area Networks for Team Training

**Yun-Chieh Fan [1,2] and Chih-Yu Wen [2,*]**

1 Simulator Systems Section, Aeronautical System Research Division, National Chung-Shan Institute of Science and Technology, Taichung 407, Taiwan; d105064101@mail.nchu.edu.tw
2 Department of Electrical Engineering, Innovation and Development Center of Sustainable Agriculture (IDCSA), National Chung Hsing University, Taichung 402, Taiwan
* Correspondence: cwen@dragon.nchu.edu.tw; Tel.: +886-04-2285-1549

Received: 16 November 2018; Accepted: 18 January 2019; Published: 22 January 2019

**Abstract:** Soldier-based simulators have been attracting increased attention recently, with the aim of making complex military tactics more effective, such that soldiers are able to respond rapidly and logically to battlespace situations and the commander's decisions in the battlefield. Moreover, body area networks (BANs) can be applied to collect the training data in order to provide greater access to soldiers' physical actions or postures as they occur in real routine training. Therefore, due to the limited physical space of training facilities, an efficient soldier-based training strategy is proposed that integrates a virtual reality (VR) simulation system with a BAN, which can capture body movements such as walking, running, shooting, and crouching in a virtual environment. The performance evaluation shows that the proposed VR simulation system is able to provide complete and substantial information throughout the training process, including detection, estimation, and monitoring capabilities.

**Keywords:** virtual reality; body area network; training simulator

## 1. Introduction

In recent years, since virtual reality (VR) training simulators allow soldiers to be trained with no risk of exposure to real situations, the development of a cost-effective virtual training environment is critical for training infantry squads [1–3]. This is because if soldiers do not develop and sustain tactical proficiency, they will not be able to react in a quickly evolving battlefield. To create a VR military simulator that integrates immersion, interaction, and imagination, the issue of how to use VR factors (e.g., VR engine, software and database, input/output devices, and users and tasks) is an important one. The VR simulator is able to integrate the terrain of any real place into the training model, and the virtual simulation trains soldiers to engage targets while working as a team [4,5].

With a head-mounted display (HMD), the key feature used in VR technology, soldiers are immersed in a complex task environment that cannot be replicated in any training areas. Visual telescopes are positioned in front of the eyes, and the movement of the head is tracked by micro electro mechanical system (MEMS) inertial sensors. Gearing up with HMDs and an omnidirectional treadmill (ODT), soldiers can perform locomotive motions without risk of injury [6–8]. Note that the above systems do not offer any posture or gesture interactions between the soldiers and the virtual environment. To address this problem, the authors of [9–17] proposed an action recognition method based on multiple cameras. To tackle the occlusion problem, the authors mostly used Microsoft Kinect to capture color and depth data from the human body, which requires multiple devices arranged in a specific pattern for the recognizing different human actions. Therefore, the above methods are not feasible in a real-time virtual environment for military training.

To overcome the limitations of the real-time training system, the dismounted soldier training system (DSTS) was developed [18–21]. In the DSTS, soldiers wear special training suits and HMDs, and stand on a rubber pad to create a virtual environment. However, the inertial sensors of training suits can only recognize simple human actions (e.g., standing and crouching). When soldiers would like to walk or run in the virtual environment, they have to control small joysticks and function keys on simulated rifles, which are not immersive for locomotion. Thus, in a training situation, soldiers will not be able to react quickly when suddenly encountering enemy fire. As a result, though the DSTS provides a multi-soldier training environment, it has weaknesses with respect to human action recognition. Therefore, in order to develop a training simulator that is capable of accurately capturing soldiers' postures and movements and is less expensive and more effective for team training, body area networks (BANs) represent an appropriate bridge for connecting physical and virtual environments [22–26], which contain inertial sensors like accelerometers, e-compasses and gyroscopes to capture human movement.

Xsens Co. have been developing inertial BANs to track human movement for several years. Although the inertial BAN, called the MVN system, is able to capture human body motions wirelessly without using optical cameras, the 60 Hz system update rate is too slow to record fast-moving human motion. The latest Xsens MVN suit solved this problem, reaching a system update rate of 240 Hz. However, the MVN suit is relatively expensive in consumer markets. In addition, it is difficult to support more than four soldiers in a room with the MVN system, since the wireless sensor nodes may have unstable connections and high latency. On the other hand, since the Xsens MVN system is not an open-source system, the crucial algorithm of the inertial sensors does not adaptively allow the acquisition by and integration with the proposed real-time program. Moreover, the MVN system is only a motion-capture BAN and is not a fully functional training simulator with inertial BANs [26].

For the above-mentioned reasons, this paper proposes a multi-player training simulator that immerses infantry soldiers in the battlefield. The paper presents a technological solution for training infantry soldiers. Fully immersive simulation training provides an effective way of imitating real situations that are generally dangerous, avoiding putting trainees at risk. To achieve this, the paper proposes an immersive system that can immediately identify human actions. Training effectiveness in this simulator is highly remarkable in several respects:

1. *Cost-Effective Design*: We designed and implemented our own sensor nodes, such that the fundamental operations of the inertial sensors could be adaptively adjusted for acquisition and integration. Therefore, it has a competitive advantage in terms of system cost.
2. *System Capacity*: Based on the proposed simulator, a six-man squad is able to conduct military exercises that are very similar to real missions. Soldiers hold mission rehearsals in the virtual environment such that leaders can conduct tactical operations, communicate with their team members, and coordinate with the chain of command.
3. *Error Analysis*: This work provides an analysis of the quaternion error and further explores the sensing measurement errors. Based on the quaternion-driven rotation, the measurement relation with the quaternion error between the earth frame and the sensor frame can be fully described.
4. *System Delay Time:* The update rate of inertial sensors is about 160 Hz (i.e., the refresh time of inertial sensors has a time delay of about 6 ms). The simulator is capable of training six men at a 60 Hz system update rate (i.e., the refresh time of the entire system needs about 16 ms), which is acceptable for human awareness with delay ($\leq$40 ms).
5. *System Feedback*: Instructors can provide feedback after mission rehearsals using the visual after action review (AAR) function in the simulator, which provides different views of portions of the action and a complete digital playback of the scenario, allowing a squad to review details of the action. Furthermore, instructors can analyze data from digital records and make improvements to overcome the shortcomings in the action (Figure 1). Accordingly, in the immersive virtual environment, soldiers and leaders improve themselves with respect to situational awareness, decision-making, and coordination skills.

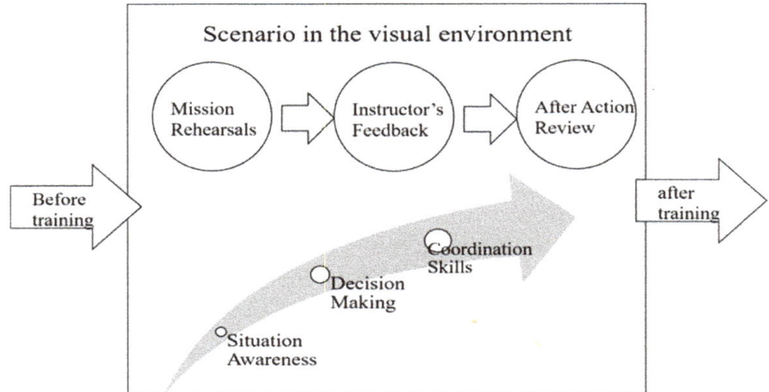


**Figure 1.** System training effectiveness process.

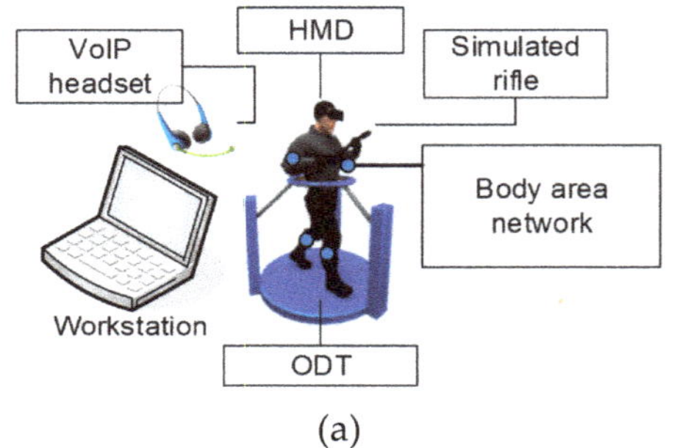


(a)

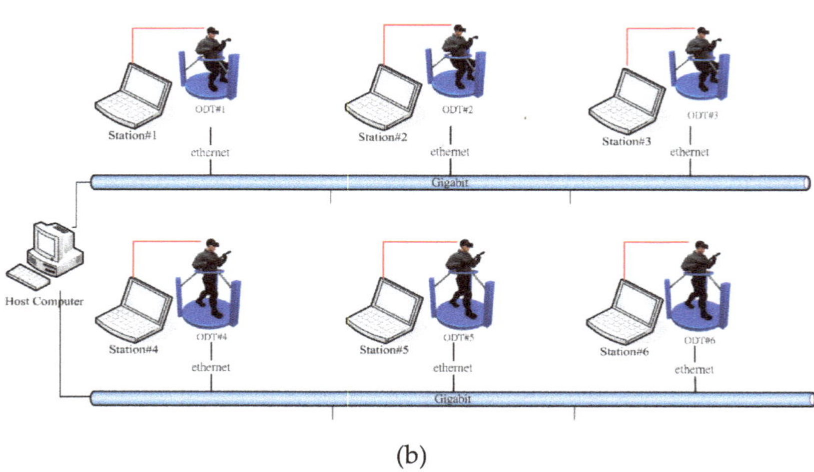


(b)

**Figure 2.** System architecture: single-soldier layout (**a**); multi-soldier network (**b**).

## 2. System Description

The proposed system, which is deployed in five-square-meter area (Figure 2), is a robust training solution that can support up to six soldiers. The virtual reality environment is implemented in C++

and DirectX 11. In addition to the virtual system, each individual soldier stands in the center of an ODT. The ODTs are customized hardware designed for the soldier training simulator, which is equipped with a high-performance workstation that generates the visual graphics for the HMD and provides voice communication functions based on VoIP technology. The dimensions of the ODT are 47" x47 "x78", 187 lbs. The computing workstation on the ODT is equipped with a 3.1 GHz Intel Core i7 processor, 8 GB RAM, and an NVidia GTX 980M graphics card. The HMD is an Oculus Rift DK2, which has two displays with 960 × 1080 resolution per eye. The VoIP software is implemented based on Asterisk open-source PBX. Soldiers are not only outfitted with HMDs, but also equipped with multiple inertial sensors, which consist of wearable sensor nodes (Figure 3) deployed over the full body (Figure 4). As depicted in Figure 3, a sensor node consists of an ARM-based microcontroller and nine axial inertial measurement units (IMUs), which are equipped with ARM Cotex-M4 microprocessors and ST-MEMS chips. Note that these tiny wireless sensor nodes can work as a group to form a wireless BAN, which uses an ultra-low power radio frequency band of 2.4G.

**Figure 3.** Top view of sensor nodes (**Left**); the wearable sensor node (**Right**).

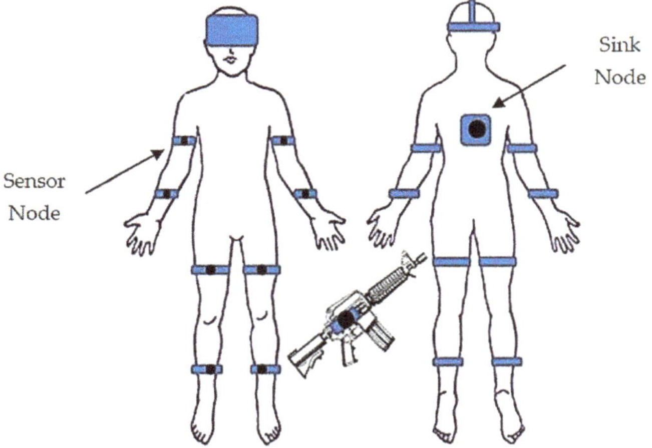


**Figure 4.** Deployment of sensor nodes and the sink node.

*2.1. Sensor Modeling*

This subsection the sensors are modeled in order to formulate the orientation estimation problem. The measurement models for the gyroscope, accelerometer, and magnetometer are briefly discussed in the following subsections [27].

*(1) Gyroscope:* The measured gyroscope signal ${}^{s}\omega_t^m$ can be represented in the sensor frame $s$ at time $t$ using

$$ {}^{s}\omega_t^m = {}^{s}\omega_t + b_{\omega,\,t} + e_{\omega,\,t} \tag{1} $$

where ${}^{s}\omega_t$ is the true angular velocity, $b_{\omega,\,t}$ is the bias caused by low-frequency offset fluctuations, and the measurement error $e_{\omega,\,t}$ is assumed to be zero-mean white noise. As shown in Figure 5, the raw data of the measured gyroscope signals in the x, y and z directions includes true angular velocity, bias and error.

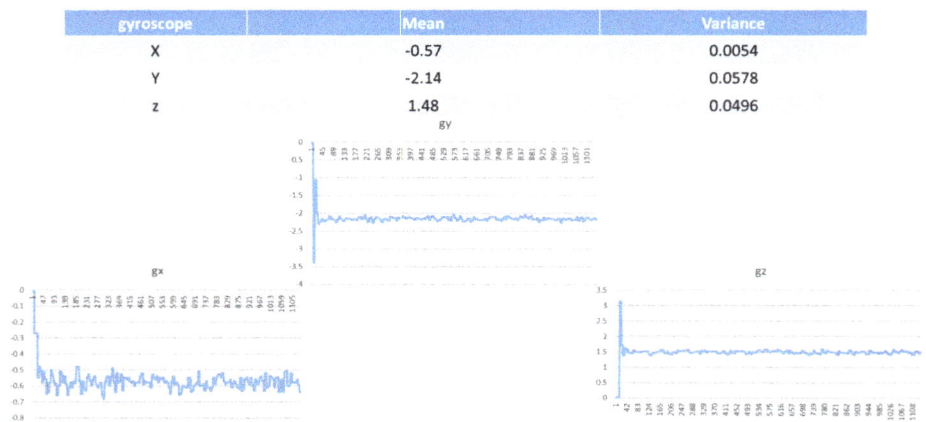

| gyroscope | Mean | Variance |
| --- | --- | --- |
| X | -0.57 | 0.0054 |
| Y | -2.14 | 0.0578 |
| z | 1.48 | 0.0496 |



**Figure 5.** Mean and variance of the gyroscope in the x, y and z directions.

*(2) Accelerometer:* Similar to the gyroscope signal, the measured accelerometer signal ${}^{s}a_t^m$ can be represented using

$$ {}^{s}a_t^m \approx {}^{s}a_{b,t} + b_{a,\,t} + e_{a,\,t} \tag{2} $$

where ${}^{s}a_{b,\,t}$ is the linear acceleration of the body after gravity compensation, $b_{a,\,t}$ is the bias caused by low-frequency offset fluctuations, and the measurement error $e_{a,\,t}$ is assumed to be zero-mean white noise. As shown in Figure 6, the raw data of the measured accelerometer signals in the x, y and z directions includes true linear acceleration, offset and error.

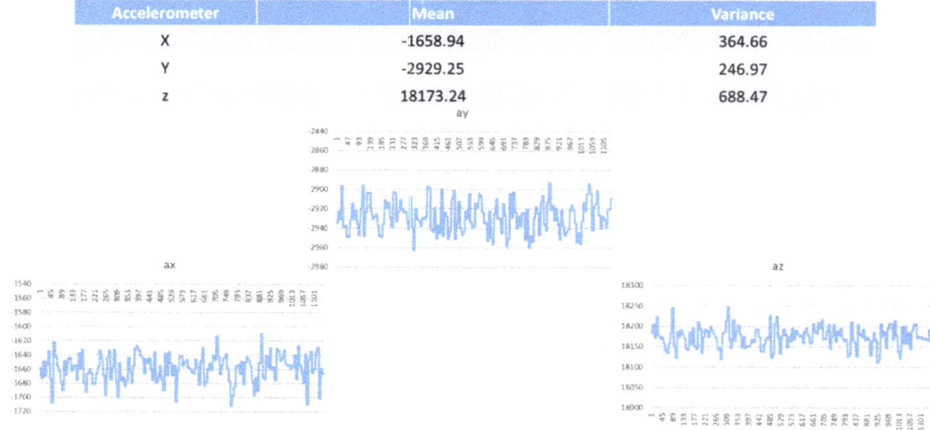

| Accelerometer | Mean | Variance |
| --- | --- | --- |
| X | -1658.94 | 364.66 |
| Y | -2929.25 | 246.97 |
| z | 18173.24 | 688.47 |



**Figure 6.** Mean and variance of the accelerometer in the x, y and z directions.

*(3) Magnetometer:* For the magnetometer, the measured signal ${}^{s}m_t^m$ is often modeled as the sum of the earth's magnetic field ${}^{s}m_t$, magnetic offset $b_{m,\,t}$, and noise $e_{m,\,t}$, which yields

$$ {}^{s}m_t^m \approx {}^{s}m_t + b_{m,\,t} + e_{m,\,t} \tag{3}$$

As shown in Figure 7, the raw data of the measured magnetometer signals in the x, y and z directions includes the earth's magnetic field, offset and error.

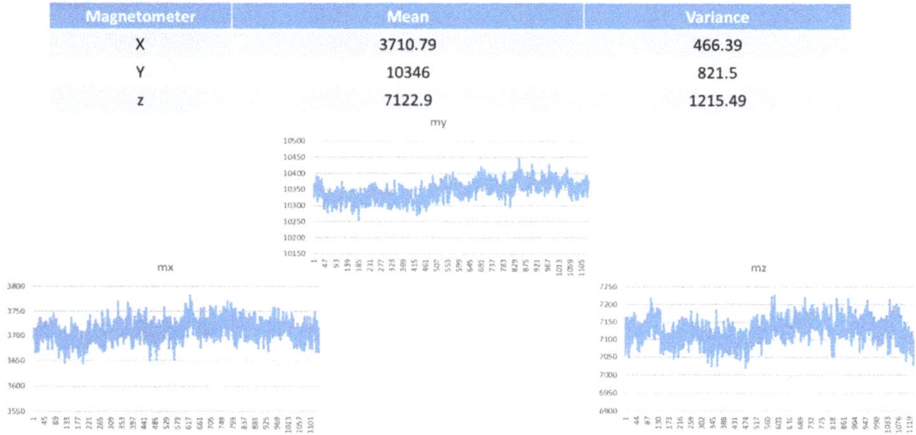

| Magnetometer | Mean | Variance |
|---|---|---|
| X | 3710.79 | 466.39 |
| Y | 10346 | 821.5 |
| Z | 7122.9 | 1215.49 |



**Figure 7.** Mean and variance of the magnetometer in the x, y and z directions.

## 2.2. Sensor Calibration

Since the accuracy of the orientation estimate depends heavily on the measurements, the null point, and the scale factor of each axis of the sensor, a calibration procedure should be performed before each practical use. To this end, the sensor calibration is described as follows:

- Step 1: Given a fixed gesture, we measure the sensing data (i.e., the raw data) and calculate the measurement offsets.
- Step 2: Remove the offset and normalize the modified raw data to the maximum resolution of the sensor's analog-to-digital converter. In this work, the calibrated results (CR) of the sensors are described by Equations (4)–(6).

Table 1 shows the means of the calibrated results in the x, y and z directions of sensor nodes, which include the accelerometer, the magnetometer, and the gyroscope. Note that the inertial sensor signals (e.g., ${}^{s}a_t^m$, ${}^{s}m_t^m$, and ${}^{s}\omega_t^m$) are measured without movement, and that a small bias or offset (e.g., $b_{a,\,t}$ or $b_{m,\,t}$ or $b_{\omega,\,t}$) in the measured signals can therefore be observed. In general, the additive offset that needs to be corrected for before calculating the estimated orientation is very small. For instance, the additive offset would be positive or negative in the signal output (e.g., the average $b_{\omega,\,t} = -0.08$ degree/sec and ${}^{s}\omega_t^m = 0$ degree/sec) without movement.

$$ \mathrm{CR}_{acc} = ({}^{s}a_t^m \pm b_{a,\,t})/\max({}^{s}a_t^m) \tag{4}$$

$$ \mathrm{CR}_{mag} = ({}^{s}m_t^m \pm b_{m,\,t})/\max({}^{s}m_t^m) \tag{5}$$

$$ \mathrm{CR}_{gyro} = ({}^{s}\omega_t^m \pm b_{\omega,\,t})/\max({}^{s}\omega_t^m) \tag{6}$$

**Table 1.** The calibrated results of the sensor nodes.

| | $CR_{acc}$ | $CR_{mag}$ | $CR_{gyro}$ |
|---|---|---|---|
| x | 0.95 | 0.98 | 0.96 |
| y | 0.97 | 0.99 | 0.96 |
| z | 0.99 | 0.98 | 0.94 |

After bias error compensation for each sensor, the outputs of the sensors represented in (1)–(3) can be rewritten as ${}^{s}\omega_t^m = {}^{s}\omega_t + e_{\omega,\,t}$, ${}^{s}a_t^m \approx {}^{s}a_{b,t} + e_{a,\,t}$, and ${}^{s}m_t^m \approx {}^{s}m_t + e_{m,\,t}$, and they are then applied as inputs for the simulation system. In particular, the proposed algorithm can be good at handling zero-mean white noise. After the completion of modeling and calibration of the sensor measurements from the newly developed MARG, all measurements are then represented in the form of quaternions as inputs of the proposed algorithm.

As shown in Figure 8, a typical run of the attitude angles of a sensor node placed on the upper arm when standing in a T-pose is recorded. In the first step, we stand with the arms naturally at the sides. Then, we stretch the arms horizontally with thumbs forward, and then we move back to the first step. This experiment shows that the sensor node is capable of tracking the motion curve directly.

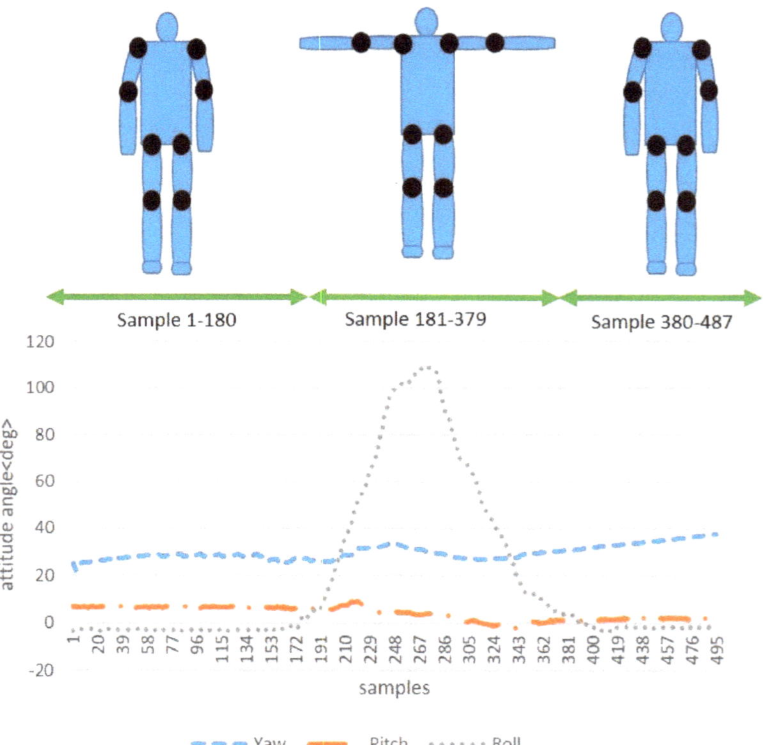


**Figure 8.** The attitude angles of a sensor node placed on the upper arm when standing in a T-pose.

### 2.3. Information Processing

As mentioned above, human action recognition is crucial to developing an immersive training simulator. The wireless BAN is able to instantaneously track the action of the skeleton of a soldier. The microcontroller of the sensor node acquires the raw data from the accelerometer, gyroscope and magnetometer through SPI and I2C interfaces. The measurements of the accelerometer, gyroscope and magnetometer are contaminated by errors of scale factor and bias, resulting in a large estimation error during sensor fusion. Thus, a procedure for normalization and error correction is needed for the sensor nodes to ensure the consistency of the nine-axis data (Figure 9). After that, the calibrated accelerometer and magnetometer data can be used to calculate the orientation and rotation. Estimation based on the accelerometer does not include the heading angle, which is perpendicular to the direction of gravity. Thus, the magnetometer is used to measure the heading angle. the Euler angles are defined as follows:

$$\begin{aligned} \text{Roll} - \Phi &: \text{ rotation about the X-axis} \\ \text{Pitch} - \theta &: \text{ rotation about the Y-axis} \\ \text{Heading} - \psi &: \text{ rotation about the Z-axis.} \end{aligned}$$

Let $X_{acc}, Y_{acc}, Z_{acc}$ be the accelerometer components. We have:

$$\text{Pitch } \theta = \tan^{-1}\left(\frac{-X_{acc}}{\sqrt{Z_{acc}{}^2 + Y_{acc}{}^2}}\right) \tag{7}$$

$$\text{Roll } \Phi = \tan^{-1}\left(\frac{Y_{acc}}{Z_{acc}}\right) \tag{8}$$

Let $X_{mag}, Y_{mag}, Z_{mag}$ be the magnetometer components. We have:

$$X_h = X_{mag}\cos(\theta) + Y_{mag}\sin(\theta) \times \sin(\Phi) + Z_{mag}\sin(\theta) \times \cos(\Phi) \tag{9}$$

$$Y_h = -Y_{mag}\cos(\Phi) + Z_{mag}\sin(\Phi) \tag{10}$$

$$\text{Heading } \psi = \tan^{-1}\left(\frac{Y_h}{X_h}\right) \tag{11}$$

Accordingly, the complementary filter output (using gyroscope) is

$$Angle_{output} = W \times (Angle_{output} + Angle_{gyro}) + (1 - W) \times Angle_{acc+mag}, \quad 0 < W < 1 \tag{12}$$

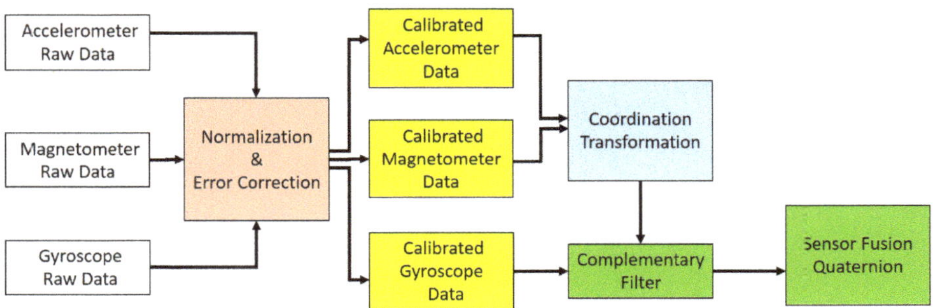


**Figure 9.** The diagram shows the process of sensor fusion.

To mitigate the accumulated drift from directly integrating the linear angular velocity of the gyroscope, the accelerometer and magnetometer are used as aiding sensors to provide the vertical and horizontal references for the Earth. Moreover, a complementary filter [28,29] is able to combine the measurement information of an accelerometer, a gyroscope and a magnetometer, offering big advantages in terms of both long-term and short-term observation. For instance, an accelerometer does not drift over the course of a long-term observation, and a gyroscope is not susceptible to small forces during a short-term observation. The W value of the complementary filter in Equation (12) is the ratio in which inertial sensor data is fused, which can then be used to compensate for gyroscope drift by using the accelerometer and magnetometer measurements. We always set the W value to 0.9, or probably even higher than 0.9. Therefore, the complementary filter provides an accurate orientation estimation.

### 2.4. Communication and Node Authentication Procedures

To tackle the problem of interference and to reduce the bit error rate (BER) of the wireless data transmission between the sensor nodes and the sink node, communication mechanisms can be applied to build a robust wireless communication system [30,31]. The communication technique employed for communication between the sensor nodes and the sink node is frequency hopping spread spectrum (FHSS). The standard FHSS modulation technique is able to avoid interference between different sensor nodes in the world-wide ISM frequency band, because there are six BANs in total, all of which are performing orientation updates in a small room. Moreover, the sensor nodes' transmission of data to the sink node is based on packet communication.

In the beginning, sensor nodes send packets, which include orientation and rotation data. When the sink node receives a packet from a sensor node, an acknowledgement packet will be transmitted immediately (Figure 10). If the acknowledgement fails to arrive, the sensor node will retransmit the packet, unless the number of retries exceeds the retransmission limit. When the sink node receives packets in operation mode, it will check the payload of the packets. As shown in Figure 11, if the payload is equal to the sensor node's unique identification information, the sink node will accept the packet and scan the next sensor node's channel during the next time slot. The ten wireless sensor node channels operate at a frequency range of 2400–2525 GHz.

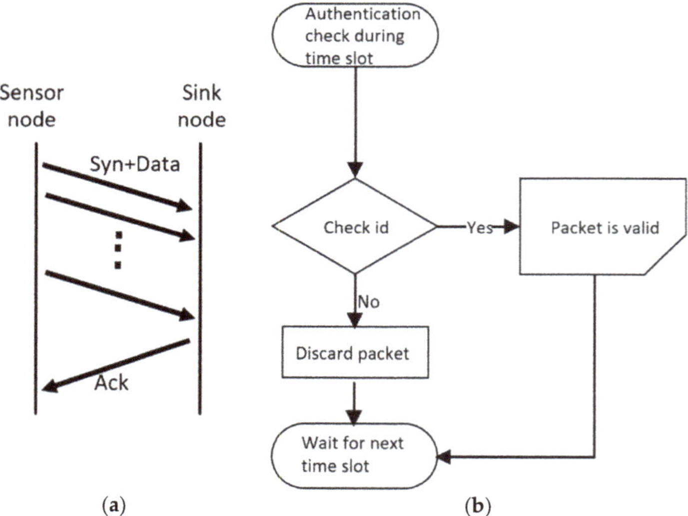


**Figure 10.** Operation mode during each time slot. (**a**) Step 1: automatic packet synchronization; (**b**) Step 2: identification check on valid packets by the sink node.

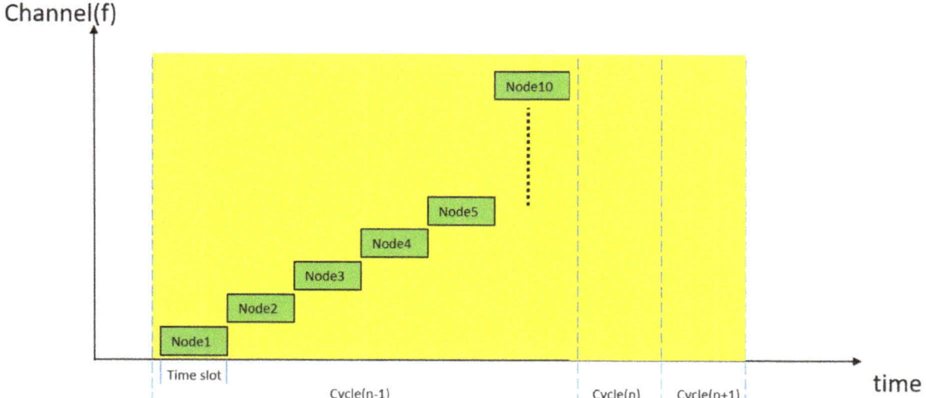


**Figure 11.** The diagram shows that sensor nodes communicate with sink node in two domains.

## 2.5. System Intialization

The sink node is deployed on the back of the body and collects the streaming data from multiple wireless sensor nodes. Afterwards, the streaming data from the sensor nodes is transmitted to the workstation via Ethernet cables, modeling the skeleton of the human body along with the structure of the firearm (Figure 12). Table 2 describes the data structure of a packet, which is sent by a sensor node to the sink node in every ODT. Table 3 shows the integrated data that describes the unique features of each soldier in the virtual environment. The description provides the appearance characteristics, locomotion, and behaviors of each soldier [32,33].

**Figure 12.** A fully equipped soldier.

**Table 2.** Data structure of a packet.

| Header | Data to Show Packet Number | 8 bits |
|---|---|---|
| Payload | Bone data size<br>Total data length<br>Soldier No.<br>T-Pose status<br>Sensor nodeID<br>Total Bones of a skeleton<br>Yaw value of the bone<br>Pitch value of the bone<br>Roll value of the bone | 144 bits |
| Tail | Data to show end of packet | 8 bits |

**Table 3.** Description of the features of each soldier.

| DATA | UNIT | RANGE |
|---|---|---|
| Soildier no. | N/A | 0~255 |
| Friend or Foe | N/A | 0~2 |
| exterbal Control | N/A | 0/1 |
| Team no. | N/A | 0~255 |
| Rank | N/A | 0~15 |
| Appearance | N/A | 0~255 |
| BMI | $Kg/Meter^2$ | 18~32 |
| Health | Percentage | 0~100 |
| Weapon | N/A | 0~28 |
| Vechicle | N/A | 0~5 |
| Vechicle Seat | N/A | 0~5 |
| Position X | Meter | N/A |
| Position Y | Meter | N/A |
| Position Z | Meter | N/A |
| Heading | Degree | −180~180 |
| Movement | N/A | 0~255 |
| Behavior | N/A | 0~255 |

For sensor nodes, the data stream reported to the sink node is interpreted as the movement-based data structures of a skeleton. Each wearable sensor node of the skeleton has a unique identification information during the initialization phase. As a result, the sink node can distinguish which sensor nodes are online. Accordingly, when turning the power on, the HMDs on the soldiers are automatically connected to the workstations, and the voice communication group setting is immediately ready. Please note that the T-pose calibration is performed in sensor node $n$ of a skeleton for initializing the root frame, which is given by

$$q_{1,n} = q_{T,n} \otimes q_{0,n} \otimes q^{\star}_{T,n}, \tag{13}$$

where $q_{T,n}$ is the reading of sensor node $n$ in the modified T-pose and $q_{1,n}$ is the new body frame from the initial root frame $q_{0,n}$.

Although sensor nodes are always worn on certain positions of a human body, the positions of sensor nodes may drift due to the movements occurring during the training process. Hence, the T-pose calibration procedure can be applied to estimate the orientation of the sensors. After that, the system is prepared to log in for simulation tasks. The system initialization flow diagram is shown in Figure 13.

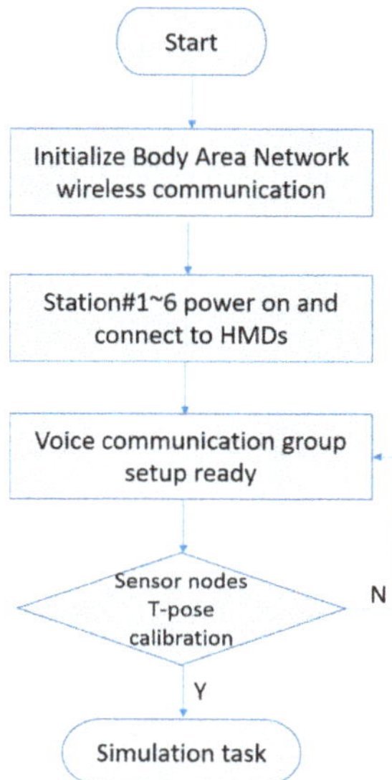


**Figure 13.** System initialization flow diagram.

## 3. Quaternion Representation

This section outlines a quaternion representation of the orientation of the sensor arrays. Quaternions provide a convenient mathematical notation for representing the orientation and rotation of 3D objects because quaternion representation is more numerically stable and efficient compared to rotation matrix and Euler angel representation. According to [34–36], a quaternion can be thought of as a vector with four components,

$$q = q_0 + q_x i + q_y j + q_z k \tag{14}$$

as a composite of a scalar and ordinary vector. The quaternion units $q_x$, $q_y$, $q_z$ are called the vector part of the quaternion, while $q_0$ is the scalar part. The quaternion can frequently be written as an ordered set of four real quantities,

$$q = [q_0, q_x, q_y, q_z]. \tag{15}$$

Denote ${}^e_s q$ as the orientation of the earth frame $\boldsymbol{u_e}$ with respect to the sensor frame $\boldsymbol{u_s}$. The conjugate of the quaternion can be used to represent an orientation by swapping the relative frame, and the sign * denotes the conjugate. Therefore, the conjugate of ${}^e_s q$ can be denoted as

$${}^e_s q^* = {}^s_e q = [q_0, -q_x, -q_y, -q_z]. \tag{16}$$

Moreover, the quaternion product $\otimes$ can be used to describe compounded orientations, and their definition is based on the Hamilton rule in [37]. For example, the compounded orientation ${}^s_h q$ can be defined by

$${}^s_h q = {}^s_e q \otimes {}^e_h q, \tag{17}$$

where ${}^e_h q$ denotes the orientation of the earth frame $\boldsymbol{u_e}$ with respect to the frame $\boldsymbol{u_h}$.

A human body model consists of a set of body segments connected by joints. For upper limbs and lower limbs, kinematic chains are modeled that branch out around the torso. The kinematic chain describes the relationship between rigid body movements and the motions of joints. A forward kinematics technique, which was introduced for the purposes of robotic control, is used to configure each pair of adjacent segments. In the system, the aim of building human kinematic chains is to determine the transformation matrix of a human body from the first to the last segments and to find the independent configuration for each joint and the relationship with the root frame.

Thus, the rotation matrix $q^n_{n-1}$ used for orientation from sensor node $n$-1 to sensor node $n$ is given by

$$q^n_{n-1} = q_{n-1} \otimes q^\star_n. \tag{18}$$

Figure 14 shows the simplified segment biomechanical model of the human body. The kinematics of segments on which no inertial sensors are attached (e.g., hand, feet, toes) are considered to be rigid connections between neighboring segments. The transformation matrix is defined as

$$Q^n_{n-1} = \begin{bmatrix} q^n_{n-1} & T^n_{n-1} \\ 0 & 1 \end{bmatrix}. \tag{19}$$

where $T^n_{n-1}$ is the translation matrix from sensor frame to body frame. According to [37,38], therefore, the transformation matrix of a human body from the first segment to the n-th segment is

$$Q^n_1 = Q^n_{n-1} Q^{n-1}_{n-2} Q^{n-2}_{n-3} \cdots Q^3_2 Q^2_1. \tag{20}$$

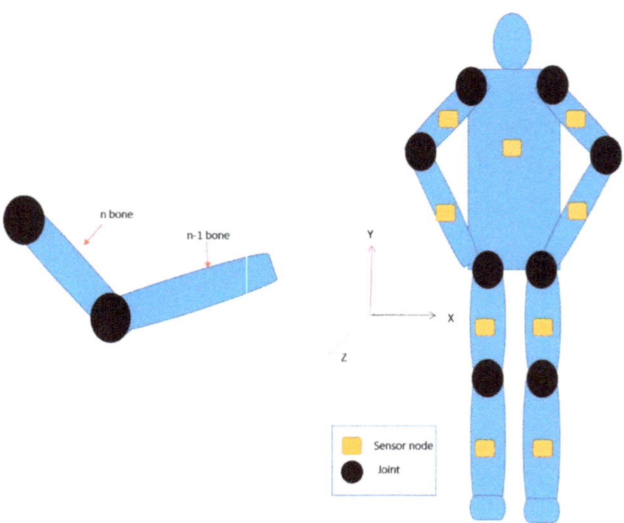


**Figure 14.** The kinematic chain of a human body.

## 4. Performance Analysis

The analysis focuses on the quaternion error and further explores the sensing measurement errors. Based on the quaternion-driven rotation, the measurement relation with the quaternion error between the earth frame and sensor frame can be further described.

### 4.1. Rotation Matrix

According to [36], given a unit quaternion $\boldsymbol{q} = q_r + q_x i + q_y j + q_z k$, the quaternion-driven rotation can be further described by the rotation matrix $\boldsymbol{R}$, which yields

$$\boldsymbol{R} = \begin{bmatrix} 1 - 2({q_y}^2 + {q_z}^2) & 2q_xq_y - 2q_rq_z & 2q_xq_z + 2q_rq_y \\ 2q_xq_y + 2q_rq_z & 1 - 2({q_x}^2 + {q_z}^2) & 2q_yq_z - 2q_rq_x \\ 2q_xq_z - 2q_rq_y & 2q_yq_z + 2q_rq_x & 1 - 2({q_x}^2 + {q_y}^2) \end{bmatrix}. \tag{21}$$

Let $\hat{q}$ be an estimate of the true attitude quaternion $q$. The small rotation from the estimated attitude, $\hat{q}$, to the true attitude is defined as $q_{err}$. The error quaternion is small but non-zero, due to errors in the various sensors. The relationship is expressed in terms of quaternion multiplication as follows:

$$q = \hat{q} \otimes q_{err}. \tag{22}$$

Assuming that the error quaternion, $q_{err}$, is to represent a small rotation, it can be approximated as follows:

$$c = \left[ q_r \; q_x^{(err)} \; q_y^{(err)} \; q_z^{(err)} \right]^T = \left[ q_r \; \vec{q}_{err} \right]^T. \tag{23}$$

Noting that the error quaternion $q_{err}$ is a perturbation of the rotation matrix, and the vector components of $\vec{q}_{err}$ are small, the perturbation of the rotation matrix R in Equation (21) can be written as:

$$R(q_{err}) \cong \begin{bmatrix} 1 & -2q_r q_z^{(err)} & 2q_r q_y^{(err)} \\ 2q_r q_z^{(err)} & 1 & -2q_r q_x^{(err)} \\ -2q_r q_y^{(err)} & 2q_r q_x^{(err)} & 1 \end{bmatrix} = I_{3x3} + 2q_r [\vec{q}_{err}]^{\times}. \tag{24}$$

Equation (22) relating $\hat{q}$ and $q$ can be written as

$$R(q) = R(\hat{q}) R(q_{err}) = R(\hat{q}) \left[ I_{3x3} + 2q_r [\vec{q}_{err}]^{\times} \right]. \tag{25}$$

$R(\hat{q})$ is the estimate of the rotation matrix or the equivalent of $\hat{q}$. Now, considering the sensor frame $u_s$ and the earth frame $u_e$, we have

$$u_e = R(\hat{q}) \left[ I_{3x3} + 2q_r [\vec{q}_{err}]^{\times} \right] u_s = \hat{u}_e + 2q_r [\vec{q}_{err}]^{\times} u_s. \tag{26}$$

Thus, the measurement relation for the quaternion error is obtained:

$$\Delta u_e \triangleq u_e - \hat{u}_e = 2q_r [\vec{q}_{err}]^{\times} u_s. \tag{27}$$

Accordingly, given the error quaternion $q_{err}$ and the sensor frame $u_s$, the perturbation of the earth frame $u_e$ can be described. The quantitative analysis of the error quaternion is detailed in Section 6.1.

### 4.2. Error Analysis

The analysis in Section 4.1 focuses on the quaternion error. Here we further explore the sensing measurement errors, which consist of the elements of the error quaternion.

Roll $- \Phi$ : rotation about the X-axis
Pitch $- \theta$ : rotation about the Y-axis
Heading $- \psi$ : rotation about the Z-axis

Now we associate a quaternion with Euler angles, which yields

$$\hat{q} = \begin{bmatrix} -\sin\frac{\Phi}{2}\sin\frac{\theta}{2}\sin\frac{\psi}{2} + \cos\frac{\Phi}{2}\cos\frac{\theta}{2}\cos\frac{\psi}{2} \\ +\sin\frac{\Phi}{2}\cos\frac{\theta}{2}\cos\frac{\psi}{2} + \cos\frac{\Phi}{2}\sin\frac{\theta}{2}\sin\frac{\psi}{2} \\ -\sin\frac{\Phi}{2}\cos\frac{\theta}{2}\sin\frac{\psi}{2} + \cos\frac{\Phi}{2}\sin\frac{\theta}{2}\cos\frac{\psi}{2} \\ +\sin\frac{\Phi}{2}\sin\frac{\theta}{2}\cos\frac{\psi}{2} + \cos\frac{\Phi}{2}\cos\frac{\theta}{2}\sin\frac{\psi}{2} \end{bmatrix} \tag{28}$$

Denote the pitch angle measurement as $\theta + \Delta\theta$, where $\theta$ is the true pitch angle information and $\Delta\theta$ is the measurement error. To simplify the error analysis, assume the rotation errors are neglected in roll angle and heading angle measurements. Let $\sin(\Phi/2) = A$ and $\sin(\psi/2) = B$. Accordingly, considering the measurement error in the pitch angle, the quaternion can be rewritten as

$$q' = \begin{bmatrix} -AB\sin\frac{\theta+\Delta\theta}{2} + \sqrt{(1-A^2)(1-B^2)}\cos\frac{\theta+\Delta\theta}{2} \\ +A\sqrt{1-B^2}\cos\frac{\theta+\Delta\theta}{2} + B\sqrt{1-A^2}\sin\frac{\theta+\Delta\theta}{2} \\ -AB\cos\frac{\theta+\Delta\theta}{2} + \sqrt{(1-A^2)(1-B^2)}\sin\frac{\theta+\Delta\theta}{2} \\ +A\sqrt{1-B^2}\sin\frac{\theta+\Delta\theta}{2} + B\sqrt{1-A^2}\cos\frac{\theta+\Delta\theta}{2} \end{bmatrix} \tag{29}$$

Assuming that the measurement error in the pitch angle is small, we obtain

$$\begin{aligned} \sin\frac{\theta+\Delta\theta}{2} &= \sin\frac{\theta}{2}\cos\frac{\Delta\theta}{2} + \cos\frac{\theta}{2}\sin\frac{\Delta\theta}{2} \\ &\simeq \sin\frac{\theta}{2} + \cos\frac{\theta}{2}\cdot\frac{\Delta\theta}{2} \end{aligned} \tag{30}$$

$$\begin{aligned} \cos\frac{\theta+\Delta\theta}{2} &= \cos\frac{\theta}{2}\cos\frac{\Delta\theta}{2} - \sin\frac{\theta}{2}\sin\frac{\Delta\theta}{2} \\ &\simeq \cos\frac{\theta}{2} - \sin\frac{\theta}{2}\cdot\frac{\Delta\theta}{2}. \end{aligned} \tag{31}$$

According to Equation (29), the quaternion with measurement error in the pitch angle can be further approximated by

$$\begin{aligned} q' &\simeq \begin{bmatrix} -AB\sin\frac{\theta}{2} + \sqrt{(1-A^2)(1-B^2)}\cos\frac{\theta}{2} - AB\cos\frac{\theta}{2}\cdot\frac{\Delta\theta}{2} - \sqrt{(1-A^2)(1-B^2)}\sin\frac{\theta}{2}\cdot\frac{\Delta\theta}{2} \\ +A\sqrt{1-B^2}\cos\frac{\theta}{2} + B\sqrt{1-A^2}\sin\frac{\theta}{2} - A\sqrt{1-B^2}\sin\frac{\theta}{2}\cdot\frac{\Delta\theta}{2} + B\sqrt{1-A^2}\cos\frac{\theta}{2}\cdot\frac{\Delta\theta}{2} \\ -AB\cos\frac{\theta}{2} + \sqrt{(1-A^2)(1-B^2)}\sin\frac{\theta}{2} + AB\sin\frac{\theta}{2}\cdot\frac{\Delta\theta}{2} + \sqrt{(1-A^2)(1-B^2)}\cos\frac{\theta}{2}\cdot\frac{\Delta\theta}{2} \\ +A\sqrt{1-B^2}\sin\frac{\theta}{2} + B\sqrt{1-A^2}\cos\frac{\theta}{2} + A\sqrt{1-B^2}\cos\frac{\theta}{2}\cdot\frac{\Delta\theta}{2} - B\sqrt{1-A^2}\sin\frac{\theta}{2}\cdot\frac{\Delta\theta}{2} \end{bmatrix} \\ &= \hat{q}\otimes q_{err} \end{aligned} \tag{32}$$

Note that, given a measurement error in the pitch angle, and the roll, pitch, and heading angle measurements, the error quaternion $\boldsymbol{q}_{err}$ can be approximately derived by Equation (32). Therefore, the measurement relation with the quaternion error between the earth frame and the sensor frame can be further described using Equation (27).

## 5. System Operating Procedures

To evaluate the effectiveness and capability of the virtual reality simulator for team training, we designed a between-subjects study. In the experiments, the impacts of three key factors on system performance are considered: training experience, the group size of the participants, and information exchange between the group members. The experimental results are detailed as follows.

### 5.1. Participants

The experiment involved 6 participants. No participants had ever played the system before. Half of the participants were volunteers who had done military training in live situations, while the other half had never done live training. The age of the participants ranged from 27 to 35.

### 5.2. Virtual Environment

The immersive environment was designed for a rescue mission in an enemy-held building (Figure 15). In addition, three hostages were being guarded on the top floor of a three-storied building which was controlled by 15 enemies. To ensure that the virtual environment was consistent with an actual training facility, we simulated a real military training site, including the urban street, the enemy-held building, and so on. All enemies controlled automatically by the system were capable of shooting, evading attacks, and team striking. When participants were immersed in the virtual environment, they could interact with other participants not only through gesture tracking, but also through VoIP communication technology.

Figure 15. Virtual environment on a HMD. (a) An indoor view. (b) An outdoor view.

As mentioned above, sensing measurement errors greatly affect the sensor nodes, which are attached to a human body. The integration of the inertial sensors, including sensor signals and drift errors, is performed on the basis of the kinematics of the human body. Therefore, sensing errors will be accumulated in quaternion form. As shown in Figure 16, the sensing measurement errors are calibrated when the T-pose calibration is performed. In the first step, we normalize the accelerometer, gyroscope and magnetometer in all of the sensor nodes and compensate for bias errors. In the following step, a complementary filter is used to mitigate the accumulated errors based on the advantages of long-term observation and short-term observation respectively. In the final step, T-pose calibration is performed to align the orientation of the sensor nodes with respect to the body segments, after which the sensor node is able to capture body movement accurately in the virtual environment.

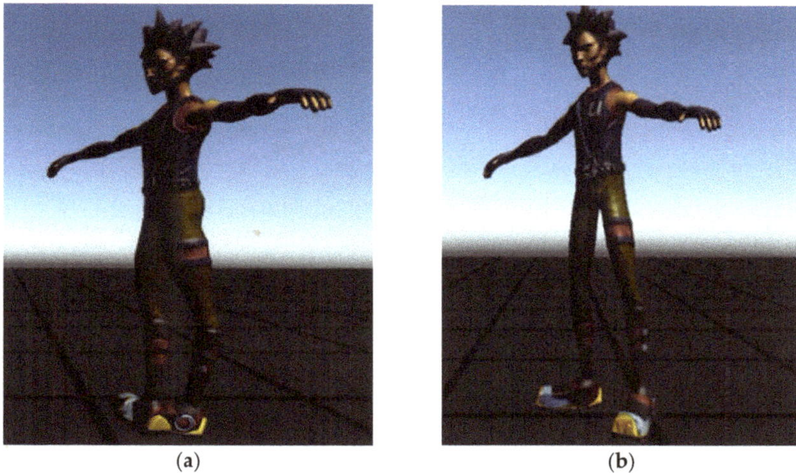

Figure 16. Sensing measurement errors of the sensor nodes were calibrated when the T-pose calibration was performed. (a) All sensor nodes were calibrated well during the T-pose procedure. (b) One sensor node attached to the right thigh was not calibrated well, and a sensing error was derived in the pitch direction.

*5.3. Procedure*

The purpose of the experiment is to evaluate the system performance. Participants follow the same path through the virtual environment in the experiment. The time-trial mission starts when the participants begin from the storm gate. Moreover, the time taken by the participants to kill all of the enemies who guard the three hostages on the top floor will be recorded. All participants control simulated M4 rifles, and the enemies control virtual AK47 rifles. All of the weapons have unlimited ammo. Under these experimental conditions, the participants' death rate (hit rate) is recorded for data analysis by the experimenters. Moreover, if all participants are killed before they complete the mission, the rescue task is terminated, and the time will not be recorded for the experiment.

## 6. Experimental Results

In order to assess the system performance, four sets of experiments were performed to explore the impact of quaternion error and the training experience on mission execution and management.

### 6.1. Error Analysis

In the first set of simulations, we explored the characteristic of the error quaternion $q_{err}$. With reference to the analysis in Section 4.2, the rotation errors are assumed to be negligible in the roll angle and heading angle measurements, and the measurement error in the pitch angle is considered to be $\Delta\theta$. With angle information (e.g., the heading angle $60^{\circ}$, the roll angle $30^{\circ}$, the pitch angle $15^{\circ}$) and $\Delta\theta$, Figure 17 presents the behavior of the error quaternion $q_{err}$ when varying the measurement error of the pitch angle. Note that, given $\Delta\theta$ ranging from $0^{\circ}$ to $0.2^{\circ}$, the vector parts of $q_{err}$ (i.e., $q_x^{(err)}$, $q_y^{(err)}$, $q_z^{(err)}$) are approximately linear with respect to the $\Delta\theta$, which can provide a sensible way of describing the error behaviors of rotation X, rotation Y, and rotation Z. According to Equation (27), given the quaternion error $\mathrm{q}_{\mathrm{err}}$ and the sensor frame $\mathrm{u_s}$, the perturbation of the earth frame $\mathrm{u_e}$ can be described. As shown in Figure 17, when the measurement error in the pitch angle is small, the small vector components of $\mathrm{q} \rightarrow_{\mathrm{err}}$ lead to a small perturbation of the earth frame $\mathrm{u_e}$. In contrast, as the measurement error in the pitch angle increases, the perturbation in the Y-axis increases, which results in a larger error component in the Y-axis (e.g., with $\Delta\theta = 0.1$, $q_x^{(err)} = 0.006$, $q_y^{(err)} = 0.05$, $q_z^{(err)} = 0.0004$).

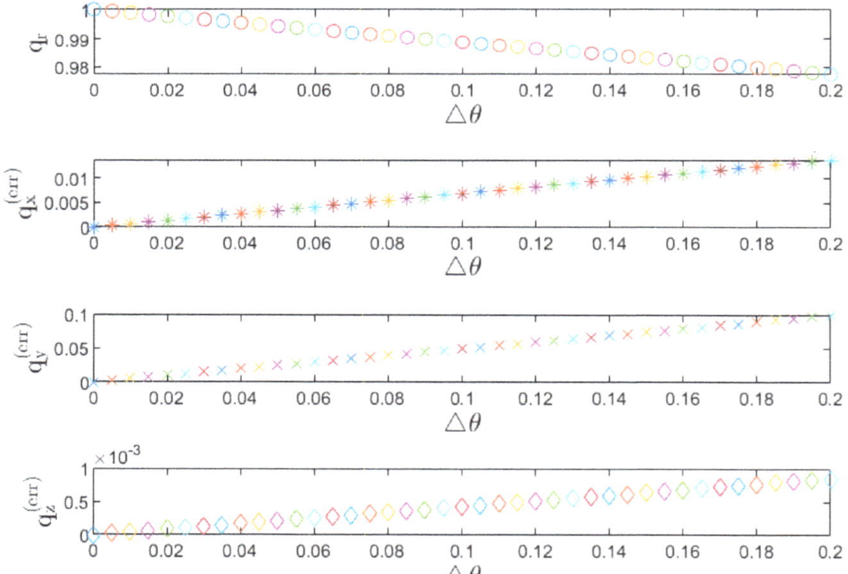


**Figure 17.** The scalar part and the vector part of the quaternion error for a small rotation of measurement error of pitch angle.

### 6.2. Simulated Training Performance

Figure 18 shows a snapshot of the proposed system. All three individual participants, who had done the same military training at the actual training site, successfully completed the rescue mission, with times of 22′16″, 25′40″, 28′51″, respectively (mean = 25′36″, standard deviation = 3′18″). However, of the three participants who had never done the same training and started the mission individually, only one participant completed the rescue mission, with a time of 36′59″, and the other two participants were killed by enemies before completing the rescue task. The three two-man teams who had been trained in the live situation completed the rescue mission with times of 11′28″, 18′19″, 16′5″, respectively (mean = 15′17″, standard deviation = 3′30″). However, the three two-man teams who had never done the same training also completed the rescue mission, with times of 25′19″, 28′33″, 26′12″, respectively (mean = 26′41″, standard deviation = 1′40″). Finally, the three-man team that had live training experience completed rescue mission with a time of 7′49″. On the other hand, the three-man team that had never done the same training completed the rescue mission with a time of 13′47″. The results of experienced and unexperienced participants' mean times in the experiment are shown in Figure 19. We also evaluated another situation, in which two subjects in the three-man groups completed the mission without the VoIP communication function. The mean time in this experiment increased by 1′26″ (Figure 20), which implies that communication and information processing can improve the performance for rescue missions.

Finally, we evaluated a six-man team of all participants in the rescue task, because standard deployment for a real live mission is a six-man entry team. The mission time decreased by 2′28″ with respect to the three-man experiment with experienced participants. The experimental results for mean times with different numbers of participants are shown in Figure 21. In addition, death rate (hit rate) revealed another difference between single and multiple participants. From the results, the mean of death rate (hit rate) was 1.5 shots/per mission when a single participant interacted with the system. However, the mean of death rate decreased to 0.38 shots/per mission when multiple participants interacted with the system.

**Figure 18.** Snapshot of the system.

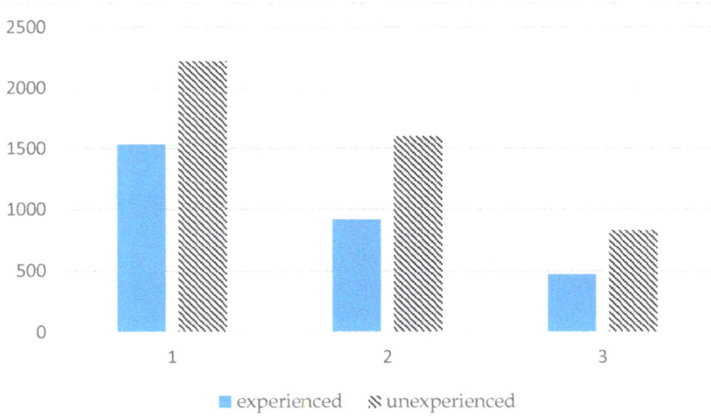


**Figure 19.** Means of experienced participants and unexperienced participants under various experimental conditions (horizontal axis: single-man, two-man, three-man; vertical axis unit: seconds).

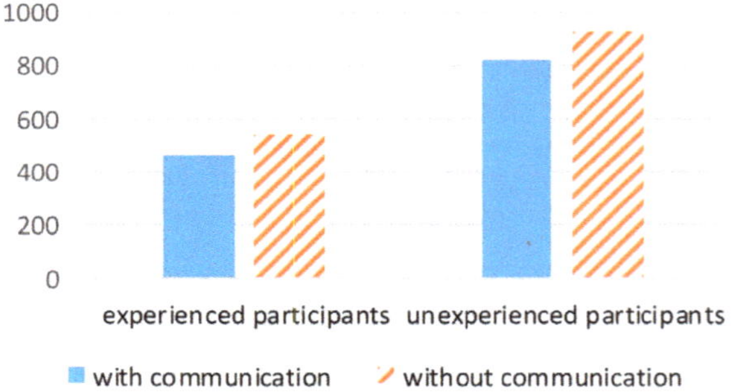


**Figure 20.** Mean of three-man teams with voice communication/without voice communication under the experimental conditions (horizontal axis: experienced/unexperienced participants; vertical axis unit: seconds).

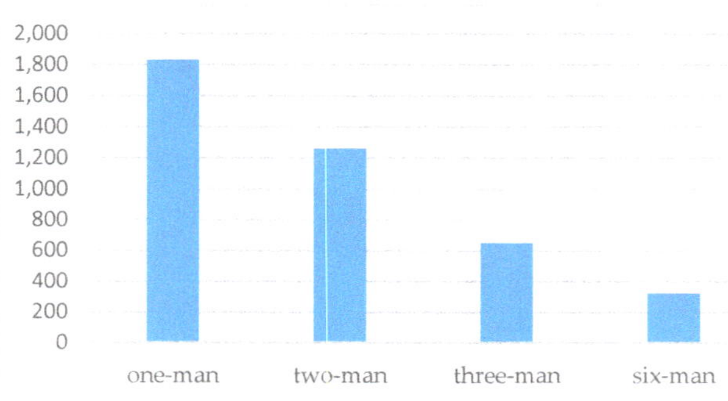


**Figure 21.** Mean times with different numbers of participants under the experimental conditions (horizontal axis: number of participants; vertical axis unit: seconds).

### 6.3. Discussion

When the participants executed the rescue mission, the activities involved in the experiment included detecting enemies, engaging enemies, moving inside the building and rescuing the hostages. The results reveal significant differences in several respects, including experience, quantity, and communication, and show that compared with the inexperienced participants, all experienced participants who had done the same training in a live situation took less time to complete the rescue mission. The wireless BANs of the participants are able to work accurately in the virtual environment for experienced participants. Tactical skills (e.g., moving through built-up areas, reconnoiter area, reacting to contact, assaulting, and clearing a room) absolutely require team work, demanding that wireless BANs interact with each other perfectly in terms of connection and accuracy. Without proper BANs, participants may feel mismatched interaction with their virtual avatars, and may feel uncomfortable or sick in the virtual environment.

The experimental results show that a larger sized group of participants took less time to complete the rescue mission than a smaller sized group of participants. Moreover, a group of multiple participants had a lower death rate compared with that of a single participant. This is due to the fact that, as the group size of participants increases, team movement is more efficient and fire power is greater in the virtual environment, which is similar to a real world mission. Furthermore, when the VoIP communication function was disabled, whether participants were experienced or not, the rescue mission time in the experiment consequently increased. As we know, in the real rescue mission, team coordination is important in the battlefield. In the system, all participants are able to interact with each other through hand signal tracking and voice communication. As a result, multiple-user training may become a key feature of the system.

## 7. Conclusions

In this paper, we have addressed problems arising when building an infantry training simulator for multiple players. The immersive approach we proposed is an appropriate solution that is able to train soldiers in several VR scenarios. The proposed simulator is capable of training six men at a system update rate of 60 Hz (i.e., the refresh time of the entire system takes about 16 ms), which is acceptable for human awareness with delay ($\leq$ 40 ms). Compared with the expensive Xsens MVN system, the proposed simulator has a competitive advantage in terms of system cost. For future work, we intend to develop improved algorithms to deal with accumulated sensing errors and environment noise on wireless BANs. Consequently, the system can develop finer gestures for military squad actions and enrich the scenario simulation for different usages in military training. The system is expected to be applied in different kind of fields and situations.

**Author Contributions:** Y.-C.F. and C.-Y.W. conceived and designed the experiments; Y.-C.F. performed the experiments; Y.-C.F. analyzed the data; Y.-C.F. and C.-Y.W. wrote the paper.

**Funding:** This research was funded by the Ministry of Science and Technology of Taiwan under grant number MOST-108-2634-F-005-002, and by the "Innovation and Development Center of Sustainable Agriculture" from The Featured Areas Research Center Program within the framework of the Higher Education Sprout Project by the Ministry of Education (MOE) in Taiwan.

**Conflicts of Interest:** The authors declare no conflict of interest.

## References

1. Dimakis, N.; Filippoupolitis, A.; Gelenbe, E. Distributed Building Evacuation Simulator for Smart Emergency Management. *Comput. J.* **2010**, *53*, 1384–1400. [CrossRef]
2. Knerr, B.W. *Immersive Simulation Training for the Dismounted Soldier*; No. ARI-SR-2007-01; Army Research Inst Field Unit: Orlando, FL, USA, 2007.
3. Lele, A. Virtual reality and its military utility. *J. Ambient Intell. Hum. Comput.* **2011**, *4*, 17–26. [CrossRef]

4. Zhang, Z.; Zhang, M.; Chang, Y.; Aziz, E.-S.; Esche, S.K.; Chassapis, C. Collaborative Virtual Laboratory Environments with Hardware in the Loop. In *Cyber-Physical Laboratories in Engineering and Science Education*; Springer: Berlin/Heidelberg, Germany, 2018; pp. 363–402.
5. Stevens, J.; Mondesire, S.C.; Maraj, C.S.; Badillo-Urquiola, K.A. Workload Analysis of Virtual World Simulation for Military Training. In Proceedings of the MODSIM World, Virginia Beach, VA, USA, 26–28 April 2016; pp. 1–11.
6. Frissen, I.; Campos, J.L.; Sreenivasa, M.; Ernst, M.O. *Enabling Unconstrained Omnidirectional Walking through Virtual Environments: An Overview of the CyberWalk Project*; Human Walking in Virtual Environments; Springer: New York, NY, USA, 2013; pp. 113–144.
7. Turchet, L. Designing presence for real locomotion in immersive virtual environments: An affordance-based experiential approach. *Virtual Real.* **2015**, *19*, 277–290. [CrossRef]
8. Park, S.Y.; Ju, H.J.; Lee, M.S.L.; Song, J.W.; Park, C.G. Pedestrian motion classification on omnidirectional treadmill. In Proceedings of the 15th International Conference on Control, Automation and Systems (ICCAS), Busan, Korea, 13–16 October 2015.
9. Papadopoulos, G.T.; Axenopoulos, A.; Daras, P. Real-Time Skeleton-Tracking-Based Human Action Recognition Using Kinect Data. In Proceedings of the MMM 2014, Dublin, Ireland, 6–10 January 2014.
10. Cheng, Z.; Qin, L.; Ye, Y.; Huang, Q.; Tian, Q. Human daily action analysis with multi-view and color-depth data. In Proceedings of the European Conference on Computer Vision, Florence, Italy, 7–13 October 2012; Springer: Berlin/Heidelberg, Germany, 2012.
11. Kitsikidis, A.; Dimitropoulos, K.; Douka, S.; Grammalidis, N. Dance analysis using multiple kinect sensors. In Proceedings of the 2014 International Conference on Computer Vision Theory and Applications (VISAPP), Lisbon, Portugal, 5–8 January 2014; Volume 2.
12. Kwon, B.; Kim, D.; Kim, J.; Lee, I.; Kim, J.; Oh, H.; Kim, H.; Lee, S. Implementation of human action recognition system using multiple Kinect sensors. In Proceedings of the Pacific Rim Conference on Multimedia, Gwangju, Korea, 16–18 September 2015.
13. Beom, K.; Kim, J.; Lee, S. An enhanced multi-view human action recognition system for virtual training simulator. In Proceedings of the 2016 Asia-Pacific Signal and Information Processing Association Annual Summit and Conference (APSIPA), Jeju, Korea, 13–16 December 2016.
14. Liu, T.; Song, Y.; Gu, Y.; Li, A. Human action recognition based on depth images from Microsoft Kinect. In Proceedings of the 2013 Fourth Global Congress on Intelligent Systems, Hong Kong, China, 3–4 December 2013.
15. Berger, K.; Ruhl, K.; Schroeder, Y.; Bruemmer, C.; Scholz, A.; Magnor, M.A. Marker-less motion capture using multiple color-depth sensors. In Proceedings of the the Vision, Modeling, and Visualization Workshop 2011, Berlin, Germany, 4–6 October 2011.
16. Kaenchan, S.; Mongkolnam, P.; Watanapa, B.; Sathienpong, S. Automatic multiple kinect cameras setting for simple walking posture analysis. In Proceedings of the 2013 International Computer Science and Engineering Conference (ICSEC), Nakorn Pathom, Thailand, 4–6 September 2013.
17. Kim, J.; Lee, I.; Kim, J.; Lee, S. Implementation of an Omnidirectional Human Motion Capture System Using Multiple Kinect Sensors. *IEICE Trans. Fundam.* **2015**, *98*, 2004–2008. [CrossRef]
18. Taylor, G.S.; Barnett, J.S. Evaluation of Wearable Simulation Interface for Military Training. *Hum Factors* **2012**, *55*, 672–690. [CrossRef] [PubMed]
19. Barnett, J.S.; Taylor, G.S. *Usability of Wearable and Desktop Game-Based Simulations: A Heuristic Evaluation*; Army Research Inst for the Behavioral and Social Sciences: Alexandria, VA, USA, 2010.
20. Bink, M.L.; Injurgio, V.J.; James, D.R.; Miller, J.T., II. *Training Capability Data for Dismounted Soldier Training System*; No. ARI-RN-1986; Army Research Inst for the Behavioral and Social Sciences: Fort Belvoir, VA, USA, 2015.
21. Cavallari, R.; Martelli, F.; Rosini, R.; Buratti, C.; Verdone, R. A Survey on Wireless Body Area Networks: Technologies and Design Challenges. *IEEE Commun. Surv. Tutor.* **2014**, *16*, 1635–1657. [CrossRef]
22. Alam, M.M.; Ben Hamida, E. Surveying wearable human assistive technology for life and safety critical applications: Standards, challenges and opportunities. *Sensors* **2014**, *14*, 9153–9209. [CrossRef] [PubMed]
23. Bukhari, S.H.R.; Rehmani, M.H.; Siraj, S. A Survey of Channel Bonding for Wireless Networks and Guidelines of Channel Bonding for Futuristic Cognitive Radio Sensor Networks. *IEEE Commun. Surv. Tutor.* **2016**, *18*, 924–948. [CrossRef]

24. Ambroziak, S.J.; Correia, L.M.; Katulski, R.J.; Mackowiak, M.; Oliveira, C.; Sadowski, J.; Turbic, K. An Off-Body Channel Model for Body Area Networks in Indoor Environments. *IEEE Trans. Antennas Propag.* **2016**, *64*, 4022–4035. [CrossRef]
25. Seo, S.; Bang, H.; Lee, H. Coloring-based scheduling for interactive game application with wireless body area networks. *J. Supercomput.* **2015**, *72*, 185–195. [CrossRef]
26. Xsens MVN System. Available online: https://www.xsens.com/products/xsens-mvn-animate/ (accessed on 21 January 2019).
27. Tian, Y.; Wei, H.X.; Tan, J.D. An Adaptive-Gain Complementary Filter for Real-Time Human Motion Tracking with MARG Sensors in Free-Living Environments. *IEEE Trans. Neural Syst. Rehabil. Eng.* **2013**, *21*, 254–264. [CrossRef] [PubMed]
28. Euston, M.; Coote, P.; Mahony, R.; Kim, J.; Hamel, T. A complementary filter for attitude estimation of a fixed-wing UAV. In Proceedings of the 2008 IEEE/RSJ International Conference on Intelligent Robots and Systems, Nice, France, 22–26 September 2008.
29. Yoo, T.S.; Hong, S.K.; Yoon, H.M.; Park, S. Gain-Scheduled Complementary Filter Design for a MEMS Based Attitude and Heading Reference System. *Sensors* **2011**, *11*, 3816–3830. [CrossRef] [PubMed]
30. Wu, Y.; Liu, K.S.; Stankovic, J.A.; He, T.; Lin, S. Efficient Multichannel Communications in Wireless Sensor Networks. *ACM Trans. Sens. Netw.* **2016**, *12*, 1–23. [CrossRef]
31. Fafoutis, X.; Marchegiani, L.; Papadopoulos, G.Z.; Piechocki, R.; Tryfonas, T.; Oikonomou, G.Z. Privacy Leakage of Physical Activity Levels in Wireless Embedded Wearable Systems. *IEEE Signal Process. Lett.* **2017**, *24*, 136–140. [CrossRef]
32. Ozcan, K.; Velipasalar, S. Wearable Camera- and Accelerometer-based Fall Detection on Portable Devices. *IEEE Embed. Syst. Lett.* **2016**, *8*, 6–9. [CrossRef]
33. Ferracani, A.; Pezzatini, D.; Bianchini, J.; Biscini, G.; Del Bimbo, A. Locomotion by Natural Gestures for Immersive Virtual Environments. In Proceedings of the 1st International Workshop on Multimedia Alternate Realities, Amsterdam, The Netherlands, 16 October 2016.
34. Kuipers, J.B. *Quaternions and Rotation Sequences*; Princeton University Press: Princeton, NJ, USA, 1999; Volume 66.
35. Karney, C.F. Quaternions in molecular modeling. *J. Mol. Graph. Model.* **2007**, *25*, 595–604. [CrossRef] [PubMed]
36. Gebre-Egziabher, D.; Elkaim, G.H.; Powell, J.D.; Parkinson, B.W. A gyro-free quaternion-based attitude determination system suitable for implementation using low cost sensors. In Proceedings of the IEEE Position Location and Navigation Symposium, San Diego, CA, USA, 13–16 March 2000.
37. Horn, B.K.P.; Hilden, H.M.; Negahdaripour, S. Closed-form solution of absolute orientation using orthonormal matrices. *JOSA A* **1988**, *5*, 1127–1135. [CrossRef]
38. Craig, J.J. *Introduction to Robotics: Mechanics and Control*; Pearson/Prentice Hall: Upper Saddle River, NJ, USA, 2005; Volume 3.

Article

# A Processing-in-Memory Architecture Programming Paradigm for Wireless Internet-of-Things Applications

**Xu Yang [1], Yumin Hou [2] and Hu He [2,*]**

1 School of Computer Science and Technology, Beijing Institute of Technology, Beijing 100081, China; yangxu@tsinghua.edu.cn

2 Institute of Microelectronics, Tsinghua University, Beijing 100084, China; hou-ym12@mails.tsinghua.edu.cn

* Correspondence: hehu@tsinghua.edu.cn; Tel.: +86-010-6279-5139

Received: 6 December 2018; Accepted: 27 December 2018; Published: 3 January 2019

**Abstract:** The widespread applications of the wireless Internet of Things (IoT) is one of the leading factors in the emerging of Big Data. Huge amounts of data need to be transferred and processed. The bandwidth and latency of data transfers have posed a new challenge for traditional computing systems. Under Big Data application scenarios, the movement of large scales of data would influence performance, power efficiency, and reliability, which are the three fundamental attributes of a computing system. Thus, changes in the computing paradigm are demanding. Processing-in- Memory (PIM), aiming at placing computation as close as possible to memory, has become of great interest to academia as well as industries. In this work, we propose a programming paradigm for PIM architecture that is suitable for wireless IoT applications. A data-transferring mechanism and middleware architecture are presented. We present our methods and experiences on simulation-platform design, as well as FPGA demo design, for PIM architecture. Typical applications in IoT, such as multimedia and MapReduce programs, are used as demonstration of our method's validity and efficiency. The programs could successfully run on the simulation platform built based on Gem5 and on the FPGA demo. Results show that our method could largely reduce power consumption and execution time for those programs, which is very beneficial in IoT applications.

**Keywords:** Processing-in-Memory; programming paradigm; Internet of Things

---

## 1. Introduction

We have entered the Era of Big Data, and the world is encountering the processing evolution of those Big Data. Existing systems used in Big Data processing are becoming less energy-efficient and fail to scale in terms of power consumption and area [1,2]. The widespread applications of the wireless Internet of Things (IoT) is one of the leading factors in the emerging of Big Data. Huge amounts of data need to be transferred and processed. Under Big Data application scenarios, the movement of large scales of data influences performance, power efficiency, and reliability, which are the three fundamental attributes of a computing system.

The trend of the ever-growing number of applications of wireless IoT is leading to changes in the computing paradigm and, in particular, to the notion of moving computation to data in what we call the Processing-in-Memory (PIM) approach. A traditional computing architecture is shown in Figure 1. Computing units may include the CPU, GPU, and DSP. Data are transferred between the computing units and the main memory through the memory-hierarchy levels. The bottleneck of data processing for a traditional computing architecture is the bandwidth and latency of data transfer, since a large amount of data are stored in the DRAM [3,4]. Although processors have large caches and an embedded memory, there is an increasing number of data stored in DRAM for high-throughput applications

(Big Data processing scenarios, as well as radar-signal processing, video/image data processing, deep learning, etc.). If we want to overcome the shortage of traditional computing architectures to better suit Big Data processing, we need to move some computation units to DRAM to exploit PIM technology. With the evolution of emerging DRAM technologies, PIM has now become of great interest to academia as well as different industries [5,6] after a period of dormancy. PIM prototypes always integrate simple processing units with DRAM arrays to minimize data movement and perform computation right at data's home. This is in contrast to the movement of data toward a CPU independent of where it resides, as it is done traditionally. It is also proposed that data computation can be performed in caches, or persistent storage (Solid State Drive—SSD) [7].

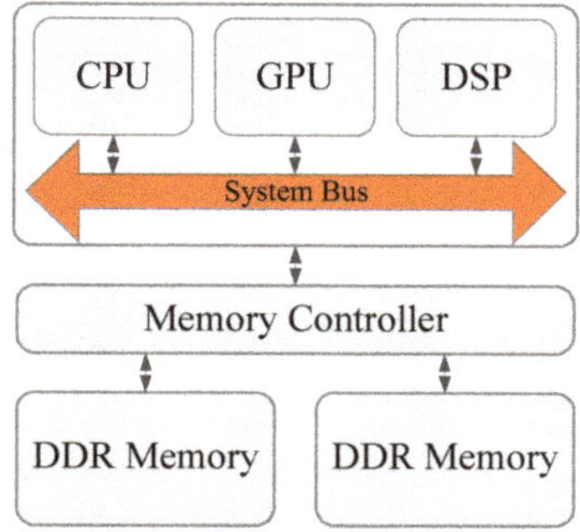


**Figure 1.** Traditional computing architecture.

Figure 2 shows the PIM concept. On the basis of a traditional computing architecture, a computing unit (we call it the PIM core in this paper) is located near DDR memory. Three-dimensional packaging technology supports the integration of DDR memory and the PIM core (the circled parts in Figure 2). Three-dimensional packaging technology stacks heterogenous layers, including DRAM dice and a logic die, in a single chip, which is called a PIM device, as shown in Figure 3. The logic layer includes the PIM core, DMA, Through-Silicon Via (TSV) interface, BUS, and DDR PHY. Companies such as Micron and Samsung are dedicated to the exploration of 3D packaging technology. Products such as Hybrid Memory Cube (HMC) [8] and High Bandwidth Memory (HBM) [9,10] have already been released to markets. Three-dimensionally packaged memory provides a new approach to the memory system architecture. Heterogeneous layers are stacked with significantly more connections. TSVs enable thousands of connections of the stacked layers. The PIM core can quickly access the DDR memory. The HMC also provides a fast connection to the host CPU.

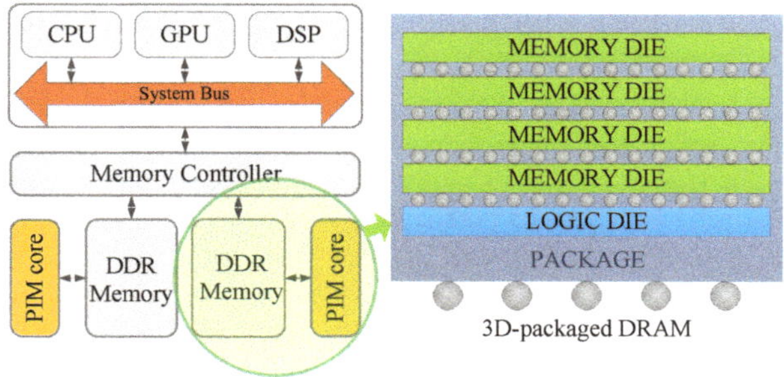


**Figure 2.** Processing-in-Memory (PIM) concept.

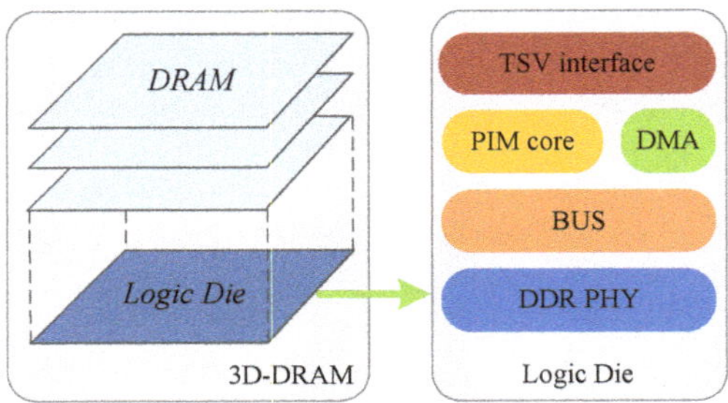


**Figure 3.** PIM device.

By adopting PIM, a PIM device becomes a small computing engine. With the PIM core fulfilling the majority of the computing tasks, the data-transfer rate between CPU and memory is largely reduced. In modern computing systems, load/store operation consumes much more power than data-processing operations. For the Intel XEON CPU, power consumed by the data transfer between CPU and memory is 19.5 times of that between CPU and the L1 cache [11]. Thus, PIM adoption largely reduces the power consumption of the computing system. By moving data-processing units to the memory, the burden of the CPU can be lightened, and the area of the CPU can also be reduced.

In typical IoT applications, huge amount of devices are connected to form a huge data transmissions and an interaction network. Data collected from end-devices might be raw, doubtful, and in large amounts. PIM could help in performing necessary end-device data preprocessing to provide a preliminary understanding of those collected data. Further compressing the amount of data needs to be transferred and exchanged, which might be very beneficial for IoT applications, where energy is very important.

In this work, we present our methods and experiences on simulation-platform design, as well as FPGA demo design, for a PIM architecture. The proposed programming paradigm was verified on both platforms. The following of this paper is organized as follows: Related works are discussed in Section 2. The target PIM architecture is introduced in Section 3. We provide details about our programming paradigm in Section 4. In Section 5, we describe our approach of the implementation of a simulator, and an FPGA demo, for the presented PIM architecture. The experiment results based on the simulation platform are given in Section 6. We show the application prospects of the PIM architecture in Section 7. Finally, we draw conclusions in Section 8.

## 2. Related Works

Many works have been done on PIM since the 1990s. EXECUBE [12], the first multinode PIM chip based on DRAM, was developed in 1992. During the same era, Intelligent RAM (IRAM) [13,14], Computational RAM (CRAM) [15], Parallel Processing RAM (PPRAM) [16], FlexRAM [5], DIVA [17], Smart Memories [18], and Intelligent Memory Manger [19] were developed. IRAM was designed for multimedia applications that include large amounts of computation. CRAM integrates many simple one-bit data-processing elements at the sense amplifiers of a standard RAM to realize the high-speed execution of applications with high parallelism. Most PIM projects of that era share the same characteristics. The PIM chip always includes several simple processing cores located near the DRAM arrays. They all realize high-speed parallel data processing.

Though promising results were witnessed at that time, no widespread commercial product emerged because producing such PIM chips was very expensive. After decades of dormancy, the interest to study PIM has been revived. With the emergence of 3D-staking technology, Big Data workloads and distributed programming paradigm, a new concept of Near Data Processing (NDP) was proposed.

Recent studies on NDP include References [1,20–23], of which References [1,22,23] all propose an ARM-like core as the PIM core integrated in 3D-DRAM. Our research most resembles Reference [1]. We both propose an ARM-like PIM core and we used gem5 [24] to realize the simulation of the PIM architecture. McPAT [25] was applied to analyze power consumption. The difference is that Reference [1] focused on PIM design-space exploration, simply simulating the PIM core based on the gem5 simulator. In our research, we simulated the PIM computing system, including both the host processor and the PIM core. Workloads were assigned between host processor and the PIM core, and communication between them was realized.

The contributions of this paper are listed as follows:

- We propose a programming paradigm for PIM architecture. Drivers and APIs were implemented. An elaborate programming example is provided.
- We simulated a complete PIM computing system, including the host CPU and PIM cores, based on the gem5 simulator. We implemented the proposed programming paradigm using system calls.
- We built a board-to-board FPGA demo for the PIM architecture. The proposed programming paradigm was verified in this demo.
- We provide a performance comparison between the PIM computing architecture and traditional architectures.
- We show the application prospects of the PIM architecture, where our programming paradigm could also be utilized.

## 3. Target Architecture

We first introduce the design philosophy of the PIM architecture. As shown in Figure 4, the PIM architecture (shown inside the large gray rectangle) should be composed of a host CPU, a PIM core, memory controller, and memory (DRAM in Figure 4). The host CPU should be a computation light processor. The host CPU, for example, can be an ARM without a NEON Out-of-Order core. According to the design methodology of PIM systems, the PIM core is integrated in the same chip with DRAM. The PIM core could be an SIMD machine, a GPU-like multithreading machine [4], a reconfigurable array, many core systems, etc. References [26,27] also states that an SIMD/VLIW/vector processor is fit for the data-processing unit in a PIM system. The host CPU and PIM cores share the same physical memory. The host CPU can access the whole memory space, making some memory space in the DRAM uncacheable. The PIM core uses the uncacheable memory space to run the program. The cached memory space is read-only for the PIM core. Users can control the PIM core through drivers. The software-development environment provides APIs for programmers.

Based on this design philosophy, we designed a PIM architecture as shown in Figure 5. The whole system consists of a host chip and a DDR chip. The host chip contains the host CPU, CPU cache, CPU TLB, internal system bus, and DDR interface. The DDR chip is a PIM device that contains the DDR memory, PIM core, and DMA. We used a simple ARM core as the PIM core, which could be configured to have one or two PIM cores. In the example presented in Figure 5, the number of PIM cores was set to two. More PIM cores can be placed in the logic layer of a 3D-packaged memory, but we limited the number to reduce simulation time. The host CPU is in charge of running the OS to control the whole system, while the PIM cores (core1 and core2) on the DDR chip focus on applications that need large amounts of computation and memory access. The CPU and the PIM cores can claim buffers in DDR memory for program execution, as shown in Figure 5. For the efficient transfer of large-scale data between the buffer of the CPU program and the buffers of PIM core program, a DMA was integrated in the DDR chip.

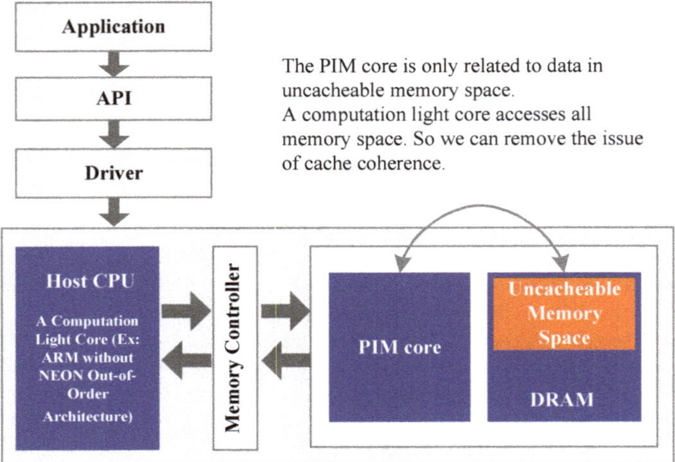


**Figure 4.** Design philosophy of the PIM architecture.

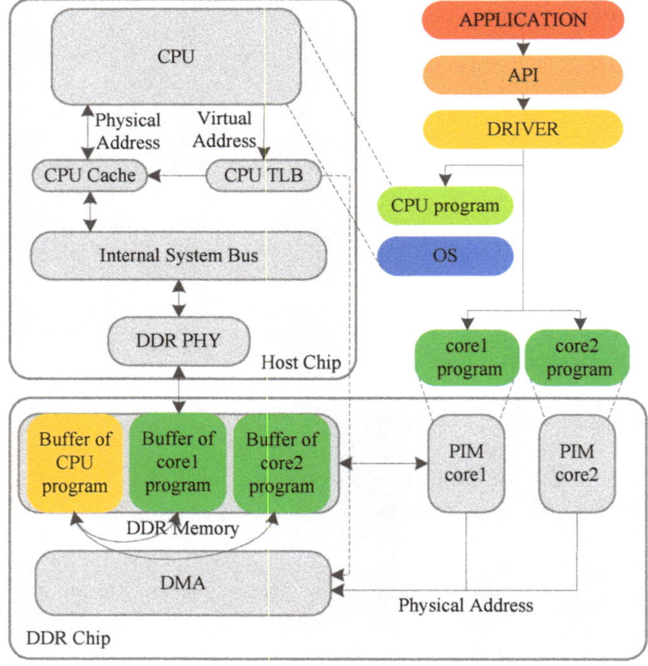


**Figure 5.** Presented PIM architecture.

## 4. Programming Paradigm

In order to fully explore the potential of the PIM computing architecture, we have designed a programming paradigm, which is discussed in detail in this section.

### 4.1. Task-Dividing Mechanism

First, a program should be analyzed to identify the behavior of each part.

As discussed before, in the presented PIM computing architecture, there is one host CPU and one PIM device. The host CPU is mainly in charge of running OS, JVM, and control-intensive tasks. The PIM cores in the PIM device are used to deal with a large scale of data processing. So, according to the results of program analysis, a program is divided into two parts: control-intensive tasks and data-intensive tasks. Control-intensive tasks are assigned to the host CPU, while data-intensive tasks are assigned to PIM cores. The division of tasks should follow predefined rules to ensure the granularity of those tasks in an appropriate level, thus reducing unnecessary frequent intercommunications between the host CPU and PIM cores.

### 4.2. Data-Transferring Mechanism

After task division, the data flow of the program between host CPU and PIM cores is clear. Then, the data-transferring mechanism for this program should be designed.

The data-exchange mechanism between the host CPU and the PIM cores is illustrated in Figure 6. In the target architecture, the CPU and the PIM cores share the same physical memory space. The CPU can access the memory by two means. Generally, data are transferred between CPU and memory through the CPU cache. The CPU can also transfer large-scale data to the PIM memory space through the DMA. The DMA does not adhere to the memory-consistent protocol, so the CPU has to perform a special process before exchanging data with the PIM cores. Before the CPU transfers data to the PIM cores, it should flush the cache data into the memory. After the PIM cores transfer data to the CPU, the CPU has to invalidate the corresponding cache line, and refresh the data from the memory. The PIM cores can directly access the DDR memory.

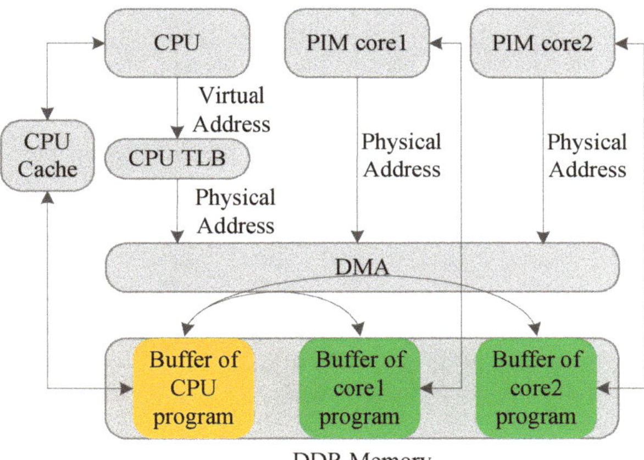


**Figure 6.** Data-transferring mechanism.

Some basic functions are designed to realize the operations stated above, as shown in Figure 7. With the help of those functions, the data-transferring mechanism between the host CPU and PIM cores is designed and decided.

| Function name | Description |
|---|---|
| settarget-memory | When executed by CPU, it creates a memory space to receive data from a PIM core. It do the same thing for PIM cores. |
| switch2cpu | Switch from current PIM core to CPU and suspend the PIM core |
| switch2dev | Switch from CPU to a specified PIM core, and suspend CPU. |
| cpu2dev | Switch from CPU to a PIM core, and transfer data to the PIM core. |
| cpu2devs | Switch from CPU to current PIM cores, and transfer data to them. |
| dev2cpu | Switch from current PIM core to CPU, and transfer data to CPU. |
| freecpu | Free CPU to run. |
| freedev | Free a specified PIM core to run. |
| freedevs | Free all PIM cores to run. |
| invalidate-cache | After a PIM core transfer data to CPU, CPU has to invalidate the cache line associate with memory space of CPU buffer. |
| suspendcpu | Suspend CPU. |
| suspenddev | Suspend PIM cores. |
| cacheflush | Before CPU transfer data to PIM core, CPU should first flush cache data into memory. |
| getdevID | Get current PIM device ID. |
| wakeupcpu | Wakeup CPU. |
| wakeupdev | Wakeup a specified PIM core. |

**Figure 7.** Basic functions supporting the proposed programming paradigm.

### 4.3. Software-Level Architecture

The software-level architecture of the PIM computing system is shown in Figure 8. This includes application, API, driver, and firmware. User application programs and API are all located in the file system. API codes are substantial library functions. Drivers run on the host CPU. Through the drivers, the CPU can interact with the firmware running on the PIM device, and control the operation the PIM device.

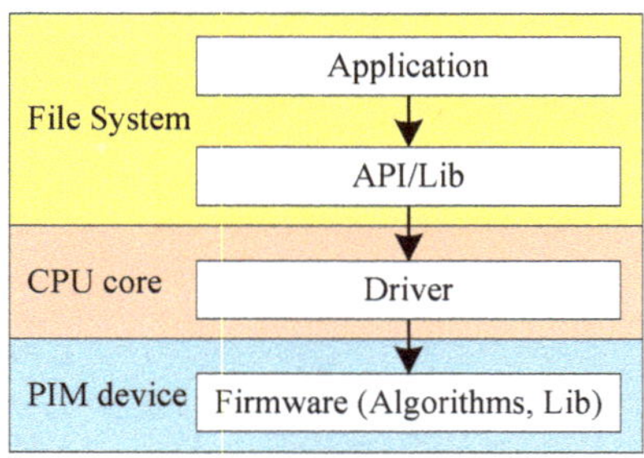


**Figure 8.** Software-level architecture of the PIM computing system.

Firmware is a microsystem running on the PIM device. Note that the PIM device does not run an operating system. The firmware allows the PIM device to interact with the user. It makes specific allocation of the whole address space. The firmware includes application-targeted algorithms. It provides plenty of functions that can be called in user applications. Users can call these functions by calling the PIM_context, which we define below. PIM_context includes three pointers. User_func_pointer points to the user function. PIM_func_pointer points to PIM firmware functions. Vars_pointer points to global variables used in these functions. PIM_context enables the PIM device to receive CPU-transferred programs and data. This helps improve users' programming efficiency. In the firmware operation process, firmware is started at first. Then, it waits until the data and algorithm are configured. Firmware uses the data to fulfil the algorithm execution. After execution is finished, it informs the driver of the end of execution, and stops working.

```
typedef struct{
    void *user_func_pointer;
    void *PIM_func_pointer;
    void *vars_pointer;
}PIM_context_t;
```

PIM drivers provide the following functions:

- **Firmware download:** The PIM device receives firmware sent by the user and downloads it to a specified location. Then, it frees the PIM device to run the firmware.
- **Data transfer:** This includes data send and receive. After firmware download is finished, data sent by a user are transferred to the PIM device and stored in a specific firmware location. After computation is finished, the specific length of the data is obtained from the specific location of the firmware, and then, the data are sent to the user.
- **Algorithm configuration and execution:** When the firmware is downloaded, the user can decide which algorithm the PIM device will run, and instruct the PIM device to start execution. The algorithm can be provided by the firmware or by the users.
- **Status check:** The user can check the status of the PIM device during execution. Only the PIM device itself can update its status. PIM device status includes PIM_start, PIM_wait_data, PIM_check_alg, PIM_running, and PIM_finish.

APIs can be divided into user-mode APIs and kernel-mode APIs. Kernel-mode APIs provide an interface to call drivers. User-mode APIs encapsulate kernel-mode APIs, and are more convenient for users to use user-mode APIs. User-mode APIs provide the following functions (A means address, indicating pointer type):

- **File operation :** to obtain file size, and read file to buffer.

  ```
  get_file_size(A file)
  read_file(A buffer, A file)
  ```

- **Function transfer :** CPU transfer user functions to PIM device. This realizes input and output buffer management for the PIM device. Since the CPU may transfer multiple functions to the PIM device, we should specify the main function running on the PIM device by the entry pointer.

  ```
  build_buf(A Obuf, A entry, A Ibuf, len)
  free_buf(A buffer)
  ```

- **Driver interaction :** The CPU obtains the PIM device information, and updates the firmware on the PIM device.

  ```
  find_device(A PIM_device)
  update_firmware(PIM_device, A buffer, len)
  ```

- **Operational configuration :** The CPU configures the PIM device to conduct computation. It chooses the algorithm on the PIM device firmware, sends and collects computation data, obtains the computation status of the PIM device, and waits for the PIM device computation to finish.

  ```
  set_algorithm(PIM_device, alg)
  get_data(PIM_device, A buffer, len)
  put_data(PIM_device, A recv_buffer, len)
  check_status(PIM_device)
  wait(PIM_device)
  ```

### *4.4. Programming Instructions*

In a program running on PIM computing architecture, the host CPU and PIM device interact with each other through PIM_context, as shown in Figure 9. When programming, users should use the PIM_func prefix to indicate the function running on PIM device, and use the PIM_vars prefix to indicate global variables used by PIM device functions. The PIM entry function and PIM variables are included in PIM_context. Through PIM_context, PIM device can call user functions, as well as firmware functions. Users should follow predefined rules to call functions and to use global variables.

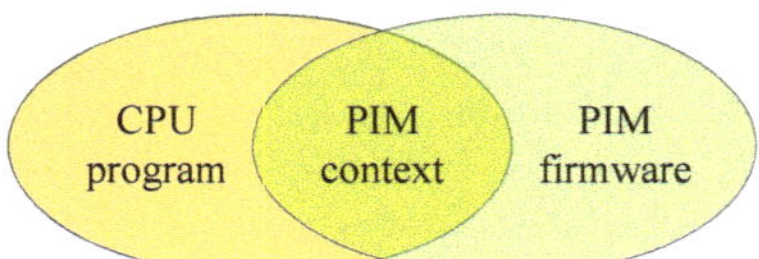


**Figure 9.** Program structure for the PIM computing architecture.

Execution of a program targeting the PIM computing architecture includes the following steps: find PIM device → send firmware function → set algorithm → send data → wait for computation finish → receive data. A program example is given below.

```
/*A program example*/
/*some parameters are omitted,
data types are omitted*/

//callee on PIM device
PIM_func int callee()

//entry function run on PIM device
PIM_func int entry(PIM_context)
{
  PIM_CALL_USER_FUNC(PIM_context, callee)
}

//firmware file
#define FIRMWARE "PIM_EXEC.bin"

int intput[n];
int output[n];
int main()
{
  //get PIM device
  find_device(&PIM_dev);
```

```
    //send program run on PIM device
    size = get_file_size(FIRMWARE);
    buffer = malloc(size);
    read_file(buffer, FIRMWARE);
    update_firmware(PIM_dev, buffer, size);
    free(buffer);

    //set algorithm
    set_algorithm(PIM_dev, ALG);

    //send function and variables
    buffer = 0;
    size = build_buf(&buffer, entry, input)
    get_data(PIM_dev, buffer, size);
    free_buf(buffer);

    //wait for computation finish
    wait(PIM_dev);

    //collect computation data
    put_data(PIM_dev, output);
}
```

## 5. Evaluation Platform Design

### 5.1. Simulator Based on Gem5

We have built a simulation platform based on gem5 [24] to evaluate the PIM computing architecture, and to verify the proposed programming paradigm. During the implementation of this platform, we experienced new challenges. Gem5 [28,29] is an open-source platform for computer-system architecture research, encompassing system-level architecture as well as processor microarchitecture. Gem5 is written in the C++ and Python languages. It has several interchangeable CPU models, and can support the simulating of multiprocessor systems. Gem5 has event-driven memory systems. However, the Gem5 simulator does not support the EPIC architecture processor models and VLIW ISA simulation.

A fast simulation methodology is crucial for exploring a sufficiently broad spectrum of applications and relevant design points. An evaluation method proposed in Reference [4] is first gathering hardware performance and power statics during execution on the current hardware. Then, the data are fed into a machine-learning model that predicts the performance and power on future PIM and host hardware configurations. However, this method is not accurate.

Recently, AMD proposed a work to explore the PIM design space [1]. They used gem5 to simulate the PIM architecture. They used a minor CPU model and a gem5 DRAM module to run MapReduce applications. In their simulation framework, the host processor was not included because gem5 does not yet support such systems.

Gem5 can now support multicore and multisystem simulations. In conjunction with full-system modeling, this feature allows the simulation of entire client–server networks. For multicore simulations, the cores are constrained to use the same CPU model. In the proposed PIM architecture, we put the PIM core near the memory, while the CPU still accessed the memory through memory hierarchy. The CPU and PIM core access memory differently. In gem5, AtomicSimpleCPU is the only CPU model that supports the fast memory (fastmem) access method, which most resembles the memory access method of the PIM core. AtomicSimpleCPU is the simplest CPU model in gem5. It finishes one instruction in one cycle. AtomicSimpleCPU can be used as the PIM core. Different CPU models should be used to simulate the host CPU. We have to modify the gem5 simulator to realize this architecture.

Figure 10 shows the gem5 simulation model we designed. The host CPU was implemented based on the O3CPU model. O3CPU is an Out-of-Order CPU model with five pipeline stages. We used the AtomicSimpleCPU as the PIM core. The number of CPUs and PIM cores can be set to an arbitrary value. The basic functions shown in Figure 7 are realized in the form of system calls on gem5 to support the programming paradigm. CPUs and PIM cores can communicate with each other using system calls.

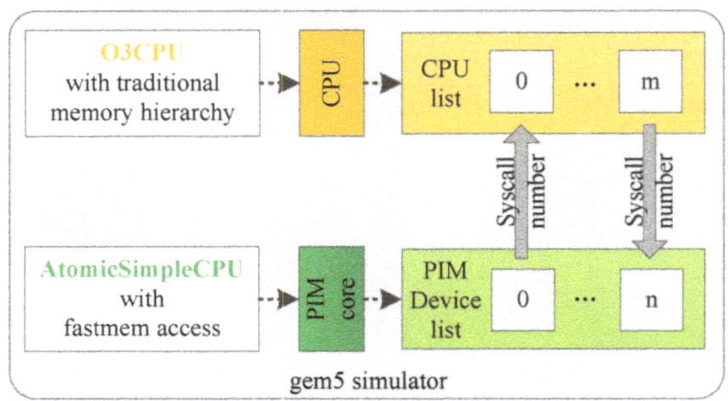


**Figure 10.** Gem5 simulator of the PIM computing architecture.

## 5.2. Board-to-Board FPGA Demo

We also built a board-to-board FPGA demo to verify the proposed programming paradigm. We used two Xilinx ZC706 evaluation boards to build the proposed PIM architecture. One of the boards worked as the master board, and the other board worked as the slave board. The Xilinx ZC706 board embraces a ZYNQ device, which integrates an ARM Cortex-A9 processor with FPGA in a single chip. We used the ARM Cortex-A9 on the master board that works as the host CPU. A self-designed ARM-compatible processor, working as the PIM core, was implemented on the FPGA of the slave board.

As shown in Figure 11, the two boards can be connected by an FPGA Mezzanine Card (FMC) Interface. Host board and slave board can communicate with each other through the chip2chip module, which is software IP supported by the ZC706 board. In the PIM computing system, the slave board can be regarded as a device of the master board. The master board can access the DDR and the control register of the slave board through the chip2chip module by accessing mapped address space. The host CPU can send control signals through the chip2chip master module to the chip2chip slave module. The PIM core receives the control signals and starts working. During execution, the PIM core can access the DDR though the AXI bus. When execution is finished, the slave board can send interrupt signals to the AXI interrupt controller, and then to the host CPU. The host CPU collects the result data from DDR.

On this FPGA demo, we were able to verify that the proposed programming paradigm is feasible and efficient.

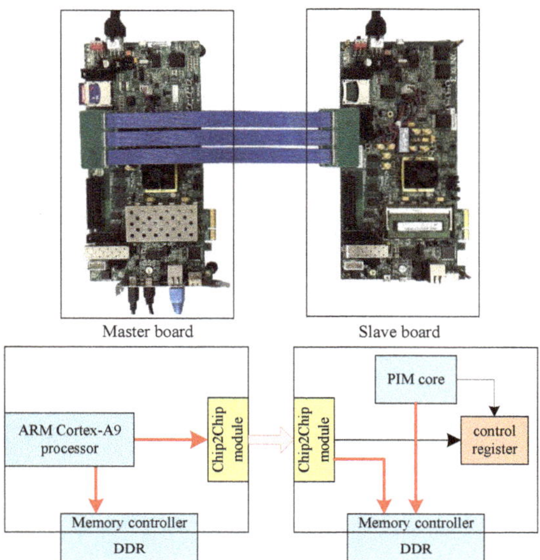


**Figure 11.** FPGA demo.

## 6. Experiments

### 6.1. Experimental Framework

Several experiments were conducted on the gem5 simulation platform. We used the Mpeg2decode and MapReduce programs to test our programming paradigm.

Mpeg2decode programs convert Mpeg2 video bit streams into uncompressed video, so there is much data-intensive work. We chose two code stream files for Mpeg2decode programs to process. The two files were centaur_1.mpg and cinedemo.m2v. Centaur_1.mpg was a black and white image, and cinedemo.m2v was a color image. The color image had a larger bitmap size, and was more time-consuming to decode. These two files are typical and illustrative enough as test files for Mpeg2decode programs.

MapReduce is a popular programming framework for parallel computing of large-scale datasets. We realized MapReduce algorithms in C language. Four testbenches, wordcount, histogram, stringmatch, and matrix-multiply, were implemented. The four MapReduce testbenches were ported to the PIM computing architecture based on the proposed programming paradigm. Note that when the dataset size was 10 MB, the matrix-multiply program was estimated to run days to finish computing on this evaluation platform. So, it was assigned to compute just 10 rows of the matrix under all datasets.

To provide a comprehensive performance comparison between the PIM architecture and other traditional architectures, four architecture models, including CPU-only, PIM, PIM2, and GPU, are referred to in the experiments. Configuration details of the four models are shown in Table 1. CPU-only is a traditional CPU architecture that is modeled by the O3CPU model in gem5. For the PIM model, the host CPU is configured the same as CPU-only model, and the PIM core is configured as the PIM core shown in the table. PIM core is modeled by AtomicSimpleCPU in gem5. For the PIM2 model, the host CPU is also configured in the same way as the CPU-only model, and it had two PIM cores. These three models were all implemented on gem5 simulator. The target ISA is ARM ISA, and the compiler we used was gcc linaro-4.7-2013.06-1. We used McPAT [25] for power analysis. McPAT is fed with the statistic generated by gem5 simulator to provide power-analysis results. Since GPU is widely used in the Big Data domain, we also provide performance comparison with GPU model. The GPU model we used is NVIDIA GeForce GTX480, based on the GPGPU-sim simulation platform. The GPUWattch model in GPGPU-sim is used for power evaluation. GPUWattch is a modified version of McPAT, dedicated for GPU architecture.

**Table 1.** Configuration detail of the test models.

| Architecture | Parameters | |
|---|---|---|
| CPU-only | Out-of-Order | |
| | L1-cache | 64 KB |
| | (64 KB Icache and 64 KB Dcache) | |
| | L2-cache | 1 MB |
| | block size | 64 B |
| | memory capacity | 2 GB |
| | Clock rate | 1 GHz |
| PIM core | in-order | |
| | L1-cache | 64 KB |
| | (64 KB Icache and 64 KB Dcache) | |
| | Clock rate | 1 GHz |
| PIM | CPU-only + one PIM core | |
| PIM2 | CPU-only + two PIM cores | |
| GPU | NVIDIA GeForce GTX480 | |
| | Fermi GPU architecture | |
| | 15 streaming multiprocessors | |
| | each containing 32 cores | |
| | virtual memory page size | 4 GB |
| | Clock rate | 700 MHz |

### *6.2. Results*

#### 6.2.1. Mpeg2decode Programs—CPU-Only vs. PIM

The first test case is Mpeg2decode. We first evaluated the performance of CPU-only and PIM architecture with different CPU cache sizes. Only one level CPU cache was set in this experiment. Two code stream files, centau_1.mpg and cinedemo.m2v, were tested on the CPU-only model and PIM model.

Figure 12 shows the execution-time comparison between CPU-only and PIM model with an increasing CPU data cache size. For the CPU-only model, performance improved with the increase of the CPU data cache size, while the performance of the PIM architecture was not affected by the size of the CPU data cache. The results show that the PIM architecture did not require a large cache size to achieve high performance. For the CPU-only model, a larger data cache brought better performance, since more data could be locally processed. This result also demonstrates the significance of processing data at the data's home. When the data cache size was 64 kB, the performance of the CPU-only model was about twice of the PIM architecture. This is because the speed of the O3CPU was twice that of the AtomicSimpleCPU in gem5.

Processor power consumption, which includes the power consumption of the processor core and L1 cache, is shown in Figure 13. The results show that the power consumption of the PIM model was much less than that of CPU-only model. On average, the PIM architecture reduced processor power consumption by 93% compared to the CPU-only model.

Comparison of cache power consumption is shown in Table 2. Comparison of bus power consumption is shown in Table 3. It is shown that the cache power consumption of the PIM model was much smaller than that of the CPU-only model. On average, cache power consumption of the CPU-only model was about $10^6$ times of that of the PIM architecture. Bus power consumption of the CPU-only model was about $10^5$ times of that of the PIM model.

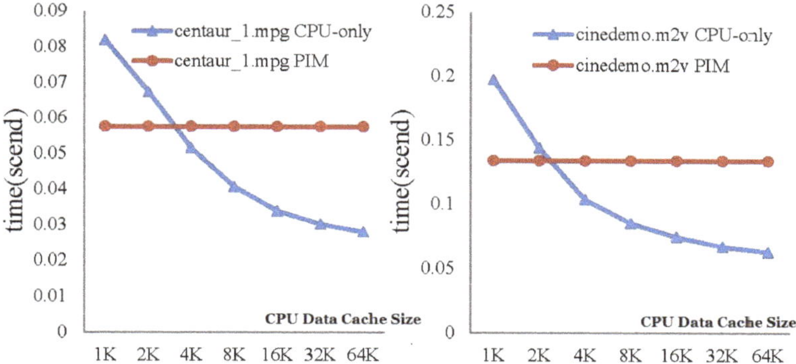


**Figure 12.** Execution-time comparison between the CPU-only and PIM models running Mpeg2decode programs, with an increasing CPU data cache size.

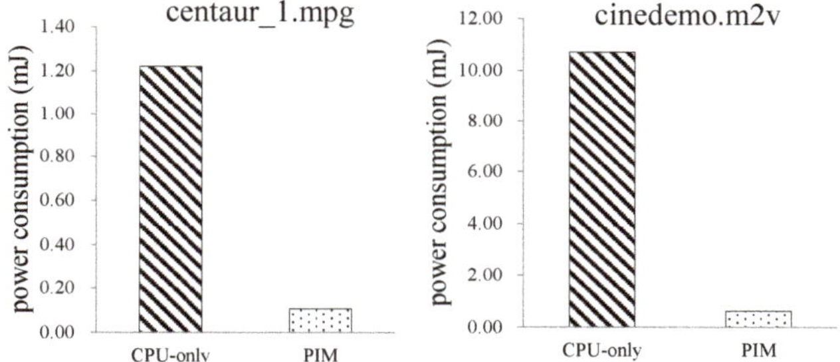


**Figure 13.** Comparison of processor power consumption between the CPU-only and PIM models running Mpeg2decode programs.

**Table 2.** Comparison of cache power consumption between the CPU-only and PIM models running Mpeg2decode programs (mJ).

| | Centaur_1.mpg | Cinedemo.m2v |
|---|---|---|
| CPU-only | $1.27 \times 10^{-2}$ | $3.74 \times 10^{-2}$ |
| PIM | $2.90 \times 10^{-8}$ | $6.72 \times 10^{-9}$ |

**Table 3.** Comparison of bus power consumption between the CPU-only and PIM models running Mpeg2decode programs (mJ).

| | Centaur_1.mpg | Cinedemo.m2v |
|---|---|---|
| CPU-only | $4.33 \times 10^{-3}$ | $1.80 \times 10^{-2}$ |
| PIM | $4.12 \times 10^{-8}$ | $4.15 \times 10^{-8}$ |

### 6.2.2. MapReduce Programs—CPU-Only vs. PIM

The second test case was the MapReduce algorithm. The dataset processed by the four MapReduce programs was about 10 MB for all.

Figure 14 shows the execution-time comparison between the CPU-only and PIM models running the four MapReduce programs. We assumed the time used by the CPU-only model was 100%. We can see from Figure 14, that for wordcount, histogram, and matrix-multiply, the PIM architecture reduced runtime by 44%, 24%, and 15%, respectively. For string-match, PIM architecture ran 30% longer than the CPU-only model. The reason is the disadvantage of the CPU-only model being long memory access delay. For the programs that require frequent memory access, the PIM architecture outperformed the CPU-only model, while for those programs that do not access memory that often, PIM might be slower than the CPU-only model. This result can also be partly attributed to the performance disparity of the AtomicSimpleCPU and O3CPU models in gem5. In other words, in the PIM model, the PIM core ran slower than the host CPU. Table 4 shows the memory access latency of the four programs in the CPU-only model. As we can see, memory access latency takes large percentage of the total runtime for wordcount, histogram, and matrix-multiply, while for string-match, memory access latency was not as significant as the other three programs. Thus, the PIM model showed performance loss when running string-match.

Comparison of processor, cache, and bus power consumption is shown in Figure 15, Tables 5 and 6 separately.

As shown in Figure 15, for the four programs, processor power consumption of the PIM model was reduced by 92.4%, 88.6%, 90.7% and 90.3%, respectively, than the CPU-only model. Table 5 shows that, for the four programs, cache power consumption of CPU-only is $10^3$–$10^4$ times of that of PIM model. We can see from Table 6 that bus power consumption of the CPU-only model was about $10^4$ times that of the PIM architecture. PIM architecture reduced processor, cache, and bus power consumption to a large extent compared to the CPU-only model.

The experimental results demonstrate the characteristics of the PIM architecture. Since the PIM core directly accesses memory, it reduces the total cache and bus access. So, cache and bus power consumption is reduced. This result shows that the PIM architecture is suitable for applications processing large datasets and requiring frequent memory access.

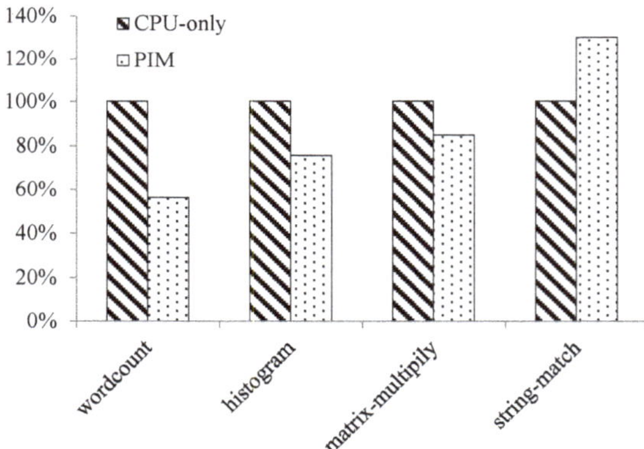


**Figure 14.** Execution-time comparison between the CPU-only model and PIM model running MapReduce programs.

**Table 4.** Memory access latency in the CPU-only model.

| | Memory Access Latency | Others |
|---|---|---|
| wordcount | 24% | 76% |
| histogram | 59% | 41% |
| matrix-multiply | 69% | 31% |
| string-match | 10% | 90% |

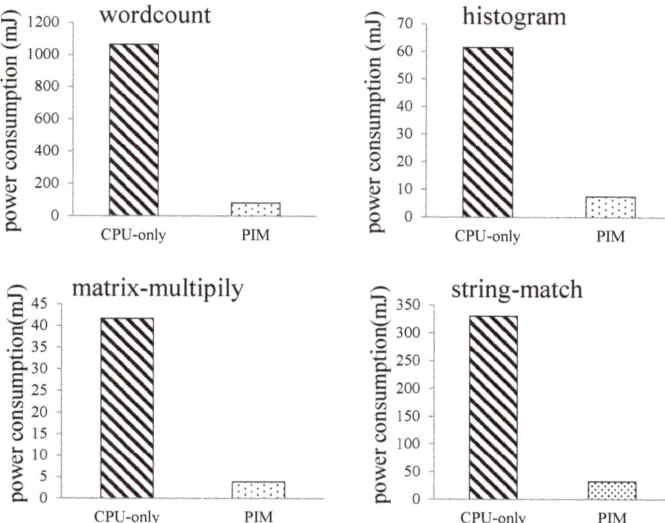


**Figure 15.** Comparison of processor power consumption between the CPU-only model and PIM model running MapReduce programs.

**Table 5.** Comparison of cache power consumption between the CPU-only model and PIM model running MapReduce programs (mJ).

| | CPU-Only | PIM |
|---|---|---|
| wordcount | 0.52 | $4.72 \times 10^{-5}$ |
| histogram | 0.16 | $3.87 \times 10^{-5}$ |
| matrix-multiply | 2.18 | $5.82 \times 10^{-5}$ |
| string-match | 0.13 | $4.77 \times 10^{-5}$ |

**Table 6.** Comparison of bus power consumption between the CPU-only model and PIM model running MapReduce programs (mJ).

| | CPU-Only | PIM |
|---|---|---|
| wordcount | 2.93 | $6.00 \times 10^{-5}$ |
| histogram | 0.40 | $5.26 \times 10^{-5}$ |
| matrix-multiply | 0.87 | $6.30 \times 10^{-5}$ |
| string-match | 0.18 | $5.92 \times 10^{-5}$ |

### 6.2.3. MapReduce Programs—CPU-Only vs. PIM vs. PIM2

In this experiment, we fed different dataset sizes to the MapReduce programs. When input data size increased from 1 to 10 MB, the performance and power consumption of the CPU-only, PIM, and PIM2 models was evaluated.

Figure 16 shows the execution-time comparison of the three models. For wordcount and histogram, the run time of the CPU-only model was always longer than the PIM and PIM2 models. The PIM2 model showed better performance than the PIM model. With the increase of the dataset size, the gap between the three models becomes larger, and the advantage of the PIM architecture becomes more evident. For matrix-multiply, the run time of the three models increased when the data size expanded. The run time of the PIM model became shorter than the CPU-only model when the dataset was larger than 4 MB. The PIM2 model always ran faster than the CPU-only model and PIM model. For string-match, the run time of the PIM model was always longer than the CPU-only model. The reason is that memory access is less frequent in this program, as shown in Table 4, while the PIM2 model showed better performance than the CPU-only model.

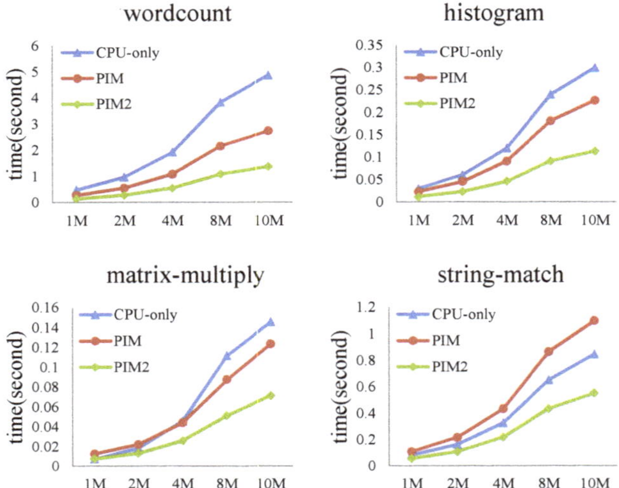


**Figure 16.** Execution-time comparison between the CPU-only, PIM, and PIM2 models running MapReduce programs with increasing datasets.

Comparison of processor power consumption is shown in Figure 17. For the four programs, processor power consumption of the PIM model and PIM2 model were almost the same, since two identical cores share the work that was previously done by a single core. The energy consumed should be approximate. Processor power consumption of the PIM model and PIM2 model slightly increased when the data size expanded. Processor power consumption of the CPU-only model running the four programs increased substantially. This result shows that PIM architecture can largely reduce processor power consumption.

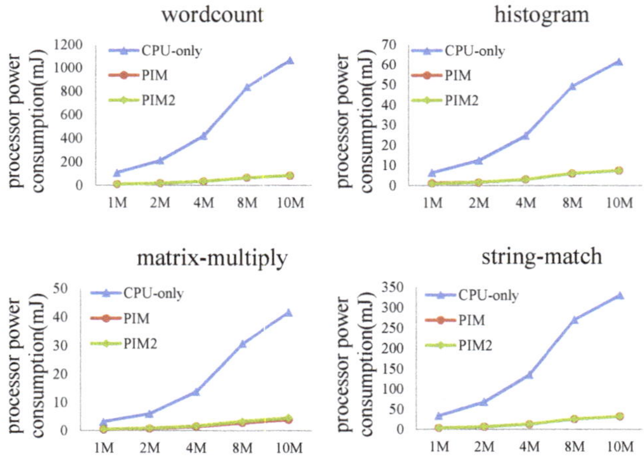


**Figure 17.** Comparison of processor power consumption between the CPU-only, PIM, and PIM2 models running MapReduce programs with increasing datasets.

Comparisons of cache and bus power consumption are shown in Figures 18 and 19, respectively. Logarithmic co-ordinates are adopted in these two figures. Cache and bus power consumption of the CPU-only model increased to a large extent for the four programs, while cache and bus power consumption of the PIM model and PIM2 model were much smaller comparatively.

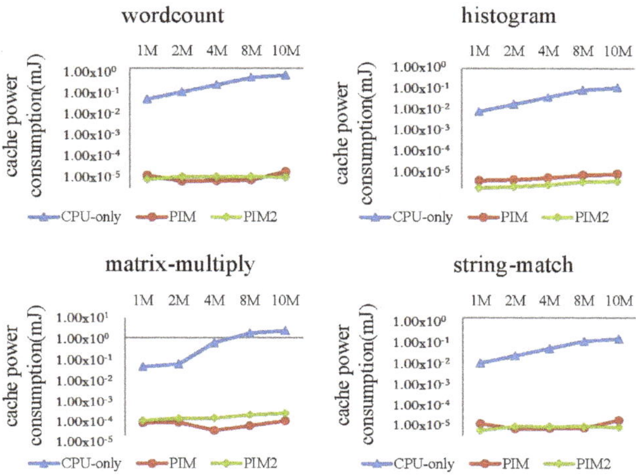


**Figure 18.** Comparison of cache power consumption between the CPU-only, PIM, and PIM2 models running MapReduce programs with increasing datasets.

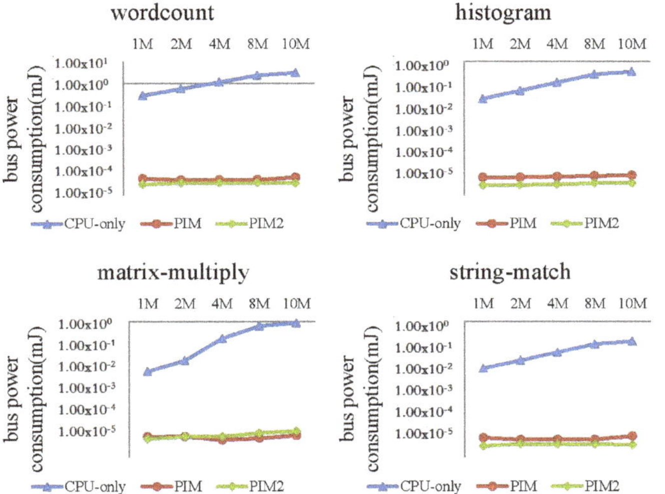


**Figure 19.** Comparison of bus power consumption between CPU-only, PIM, and PIM2 models running MapReduce programs with increasing datasets.

Figure 20 shows the performance per Joule of the three models. Logarithmic co-ordinates were adopted in this figure. The result is the ratio of reciprocal value of total runtime and average energy. We can see from the figure that performance per Joule of the three models decreased with increasing input data size. However, the PIM and PIM2 models showed about one order of magnitude higher performance per Joule than the CPU-only model. The PIM2 model showed better performance per Joule than PIM model.

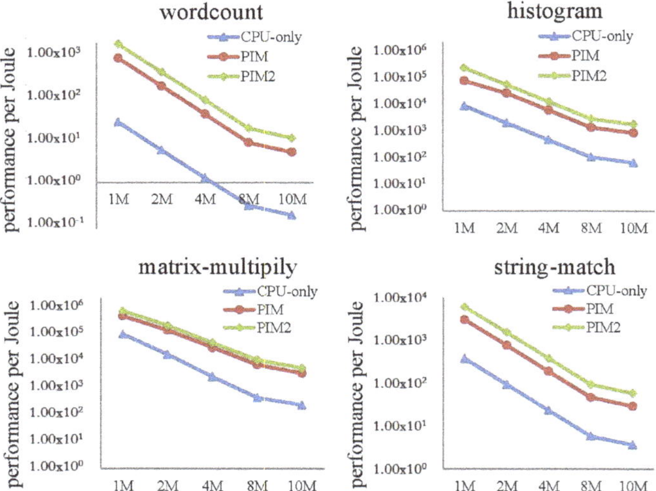


**Figure 20.** Performance per Joule comparison between the CPU-only, PIM, and PIM2 models running MapReduce programs with increasing datasets.

From the experimental results, we can conclude that the run time of the PIM architecture is largely affected by the nature of the application and input dataset size. For the programs requiring frequent memory access, the PIM architecture can improve program performance. For all the tested applications, PIM architecture reduces processor, cache, and bus power consumption to a large extent. When the input dataset size increased, the advantage of the PIM architecture became more noticeable. Performance per Joule of the PIM and PIM2 models was also much higher than the CPU-only model. The PIM2 model could further improve the performance of the PIM architecture, and also showed better performance per Joule.

#### 6.2.4. MapReduce Programs —CPU-Only vs. PIM2 vs. GPU

In this experiment, we ran the four MapReduce programs on the CPU-only, PIM2, and GPU models, with the data size increasing from 1 to 10 MB. Run time and performance per Joule were evaluated.

Figure 21 shows the execution time of the four MapReduce programs on CPU-only model, PIM2 model and GPU model. We can see from the result that the GPU run much faster than CPU-only model and PIM2 model. For the wordcount program, the advantage of GPU was less evident than the other three programs. GPU showed the best performance running the matrix-multiply program. Matrix-multiply is quite computing-intensive, and there are many approaches to accelerate the algorithm on GPU. We adopted one of the approaches in our experiment, while the wordcount program was less suitable to run on GPU. For applications running on GPU, the assignment was divided into many small parts, with each part running on a thread of the GPU. However, for wordcount, we had to assign enough workload to each thread to simplify the final data-collecting work to the CPU.

Figure 22 shows the performance per Joule of the three models. For matrix-multiply, GPU showed the best performance per Joule among the three models. Since this program quite suitable to run on GPU, and the performance advantage is evident enough to hide the high power of GPU. For wordcount, GPU showed the worst performance per Joule. Since the run time of GPU was close to the PIM2 model, as shown in Figure 21, and the power of the GPU was much higher than the CPU-only model and PIM2 model. For histogram and string-match, performance per Joule of the GPU was between the

CPU-only model and PIM2 model. This result shows that the PIM architecture has an advantage over GPU for the applications that are comparatively less computing-intensive. The reason is that the PIM core is a general-purpose processor. By analyzing all the experimental results above, it can be predicted that the PIM architecture can achieve better performance for more computing-intensive tasks if the PIM core is replaced with specialized computing units.

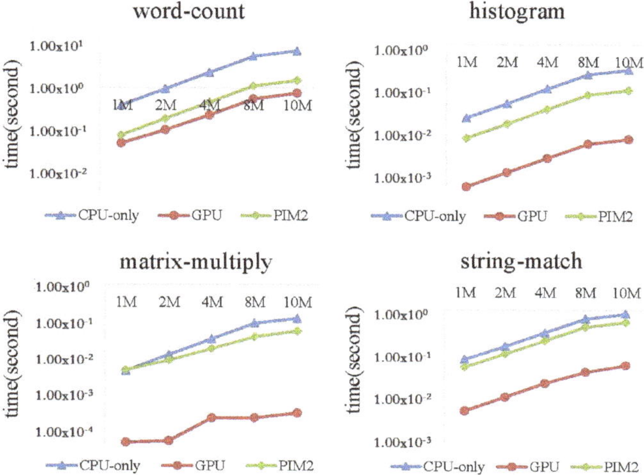


**Figure 21.** Execution-time comparison between the CPU-only, PIM2, and GPU models running MapReduce programs with increasing datasets.

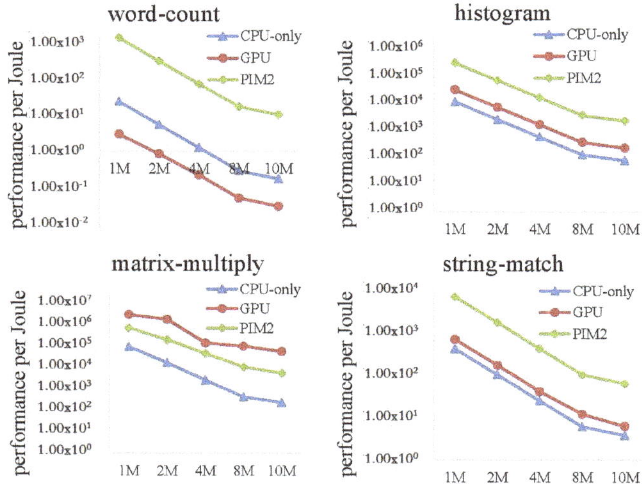


**Figure 22.** Performance per Joule comparison between CPU-only, PIM2, and GPU models running MapReduce programs with increasing datasets.

## 7. Application Prospect

In this paper, we focused on studying a PIM computing architecture with one host CPU and one PIM device. A general-purpose ARM processor was used as the PIM core integrated in the PIM device.

This architecture can be extended by allowing the host CPU access many PIM devices, as shown in Figure 23. The host CPU and the PIM devices can be connected via PCIe and SerDes. This computing architecture can be adopted by future servers to provide instant response service. The ARM processor can be replaced by other computing units according to the target application domain. The proposed programming paradigm can be also applied to the PIM computing architecture with multiple PIM devices. If a processor with embedded flash or ROM is used as the PIM core, firmware can be stored in the embedded flash/ROM to eliminate the operation of sending firmware to PIM devices every time. Users only need to update firmware when necessary. The PIM core can even be designed to run a simple operating system.

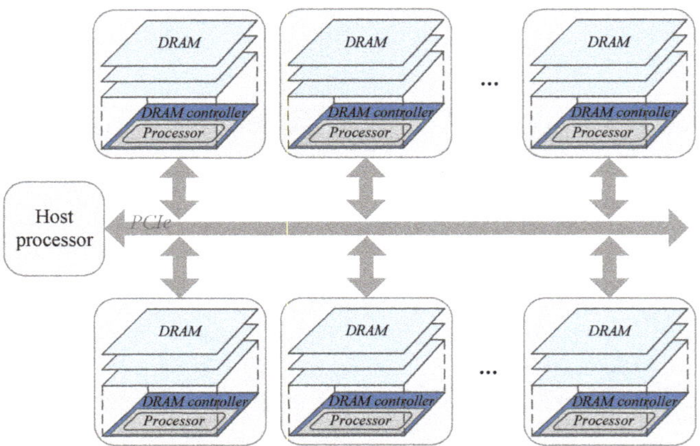


**Figure 23.** Server demo based on PIM computing architecture.

We believe that this programming paradigm might be very helpful for wireless IoT applications. IoT applications now encompass a lot of different domains, such as medicine, surveillance, transportation, and environmental protection. In those IoT applications, a lot of data might be collected in the end-devices. However, not all data, or not all raw data, need to be transferred or exchanged. Thus, PIM could be very beneficial in improving data-transmission and energy efficiency for those end-devices.

## 8. Conclusions

IoT applications are very popular today. It brings alive the Big Data scenario. Under such a scenario, huge amounts of data would be collected, transferred, and exchanged. In this paper, we proposed a programming paradigm for a PIM architecture suitable for wireless IoT applications. This programming paradigm with PIM could help perform data processing in the end-device near where the raw data are collected. Thus, preliminary understanding of the data could be given, and the amount of data needing to be transferred could be largely reduced and, hence, the required energy for IoT devices in data transmission. We have implemented several typical programs that are popular in wireless IoT applications based on the proposed programming paradigm. We ran these programs on a simulation platform, and on an FPGA demo. The proposed programming paradigm was proven to be feasible and efficient. The evaluation results based on the simulation platform were collected. The results show that, by adopting the proposed programming paradigm, we could exploit the benefits coming with a PIM architecture to largely improve data-processing performance and energy efficiency compared to traditional computing architectures. The proposed programming paradigm could also be used in future PIM computing architectures.

**Author Contributions:** Conceptualization, H.H. and Y.H.; Methodology, Y.H.; Software, Y.H.; Validation, X.Y., Y.H. and H.H.; Formal Analysis, Y.H.; Data Curation, Y.H.; Writing-Original Draft Preparation, X.Y.; Writing-Review & Editing, X.Y.; Visualization, X.Y.; Supervision, X.Y. and H.H.; Funding Acquisition, X.Y. and H.H.

**Funding:** This work was supported by the National Natural Science Foundation of China under Grant No. 91846303, the National Natural Science Foundation of China under Grant No. 61502032, and the Tsinghua and Samsung Joint Laboratory.

**Conflicts of Interest:** The authors declare no conflict of interest.

## References

1. Scrbak, M.; Islam, M.; Kavi, K.M.; Ignatowski, M.; Jayasena, N. Processing-in-memory: Exploring the design space. In *Proceedings of the 28th International Conference on Architecture of Computing Systems (ARCS 2015)*; Springer International Publishing: Cham, Switzerland, 2015; pp. 43–54.
2. Ferdman, M.; Adileh, A.; Kocberber, O.; Volos, S.; Alisafaee, M.; Jevdjic, D.; Kaynak, C.; Popescu, A.D.; Ailamaki, A.; Falsafi, B. A case for specialized processors for scale-out workloads. *IEEE Micro* **2014**, *34*, 31–42. [CrossRef]
3. Hennessy, J.L.; Patterson, D.A. *Computer Architecture: A Quantitative Approach*; Elsevier: Amsterdam, The Netherlands, 2011.
4. Zhang, D.; Jayasena, N.; Lyashevsky, A.; Greathouse, J.L.; Xu, L.; Ignatowski, M. Top-pim: Throughput-oriented programmable processing in memory. In Proceedings of the International Symposium on High-performance Parallel and Distributed Computing, Vancouver, BC, Canada, 23–27 June 2014; pp. 85–98.
5. Torrellas, J. Flexram: Toward an advanced intelligent memory system: A retrospective paper. In Proceedings of the 2012 IEEE 30th International Conference on Computer Design (ICCD), Montreal, QC, Canada, 30 September–3 October 2012; pp. 3–4.
6. Zhang, D.; Jayasena, N.; Lyashevsky, A.; Greathouse, J.; Meswani, M.; Nutter, M.; Ignatowski, M. A new perspective on processing-in-memory architecture design. In Proceedings of the ACM SIGPLAN Workshop on Memory Systems Performance and Correctness, Seattle, WA, USA, 16–19 June 2013; pp. 1–3.
7. Balasubramonian, R.; Chang, J.; Manning, T.; Moreno, J.H.; Murphy, R.; Nair, R.; Swanson, S. Near-data processing: Insights from a micro-46 workshop. *IEEE Micro* **2014**, *34*, 36–42. [CrossRef]
8. Jeddeloh, J.; Keeth, B. Hybrid memory cube new dram architecture increases density and performance. In Proceedings of the Symposium on VLSI Technology, Honolulu, HI, USA, 12–14 June 2012; pp. 87–88.
9. Lee, D.U.; Kim, K.W.; Kim, K.W.; Lee, K.S.; Byeon, S.J.; Kim, J.H.; Cho, J.H.; Lee, J.; Chun, J.H. A 1.2v 8 gb 8-channel 128 gb/s high-bandwidth memory (hbm) stacked dram with effective i/o test circuits. *IEEE J. Solid-State Circ.* **2015**, *50*, 191–203. [CrossRef]
10. Jun, H.; Cho, J.; Lee, K.; Son, H.Y.; Kim, K.; Jin, H.; Kim, K. Hbm (high bandwidth memory) dram technology and architecture. In Proceedings of the Memory Workshop (IMW), Monterey, CA, USA, 14–17 May 2017; pp. 1–4.
11. Molka, D.; Hackenberg, D.; Schone, R.; Muller, M.S. Characterizing the energy consumption of data transfers and arithmetic operations on x86-64 processors. In Proceedings of the International Conference on Green Computing, Hangzhou, China, 30 October–1 November 2010; pp. 123–133.
12. Kogge, P.M. Execube—A new architecture for scaleable mpps. In Proceedings of the International Conference on Parallel Processing, Raleigh,NC, USA, 15–19 August 1994; pp. 77–84.
13. Patterson, D.; Anderson, T.; Cardwell, N.; Fromm, R.; Keeton, K.; Kozyrakis, C.; Thomas, R.; Yelick, K. Intelligent RAM (IRAM): Chips that remember and compute. In Proceedings of the IEEE International Solid-State Circuits Conference (43rd ISSCC), San Francisco, CA, USA, 8 February 1997; pp. 224–225.
14. Patterson, D.; Anderson, T.; Cardwell, N.; Fromm, R.; Keeton, K.; Kozyrakis, C.; Thomas, R.; Yelick, K. A case for intelligent RAM: IRAM. *IEEE Micro* **1997**, *17*, 34–44. [CrossRef]
15. Nyasulu, P.M. System Design for a Computational-RAM Logic-In-Memory Parallel-Processing Machine. Ph.D. Thesis, Carleton University, Ottawa, ON, Canada, 1999,
16. Murakami, K.; Inoue, K.; Miyajima, H. Parallel processing ram (ppram). *Comp. Biochem. Physiol. Part A Physiol.* **1997**, *94*, 347–349. [CrossRef]

17. Draper, J.; Chame, J.; Hall, M.; Steele, C.; Barrett, T.; LaCoss, J.; Granacki, J.; Shin, J.; Chen, C.; Kang, C.W.; et al. The architecture of the diva processing-in-memory chip. In Proceedings of the International Conference on Supercomputing, New York, NY, USA, 22–26 June 2002; pp. 26–37.
18. Mai, K.; Paaske, T.; Jayasena, N.; Ho, R.; Dally, W.J.; Horowitz, M. Smart memories: A modular reconfigurable architecture. In Proceedings of the 27th International Symposium on Computer Architecture, Vancouver, BC, Canada, 10–14 June 2000; pp. 161–171.
19. Rezaei, M.; Kavi, K.M. Intelligent memory manager: Reducing cache pollution due to memory management functions. *J. Syst. Archit.* **2006**, *52*, 41–55. [CrossRef]
20. Tseng, H.W.; Tullsen, D.M. Data-triggered multithreading for near-data processing. In Proceedings of the Workshop on Near-Data Processing, Waikiki, HI, USA, 5 December 2003.
21. Chu, M.L.; Jayasena, N.; Zhang, D.P.; Ignatowski, M. High-level programming model abstractions for processing in memory. In Proceedings of the Workshop on Near-Data Processing, Waikiki, HI, USA, 5 December 2003.
22. Pugsley, S.H.; Jestes, J.; Zhang, H.; Balasubramonian, R.; Srinivasan, V.; Buyuktosunoglu, A.; Davis, A.; Li, F. NDC: Analyzing the impact of 3d-stacked memory+logic devices on mapreduce workloads. In Proceedings of the IEEE International Symposium on Performance Analysis of Systems and Software (ISPASS), Monterey, CA, USA, 23–25 March 2014; pp. 190–200.
23. Islam, M.; Scrbak, M.; Kavi, K.M.; Ignatowski, M.; Jayasena, N. *Improving Node-Level MapReduce Performance Using Processing-in-Memory Technologies*; Springer International Publishing: Cham, Switzerland, 2014.
24. Binkert, N.; Beckmann, B.; Black, G.; Reinhardt, S.K.; Saidi, A.; Basu, A.; Hestness, J.; Hower, D.R.; Krishna, T.; Sardashti, S. The gem5 simulator. *ACM SIGARCH Comput. Archit. News* **2011**, *39*, 1–7. [CrossRef]
25. Li, S.; Ahn, J.H.; Strong, R.D.; Brockman, J.B.; Tullsen, D.M.; Jouppi, N.P. Mcpat: An integrated power, area, and timing modeling framework for multicore and manycore architectures. In Proceedings of the 42nd Annual IEEE/ACM International Symposium on Microarchitecture, New York, NY, USA, 12–16 December 2009; pp. 469–480.
26. Ahn, J.; Hong, S.; Yoo, S.; Mutlu, O.; Choi, K. A scalable processing-in-memory accelerator for parallel graph processing. In Proceedings of the ACM/IEEE 42nd Annual International Symposium on Computer Architecture (ISCA), Portland, OR, USA, 13–17 June 2015; pp. 105–117.
27. Morad, A.; Yavits, L.; Ginosar, R. GP-SIMD processing-in-memory. *ACM Trans. Archit. Code Optim.* **2015**, *11*, 53. [CrossRef]
28. The gem5 Simulator. Available online: http://gem5.org/Main_Page (accessed on 30 December 2018).
29. Binkert, N.L.; Dreslinski, R.G.; Hsu, L.R.; Lim, K.T.; Saidi, A.G.; Reinhardt, S.K. The M5 Simulator: Modeling Networked Systems. *IEEE Micro* **2006**, *26*, 52–60. [CrossRef]

Article

# A Semantic-Enabled Platform for Realizing an Interoperable Web of Things

**Jorge Lanza [1], Luis Sánchez [1,*], David Gómez [2], Juan Ramón Santana [1] and Pablo Sotres [1]**

1 Network Planning and Mobile Communications Lab, University of Cantabria, 39012 Santander, Spain; jlanza@tlmat.unican.es (J.L.); jrsantana@tlmat.unican.es (J.R.S.); psotres@tlmat.unican.es (P.S.)
2 Atos Research & Innovation, C/ Albarracín 25, 28037 Madrid, Spain; david.gomez@atos.net
* Correspondence: lsanchez@tlmat.unican.es; Tel.: +34-942-203-940

Received: 15 January 2019; Accepted: 14 February 2019; Published: 19 February 2019

**Abstract:** Nowadays, the Internet of Things (IoT) ecosystem is experiencing a lack of interoperability across the multiple competing platforms that are available. Consequently, service providers can only access vertical data silos that imply high costs and jeopardize their solutions market potential. It is necessary to transform the current situation with competing non-interoperable IoT platforms into a common ecosystem enabling the emergence of cross-platform, cross-standard, and cross-domain IoT services and applications. This paper presents a platform that has been implemented for realizing this vision. It leverages semantic web technologies to address the two key challenges in expanding the IoT beyond product silos into web-scale open ecosystems: data interoperability and resources identification and discovery. The paper provides extensive description of the proposed solution and its implementation details. Regarding the implementation details, it is important to highlight that the platform described in this paper is currently supporting the federation of eleven IoT deployments (from heterogeneous application domains) with over 10,000 IoT devices overall which produce hundreds of thousands of observations per day.

**Keywords:** interoperability; Web-of-Things; semantics; Internet-of-Things; registry

## 1. Introduction

The Internet of Things (IoT) is unanimously identified as one of the main technology enablers for the development of future intelligent environments [1]. It is driving the digital transformation of many different domains (e.g., mobility, environment, industry, healthcare, etc.) of our everyday life. The IoT concept has attracted a lot of attention from the research and innovation community for a number of years already [2–4]. One of the key drivers for this hype towards the IoT is its applicability to a plethora of different application domains [5], like smart cities [6,7], e-health [8,9], smart-environment [10,11], smart-home [12], or Industry 4.0 [13]. This is happening by realizing the paradigm of more instrumented, interconnected, and intelligent scenarios, which are instrumented through low-cost smart sensors and mobile devices that turn the workings of the physical world into massive amounts of data points that can be measured. Interconnected so that different parts of a core system, like networks, applications, and data centers, are joined and "speak" to each other, turning data into information. Finally intelligent, with information being transformed into real-time actionable insights at massive scale through the application of advanced analytics.

"Today, there are roughly 1.5 billion Internet-enabled PCs and over 1 billion Internet-enabled smartphones. The present 'Internet of PCs' will move towards an 'Internet of Things' in which 50 to 100 billion devices will be connected to the Internet by 2020. Some estimations point to the fact that in the same year, the amount of machine sessions will be 30 times higher than the number of mobile person sessions" [14]. The IoT has drastically changed some of the key challenges that Future Internet will have to address. Until recently, researchers have focused on devices and the communication technologies used to access them. However, as it happened with nowadays Internet, most of the revenue is projected to come from the services that can be provided using all these devices and communication networks as a basis. While billions of devices are connecting to the Internet, as it happened with "Internet of PCs", the Web has emerged again as the paradigm to exploit the potential of the myriad of connected devices in so-called cyber physical environments. The Web of Things (WoT) paradigm is emerging with virtual representations of physical or abstract realities increasingly accessible via web technologies. Achieving a new phase of exponential growth, comparable to the earliest days of the Web, requires open markets, open standards, and the vision to imagine the potential for this expanding WoT.

This paper describes the design and implementation of a solution to enable the WoT paradigm. The so-called Semantic IoT Registry (IoT Registry at GitHub (https://github.com/tlmat-unican/fiesta-iot-iot-registry)) consists on a semantic web-enabled repository that provides platform independent APIs for application developers as well as the baseline for different platforms to interoperate with one another. The approach taken is based upon semantically enriched metadata that describes the data and interaction models exposed to applications. In this sense, it enables platforms to share the same meaning when they share their data through the IoT Registry. This data is available as Linked Data on Resource Description Framework (RDF) format through RESTful APIs. In addition to the support of semantically enriched data, which is the basis for fulfilling the data interoperability requirement, the IoT Registry enables the use of Uniform Resource Identifiers (URIs) as addresses for things serving as proxies for physical and abstract entities. This latter aspect is of utmost importance for realizing the infrastructure-agnostic scenario that a WoT comprising cross-domain platforms requires.

The IoT Registry is the base component underpinning the application development across federated IoT data sources through the provision of Application Programming Interfaces (APIs) to get and push data from any of the underlying IoT infrastructures. Thus, it is important to highlight that the implementation details presented in this paper are the result of overcoming the challenges, mainly in terms of heterogeneity and scalability, that the federation of eleven IoT deployments (from heterogeneous application domains), with over 10,000 IoT devices overall which produce hundreds of thousands of observations per day [15], poses on the implemented system.

The remainder of the article is structured as follows. Section 2 makes a non-exhaustive review of relevant related work in the key research topics that underpin the IoT Registry. Section 3 introduces the key design considerations that have been taken into account for the development of the IoT Registry. The detailed description of the IoT Registry architecture and its functional building blocks is presented in Section 4. Section 5 goes a level deeper and focus on the implementation details for the storage and distribution of semantic data streams paying special attention to the solutions used to address the challenges of scalability and resource abstraction. Finally, Section 6 concludes the document and briefly discusses those open issues that shall be addressed in the future.

## 2. Related Work

### 2.1. Semantic IoT Interoperability

Currently fragmented IoT ecosystem is jeopardizing the development of global solutions. The existing multiple parallel IoT platforms have to converge towards offering seamless, global, and linked services to their users. It is necessary to implement solutions that are able to make the already existing IoT infrastructures to collaborate in providing a common and portable way of offering their data services. One of the aims of the platform described in this article is to support the automation of the deployment of services/applications over heterogeneous IoT domains.

Semantic technologies will play a key role to align the descriptions of the various IoT entities from various platforms. However, defining the abstraction level required for an IoT ontology is challenging. Nowadays, there are plenty of initiatives that are specifying the models to represent IoT devices and the ecosystem around them. Nonetheless, it is still difficult to find one that addresses all the requirements.

Probably, the most widely used ontology is the Semantic Sensor Network (SSN) Ontology [16] that covers sensing, but does not take actuating or other realms of IoT into account. Moreover, this ontology is very complex to use at its full extension and is typically used as a baseline reference. The IoT-Lite ontology [17] uses the SSN as a basis and adds Architectural Reference Model (ARM) [18] key concepts to provide a more holistic IoT model. The adopted solution within FIESTA-IoT has been to reuse these generic ontologies and extend them wherever required to meet the requirements identified for the federation of IoT testbeds.

Other works that are pursuing parallel objectives and are worth mentioning are OneM2M [19] or IoT-O [20]. OneM2M, as an international partnership project of standardization bodies, is defining a standard for M2M/IoT-communications. The actual release is lacking semantic description of resources that will be addressed as one major point in the next release. The IoT-O ontology aims to unify different ontologies in the IoT landscape and consists of eight different ontologies. It is actively maintained and connected to other standardizations like OneM2M. While many similarities can be found with the FIESTA-IoT ontology, geolocation of resources and observations is not addressed and the virtual entity concept that is central for the IoT-A ARM is not properly covered. Moreover, to the best of our knowledge no major IoT platforms have already adopted it for semantically handling its resources and observations. Other ontologies are focusing on either specific subdomains, like the Sensor Web for Autonomous Mission Operations (SWAMO) ontology [21], which concentrates on marine interoperability or not specifically defined for the IoT domain like GoodRelations [22] that is dealing with products but can be taken into account in the industrial IoT area. The Smart Appliance REFerence (SAREF) ontology [23] is another initiative which is raising much attention as enables semantic interoperability for smart appliances. It has been created in close interaction with the industry, thus, fostering the future take up of the modeling defined. Still, it is centered on home appliances that might make it fall short in other domains.

### 2.2. RDF Description and Validation

Semantics define a powerful take on how to shape, allot, catalogue, and share information. Moreover, it is possible to construct a distributed database accessible from any Internet-connected device by using URIs as identifiers and locators for the different pieces of information. However, it is necessary to guarantee that all the information stored in this database complies with the rules established in the ontology that underpins the semantic nature of that data. Otherwise, it would not be of actual value, as it would be weighed down by issues related to data quality and consistency.

Recent approaches to validate RDF—the most common format for representing semantic data—have been proposed. ShEx [24] and SHACL [25] are languages for describing RDF graph structures so that it is possible to identify predicates and their associated cardinalities and data types. Both of them have the same objective, to enable validating RDF data by using a high-level language. Both define the notion of a shape, as the artifact to specify constraints on the topology of RDF nodes. SHACL shapes are analogous to ShEx shape expressions, with the difference that links to data nodes are expressed in SHACL by target declarations and in ShEx by shape maps. In most of the common cases, it is possible to translate between ShEx and SHACL. The triples involving nodes in the graph that must be validated have to comply with the defined shapes. If not matching, the RDF graph would not be consistent with the semantic model that wants to be used.

The IoT registry imposes RDF validation before storing the datasets that arrives from the underlying IoT platforms. The solution adopted in the reference implementation of the IoT registry did not make use of any of the aforementioned languages due to some practical reasons. Firstly, SHACL recommendation only appeared on July 2017 [25], but the implementation of the IoT registry started much before. The first works started on 2015, but they were only working drafts. Moreover, available implementations, both for SHACL and ShEx, are under development. Secondly, the FIESTA-IoT ontology, which was the one that we used for our reference implementation, did not have the shapes graph that is necessary to apply SHACL or ShEx validators. Finally, yet critically important, we preferred to address the validation process by directly applying SPARQL filtering in order to avoid further delays, associated with the execution of shapes validation procedure and the evaluation of its validation report, in the processing of incoming data.

### 2.3. Web of Things

The WoT emerges from applying web technologies to the IoT to access information and services of physical objects. In WoT, each physical object possesses a digital counterpart. These objects are built according to Representational state transfer (REST) architecture and accessed with HTTP protocol via RESTful API. A Web Thing can have an HTML or JSON representation, REST API to access its properties and actions, and an OWL-based semantic description.

W3C has recently launched the Web of Things Working Group [26] to develop initial standards for the Web of Things. Its main aim is "to counter the fragmentation of the IoT". They are still working on defining the WoT architecture and the description of the WoT Thing, which should define a model and representation for describing the metadata and interfaces of Things, where a Thing is the virtualization of a physical entity that provides interactions to and participates in the WoT.

In parallel to this standardization effort, several projects and platforms have been developed targeting the support of service provision based on the WoT paradigm. Paganelli et al. [27] present their WoT-based platform for the development and provision of smart city services. Precision agriculture is the application domain that benefits from the platform described by Foughali et al. [28]. While they provide some of the solutions promised by the WoT, still do not address the IoT fragmentation as they rely on proprietary modeling. Other works [29–31], instead, leverage semantic technologies to fulfill the extendable modeling requirement. As we are proposing in this paper, we believe that this is the necessary combination in order to fully develop the WoT concept into a running system. The key novelty from the work presented in this paper is that previous works have not been implemented and proven over real-world scenarios with federation of heterogeneous IoT infrastructures, as it is the case of the platform presented in this paper.

### 2.4. RDF Streams

Annotating data following RDF principles is the first step for enabling interoperability. However, it still necessary to access to this data so that services can be provided on the basis of the context knowledge that it enables. IoT data is typically associated to the Big Data paradigm. However, it is not only large in volume. An important difference between the IoT datasets compared to the conventional ones is the quick changes in data and dynamicity of the environment.

Different RDF Stream Processing (RSP) systems have been proposed to enable querying over RDF streams, in opposition to the typical static querying. Most of them are extensions of SPARQL that take into account the dynamicity of the data that is being queried [32].

The solution implemented in the IoT Registry has some similarities with RSP but the key difference is that IoT registry design considerations make it to be settled somehow in between the persistent and transient data. Most of the time, only the latest knowledge about a dynamic system that is being monitored is important. However, access to history is also necessary for some applications. IoT Registry tries to provide a solution for both demands. This is, facilitate the access to the most recent data but also enable access to relevant information in the past.

### 2.5. IoT Platforms Federation

The vision of integrating IoT platforms and associated silo applications is bound to several scientific challenges, such as the need to aggregate and ensure the interoperability of data streams stemming from them. The convergence of IoT with cloud computing is a key enabler for this integration and interoperability. It facilitates the aggregation of multiple IoT data streams so that it is possible to develop and deploy scalable, elastic, and reliable applications that are delivered on demand according to a pay-as-you-go model. During the last 4–5 years several efforts towards IoT/Cloud integration have been proposed [33,34] and a wide range of commercial systems (e.g., Xively [35] and ThingsSpeak [36]) are available. These cloud infrastructures provide the means for aggregating data streams and services from multiple IoT platforms. Moreover, other initiatives, such as Next Generation Services Interface (NGSI) [37] promoted by Open Mobile Alliance (OMA) or Hyper/Cat [38], focus on the homogenization of interfaces enabling web access to data produced by IoT deployments. However, they are not sufficient for alleviating the fragmentation of IoT platforms. This is because they emphasize on the syntactic interoperability (i.e., homogenizing data sources, interfaces, and formats) rather than on the semantic interoperability of diverse IoT platforms, services, and data streams. They intentionally enforce no rules about metadata naming. Use of a reference ontology is one of the currently explored possibilities for having interoperable systems.

Some open source initiatives like Project Haystack [39], which aims at streamlining the access to data from the IoT, or the IoT Toolkit [40], which has developed a set of tools for building multiprotocol Internet of Things gateways and service gateways, are working on standardization of semantic data models and web services. Their goal is making it easier to unlock value from the vast quantity of data being generated by the smart devices that permeate our homes, buildings, factories, and cities. In this sense, the differential aspect of the platform proposed in this paper is that it is taking a holistic approach, considering the semantic federation of testbeds as a whole, vertically and horizontally integrated, and independently of the environment (smart city, health, vehicles, etc.) they are controlling.

## 3. Semantic IoT Registry Design Considerations

In this section, we are reviewing the key requirements that have defined the main design considerations that we have observed during the development of the Semantic IoT Registry. This review is important in order to understand the design choices and the solution's architecture and building blocks, which are described in the next sections.

Interoperability: Frequently denoted as interworking, it characterizes the capacity of systems to connect and interchange data. In the IoT domain, where many vendors will coexist, access to information from diverse platforms is essential in order to not be trapped in vendor lock-in situations. Interoperability entails for conversion between systems and sharing data via well-known interfaces in order to facilitate its exploitation. Two, among the different options that can be considered, are particularly interesting:

- To replicate IoT data, leveraging the existing knowledge on the existing platforms. With this option, the response time would not be affected if new nodes were added. However, this solution requires higher capacity, which could lead to scalability problems if the data set grows in an unbounded way.
- To discover and translate the IoT data in real time. In this option, the home IoT platforms are used to retrieve the information, thus avoiding duplication, hence adapting not only the underlying IoT data but also the operations allowed by the different interfaces. On the cons side, this would introduce extra complexity and the increase in the overall system response time.

IoT data abstraction and integration: Data abstraction in IoT is related to the way in which the physical world is represented and managed. It is necessary to have a set of concepts to properly describe not only the IoT device itself but also the sensor observation and measurement data. Using semantic descriptions, the IoT data can be homogeneously characterized. However, it is necessary to employ a generic model that is capable of supporting the mapping among the used semantic models that might exist in a cross-platform IoT integration scenario. Thus, the solution has to focus on a canonical set of concepts that all IoT platforms can easily adopt. We opted for taking as reference the IoT Architecture Reference Model (ARM) as defined in the IoT-A project [18]. The foremost aspect that this choice implies is that the ontology that is used to regulate the semantic annotation of the testbeds' resources is only bound by the core concepts that compose the aforementioned ARM Domain and Information Models. These core concepts are

- A Resource is a "Computational element that gives access to information about or actuation capabilities on a Physical Entity" [18].
- An IoT service is a "Software component enabling interaction with IoT resources through a well-defined interface" [18].

These concepts conform the baseline for representing the devices and overall IoT infrastructure. However, there is still a major concept that is not tackled within the ARM models. This concept relates to the actual data that is gathered by the devices and offered through the services that expose them. It is called the Observation concept:

- An Observation is a "piece of information obtained after a sensing method has been used to estimate or calculate a value of a physical property related to a Physical Entity."

Data semantization: Besides the actual technologies used for exporting the data services, the main feature that underpins our solution is the fact that the information is exchanged in a semantically annotated format. The IoT Registry is not bound to any ontology in particular so its design is fundamentally reusable and extendable. This means that while using a common ontology is necessary to make it possible to seamlessly deal with data from different sources, the design of the IoT Registry does not specify the ontology that has to be used. Additionally, from the data representation viewpoint, a specific RDF representation formats (e.g., RDF/XML, JSON-LD, Turtle, etc.) should not be imposed. Data consumers should be able to select their preferred format in every request. Thus, all of them must be supported.

Cyberphysical Systems as a Service (CPSaaS): The value of IoT is on the ability to ease the access to the services that the IoT devices embedded in our surroundings export. IoT is typically reduced to the data that sensors can produce but it is not only that. IoT is an integral part of Cyberphysical Systems (CPS) and vice versa. One of the key design considerations of the IoT Registry is that not only data but also services have to be interoperable. The CPSaaS consideration relates precisely with the ability to access the services exported by IoT and/or CPS platforms in an interoperable and platform agnostic manner. Thus, it is necessary to guarantee that application developers, which are the ones that at the end will consume those services through their applications, can use these services in the most user-friendly manner. Nowadays, REST APIs are the de facto standard for exporting services. Thus, it is required that access to datasets, data streams, and IoT Services in general is orchestrated through REST-based interfaces. Additionally, the heterogeneity of the underlying infrastructure should be hidden to the final data and service consumers. This way, the interoperability is something that the platform has to guarantee but, at the same time, has to be simply offered to the end user. That is, they will consume data and services based on their interest and necessities but without even knowing if it comes from one or many different platforms.

Services Exportability: The challenges that IoT-enabled smart solutions face are very similar among them. Hence, the context-aware, intelligent services that are meant to optimize the scenarios to which they are applied are potentially beneficial to any of them regardless of its location (e.g., services for improving the efficiency of the city and the well-being of its citizens are applicable to any city). This is one of the key motivating factors for service providers and application developers to invest resources in designing and implementing such services. Nonetheless, if deploying the same service in different cities is not straightforward, the profit margins are put at risk due to the cost of tailoring the service to the infrastructure and platform that is deployed in each city. Only if the same service or application can be seamlessly used across cities, the full potential of IoT-based services will be reached.

Access to history: IoT is typically assumed to have just a stream nature. That is, only the current situation is important. As new observations from one sensor arrive, the previous measurements can be discarded. This is true in most of the cases but the value of accessing historical information cannot be neglected. Combination of IoT datasets and BigData algorithms are used for critical applications that are based on the identification of patterns, for which storage of historical records is fundamental.

Discoverability: It is not enough with setting the mechanisms to make data interoperable among IoT Platforms, but it is also necessary to fetch this data from the repositories were it is stored. Due to the huge variety of information and domains that can coexist, it is necessary to have an elastic discovery environment that can exploit the richness of the semantic-enabled information stored.

Programmability: Creating the ecosystem is the necessary but not sufficient condition. A fundamental risk of such ecosystem is that developers and platform providers might find the ecosystem's features unattractive. For the ecosystem to grow, developers must find it easy to interact with the interfaces offered. Enabling REST-based interfaces guarantee that in both cases interactions are straightforward.

## 4. Semantic IoT Registry Architecture

The IoT Registry is responsible of storing and managing the semantic descriptions that underlying IoT infrastructures make available. Considering that aggregated stored data can be of a different nature, the component must handle their differences while processing it homogeneously, to satisfy the needs of external users requesting information. In that sense, we mainly consider two tightly bound realms: the one related to the description of the resources belonging to the testbeds and the one related to the observations or measurements constantly gathered by those devices. Both of them are filled with semantic documents using the well-known RDF [41] serialization format. Thus, the core of the IoT registry component is composed of a Triplestore Database (TDB) and a fully-fledged API that allows the interplay with users.

In addition to the TDB, the IoT registry is composed of a data query endpoint supported by a SPARQL engine, a manager that supervises IoT data registration and exposes the stored information to authorized users using a REST API, and an access broker that securely proxies direct access to registered resources. Figure 1 shows the internal architecture and the relationships between the different building blocks.

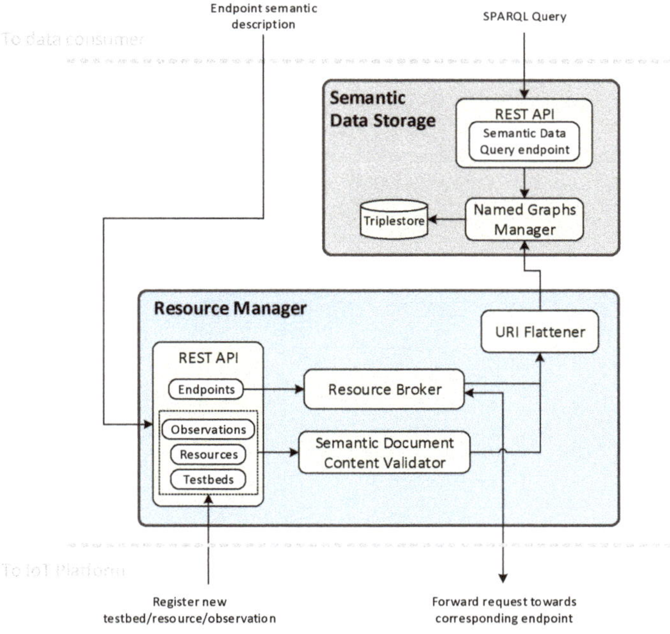


**Figure 1.** Internet of Things (IoT) registry functional architecture.

Although IoT Registry can run integrated within a more complex architecture [42], it can operate in standalone mode, providing basic authenticated access.

In the following sections, we describe in detail the main insights achieved from the implementation of these functional components, paying special attention to those features that make them suitable for addressing the challenges and design consideration brought about previously.

## 4.1. Semantic Data Storage

Because of the canonical concepts that underpin the model for data abstraction and semantization, the data stored at the IoT registry can be catalogued in two different, but closely interrelated, domains: Firstly, the descriptions of the IoT devices that are deployed in the field and, on the other hand, the measurements that they generate. The internal structure of the TDB follows a similar approach.

Before delving deeper into the TDB description, it is important to introduce the concept of Graph, which is basic for semantics and key to understand the remainder of the paper. A semantic graph is a network that represents semantic relationships between concepts, where the nodes of the network are the vertices of the graph, which represent the concepts, and the links of the network are the edges of the graph, which represent semantic relations between concepts. TDBs are said to store graphs since they store RDF triplets consisting of two nodes (concepts) and one link (the relation among them). When triplets containing nodes that were present in previously saved triplets are stored at the TDB they extend the graph whether creating new relations (i.e., the two nodes in the triplet were already in the graph) or introducing new concepts (i.e., one of the two nodes was not in the graph before). Additionally, inner graphs can be configured in TDBs in order to optimize its performance. These graphs focus on specific preconfigured concepts. This way, the information can be structured on different realms so that operations (e.g., store and fetch) that only affect one of these realms can be handled within the inner graph.

Based on the two identified realms, testbeds and its resources, and the observations collected in Figure 2, we introduced the high-level organization of the TDB. The approach followed for the IoT registry TDB considers two graphs, one for resources and another for observations' descriptions. This provides a basic structure to the information stored, making it possible to focus the target when requesting or storing data. The dataset is also configured so that the default graph (named as Global in Figure 2) is defined to be the union of the two. The identifier of the sensor that made the observation is the link between the two realms. From the semantic point of view, the subjects and objects related via the properties *ssn:madeObservation* and/or *ssn:observedBy* are *ssn:Observation* and *ssn:Sensor*. As the sensor is referenced in both realms, the inference engine can establish the required relationships. As it has been previously mentioned, the IoT Registry design is not bound to any ontology; however, for the sake of clarity in the description we are using the concepts and relations from the FIESTA-IoT ontology [43], which is the one that we employed for the reference implementation that we have developed.

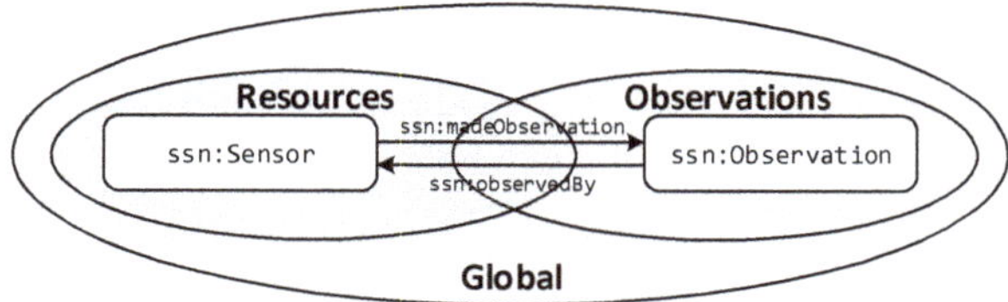


**Figure 2.** IoT registry TDB internal structure.

Although this was a design decision, the experience from real usage of our reference implementation has proven it was appropriate. Even when the users could always use the Global graph so that both graphs are used seamlessly on-the-fly, it was quite straightforward for them to direct their queries to only one domain. For example, they request either for a particular device's property (i.e., location, physical phenomenon, or just metadata) inquiring only the Resources graph, or for the value and unit of measurement of an observation requiring information just from the Observations graph. The solution adopted offers flexibility and optimization. Performance-wise, shrinking the size of the targeted information reduces the expected response time. Consequently, queries should be adapted and run on either graphs based on the required information, resulting on a better user experience.

Last but not least, it is important to highlight that the TDB can be deployed in a distributed manner. This way, when queries are sent to the central IoT Registry, they will not only handle the request with its local TDB, but also forward it to the remote semantic query endpoint. This distributed mechanism avoids data replication and enables potential query parallelization, leading to a better response time. It is important to note that all the semantic databases should have the information modeled following the same ontology.

### 4.2. Resource Manager

The Resource Manager (RM) main objective is to supervise how IoT data is registered, that is, how the underlying IoT platforms push their resources' descriptions and the generated observations. RM entry point is tightly coupled with the underlying IoT Registry semantic storage database structure, as depending on the origin and nature of the data, the storage reference differs. Initially the RM has to only choose between the resources or observations graphs.

RM registration endpoint is defined to be complaint with SPARQL 1.1 Graph Store HTTP Protocol [44]. RDF documents structure is open and may include a great variety of information (e.g., multiple resources' and observations' annotated descriptions, additional metadata like frequency, availability, quality-related features, etc.). Thus, the first duty for the RM is to guarantee that documents suit the minimum requirements. Therefore, RM must analyze and validate the annotated descriptions. Firstly, the RM checks the compliance with the ontology that is used. It inspects the semantic content to determine the type of data that is being registered and it verifies whether it includes the necessary concepts according to the cardinality expressed in the ontology. In order to avoid delays, due to internal queries, and to ease the process, the registration endpoint is defined as a REST API that replicates the basic IoT Registry database structure. Thus, underlying platforms have to use the proper endpoint when pushing new resources or observations.

Secondly, the RM substitutes the URIs of the triplets that arrive from the underlying platforms replacing them by other URIs that are under a common namespace. These new URIs are also valid URLs as they will be using the IoT Registry namespace. This way, the two conditions for the Cyberphysical Systems as a Service design consideration are fulfilled. On the one hand, the RM detaches the information from its originating platform, thus achieving the platform agnostic paradigm. On the other hand, the WoT-paradigm is also enabled, as every concept in the stored graphs will have a dereferenceable URL.

RM does not only forward the RDF documents submitted by the underlying platforms towards the IoT Registry's TDB but also exposes read endpoints for each of the semantic subjects within these documents. On each URL, the semantic model of that node can be read. This model will contain the nodes to which that particular node is connected and the relations to each of them. Following the links referenced in the URIs of those neighbor nodes, it would be possible to progressively browse through the entire semantic graph. This behavior extends the HATEOAS (Hypermedia as the Engine of Application State) constraint of a REST API, but in this case applied to semantically described data.

Besides, IoT Registry, through the RM, also provides generic reading endpoints for retrieving the list of IoT resources and observations IRIs. Furthermore, using query parameters it is possible to filter out specific information (i.e., by phenomenon or unit of measurement). This RM endpoint aims at enabling access to semantic information in a more familiar way for traditional web developers, hiding the complexity of dealing with SPARQL queries.

Summarizing, the RM makes it possible to use standard HTTP and REST API procedures to access the semantically annotated documents stored in IoT Registry, wrapping the internals and complex semantic requests used between the RM and the TDB.

### 4.3. Semantic Data Query Endpoint

Taking into account that the IoT registry stores semantically annotated documents into a semantic graph, it is essential that it provides an interface to make semantic queries. SPARQL is known to be the most common and widely used RDF query language. For this reason, we chose to export the semantic query functionality by enabling a direct SPARQL endpoint conformant with SPARQL protocol for RDF [45].

The default endpoint operates on the Global graph, which merges both resources and their gathered observations. Nevertheless, SPARQL queries can be limited to only one of the underlying graphs by using other specific endpoints provided. The underlying methodology is based on the use of FROM clause that allows to reference specific RDF datasets, instructing the SPARQL processor to take the union of the triples from all referenced graphs as the scope of the SPARQL sentence execution. As internal graphs structure and naming are not made publicly available, external users must use the defined interfaces in order to successfully access the whole datasets.

IoT Registry restricts the execution of SPARQL queries willing to modify the stored content (INSERT, UPDATE, or DELETE). As this endpoint is foreseen to be used by context information consumers, this behavior does not limit the normal operation.

The IoT registry implements an additional functionality related to semantic data query. It provides a repository of SPARQL queries. The queries stored can be static or they can be templates that allow dynamic behavior of the query by assigning values to predefined variables. This repository is designed with the purpose of sharing knowledge between users and smoothing the learning curve of using the platform as a whole. In order to keep the system secure, protection to injection attacks has been implemented.

As it has been described, the semantic data query endpoint extends the functionalities of the RM, providing a more flexible interface towards IoT Registry's TDB.

### 4.4. Resource Broker

The RDF datasets stored in IoT registry are mainly static semantic descriptions. SPARQL update request are required in order to modify the triples. However, these documents can also include metadata or references to external endpoints that complement the basic information or provide a more updated version.

As it has been already introduced, an important aspect of the IoT modeling that we have followed is that IoT devices can directly export their services. Every device can define multiple instances of the class *iot-lite:Service* (iot-lite namespace is defined as http://purl.oclc.org/NET/UNIS/fiware/iot-lite#), whose properties include a service endpoint (URL). For instance, this service can refer to the last observation of the sensor, which will be directly accessible through the URL provided, or, for the case of an actuator, it can refer to how to enable one of its functionalities. The Resource Broker (RB) is the component in charge of enabling the access to IoT devices' services while keeping the required platform agnostic nature and homogenizing the way of accessing them for the end user also.

The RB is also relevant from a security point of view. Underlying platforms will delegate access control to their sensors and actuators to the IoT Registry who, based on its own registered users and groups, and the profiles and policies defined for them by agreement with the platform owner, will in turn enable the path to the end device. The namespace transformation implemented by the RM also applies to the services exported by IoT devices. Thus, RB will be acting like a proxy that intercepts any request made to any IoT Registry endpoint URL, translates it to its original value and finally forwards it to the corresponding platform endpoint. The RB will also transform the URIs and other relevant values included in the reply. The process is carried out internally, so that it is completely transparent for the end user.

## 5. Semantic IoT Registry Implementation Details

As it has been previously introduced, the IoT Registry is the cornerstone component within a platform that is currently supporting the federation of eleven IoT different platforms [15]. In this sense, besides the design considerations, it is critically important to describe in detail the implementation path followed for some of the aforementioned design features. Moreover, the implementation details depicted in this section are the result of challenging the developed system against scalability and heterogeneity issues that only a federation with over 10,000 IoT devices that generates some hundreds of thousands observations daily can demand.

### 5.1. URI Flatten Process

The Semantic Web is all about making links between datasets understandable not only by humans but also by machines. When semantically describing entities, a unique identifier in the form of an URI is usually assigned. However, it is not mandatory that this identifier is dereferenceable on the web.

This situation applies to the data supplied to the IoT Registry by the federated platforms. To address this issue, the RM transforms original platforms' URIs into URLs associated to a common and dereferenceable namespace that can be referenced by any web application. We called this process URI flattening. The procedure pursues a twofold objective: on the one hand, enabling web references to semantic entities; and on the other hand, fulfilling the platform agnostic paradigm. The latter is achieved by hiding the binding with the source platform when renaming the entity.

For the RB to properly proxy the queries to the services exported by the underlying IoT devices, it was necessary for the process to be reversible. This is, to be able to go from the flattened URI to the original one and vice versa. In order to avoid the potential delays when accessing a big look-up table storing all the mappings, we implemented an algorithm based on symmetric cryptography. The flattened URI results from ciphering the original URI for generating its corresponding flattened URL. This solution allows us to quickly go from and back to the original URI by just knowing the secret key.

Figure 3 shows the process implemented within the RM for the transformation of the URIs for all the nodes stored at the TDB graph database.

1. The original URI is prepended by a Cyclic Redundancy Check (CRC) or a short summary of any hash function made over the original URI and an integer which represents the entity; and the entity type, which is an identifier of the entity (e.g., testbed entities are represented with the NodeType 0x03).
2. The resulting string is cyphered using AES-128 block-cypher and Base64 URL safe encoded.
3. The resulting string is appended to the corresponding common IoT namespace resulting in the corresponding transformed URI.

The values prepended to the original URI are used not only for integrity check, but also to randomize the beginning of the resource URL, so it is harder to determine the source testbed. Otherwise, as we are using a block cypher mechanism, entities with the same namespace will have similar URLs. In order to increase the randomness of the resulting URL, it is possible to include a salt also.

Table 1 includes one example of the transformation process. The procedure is applied before storing triples in the TDB to instances of classes, either subjects or objects of the RDF statements, but also to some specific literals, based on the related property. In this sense, we mainly consider the modification of literals whose value is a URL or a direct reference to an exported service. For example, instances of *iot-lite:Service* class can define a property *iot-lite:endpoint*, which is defined as *xsd:anyURI*, and usually it is a URL where the corresponding service is available. This is closely related to the RB functionality introduced above.

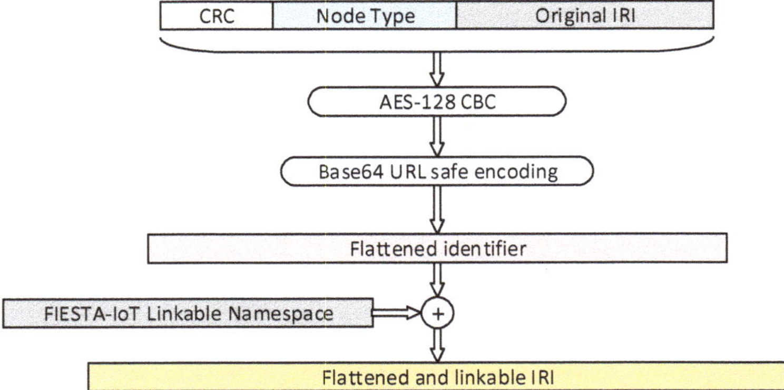


**Figure 3.** Uniform resource identifier (URI) transformation algorithm.

**Table 1.** URI flattening process example.

| Testbed | FIESTA-IoT URL |
|---|---|
| Original IRI | http://api.smartsantander.eu#SmartSantanderTestbed |
| Entity type | 0x03 (Testbed) |
| CRC | 0x50A32758 |
| IoT Registry URL | https://platform.fiesta-iot.eu/iotregistry/api/testbeds/ kscYbDJBhbywRuRSSOsucfEhrY1lTb5LF6bYBh36pTbvKDqUIfDkS7WeB9ryaC7l-C9ZExZYLwiyuw8wAKjZpQ== |

Therefore, the RM has to analyze all the triples posted by testbeds in order to identify classes instances and, for FIESTA-IoT specific case, *iot-lite:endpoint* references. Implementation-wise, we can rely on several SPARQL queries and the subsequent generation of the new semantic model. However, we have opted to take advantage of Jena functionalities, especially those related with OntModel, as it provides a better integration and coding experience. Figure 4 shows the pseudocode of the procedure.

```
procedure AdaptEntity(rdfInput)
      ontModel ← new model based on FIESTA-IoT ontology and rdfInput
      newModel ← new empty plain model
      foreach statement in input
            s ← subject from statement
            p ← predicate from statement
            o ← object from statement

            if s is class then
                  next statement
            end if

            ns ← flatten and anonymize s
            if (o is instance or (o is literal and o is xsd:anyURI)) then
                  no ← flatten and anonymize o
            end if

            newStatement ← new statement from (ns, p, no)
            add newStatement to newModel
      end foreach
      return newModel
end procedure
```

**Figure 4.** Transformation procedure pseudocode.

The main premise of the procedure is that every subject has to be dereferenceable unless it is a class definition. It does not matter whether it is blank or named node. Then, when we iterate over each RDF statement of the semantic document, we transform every subject following the previous premise. Besides, for every object we check whether the value is a reference to a class, to a class instance or to a URL literal and perform the same operation.

In order to be able to achieve this, we initially generate an OntModel from the original posted RDF document. We use this new model to check or infer the nature of each RDF statement subject and object. We also filter some properties and cache IRIs in order to reduce the inference request and speed up the process. The properties that are filtered are those

*5.2. Semantic Document Content Validation*

A particularly interesting aspect of using semantics is the ability to validate not only syntactically but also semantically the data [46]. However, even if the IoT registry could be guaranteed that documents posted by the underlying IoT infrastructures to respect the ontology employed, this is not enough to completely prevent from the injection of graph inconsistencies that might lead to the storage of loose data. RDF documents are quite flexible in terms of the data they include. For instance, registering a resource description that does not include the proper bond with its associated platform, or an observation not including a reference to the node that was generated it or its value and unit of measurement, will make the information provided not fully useful. Syntactically and semantically, these two examples can perfectly pass the filter but they would still lead to nodes in the graph that are not properly bound to the WoT. Even if the generator is willing to provide the full description, but as two separate and independent RDF documents, IoT registry cannot take the risk of not receiving one of the pieces.

In order to avoid this, the IoT registry carries out an additional validation step, in this case, for making sure that the resource(s) or observation(s) do carry all the information required for describing a resource or recording a measurement from a sensor appropriately. To do this, the module runs internally several queries that will check not only that it contains all the mandatory nodes for a resource description or an observation graph, but also that the required properties are present (as stated by the properties' cardinality at the ontology definition).

As a single RDF document can contain more than one resource or observation, it is important that the process implemented checks for every resource or observation reference in the RDF description.

As it has been previously mentioned, we have used the FIESTA-IoT ontology [43] as the basis for the reference implementation that we have developed. In this sense, we consider that a resource description must include a bond to the IoT platform to which it belongs, its location, and the measured phenomenon along with its unit of measurement. Similarly, an observation description must consist of, at least, its value, unit and quantity kind, the location, and timestamp when it was generated and a link to the sensor. Additional information might be interesting to have, but this enumeration is considered the bare minimum to accept a RDF description.

Figure 5 shows an example of a minimal document validation SPARQL. This SPARQL query is applied to the RDF document posted, and no interaction with IoT registry's TDB is required. The outcome of the execution of the SPARQL sentence is a list of resources whose description meets the minimum set of information as defined by the FIESTA-IoT ontology. Besides, another SPARQL query extracts the list of linked resources within the RDF document; both lists must match. Otherwise, it would mean that the RDF document is including invalid descriptions. Once this first check is passed, it is verified that the deployments or IoT platforms associated are already registered in the TDB. Initially, and based on the semantic approach taken, it could seem that the simplest way this can be done is through the execution of another SPARQL request on the TDB. However, in order to minimize the amount of time spent on the validation procedure and reduce the amount of tasks run on the TDB, we have taken an alternative path. As the list of underlying platforms registered can be considered quite static and not very large, we can keep it either in memory or in a relational database. Then, matching the presence of platforms in both lists is straightforward.

```
PREFIX ssn: <http://purl.oclc.org/NET/ssnx/ssn#>
PREFIX iot-lite: <http://purl.oclc.org/NET/UNIS/fiware/iot-lite#>
PREFIX geo: <http://www.w3.org/2003/01/geo/wgs84_pos#>

SELECT ?entity ?deployment
WHERE {
  ?platform geo:location ?location .
  ?location geo:lat ?lat .
  ?location geo:long ?long .

  # ssn:Devices
  {
    ?entity ssn:hasDeployment ?deployment .
    ?entity ssn:onPlatform ?platform .
  }
  UNION
  # ssn:SensingDevices
  {
    ?device ssn:hasDeployment ?deployment .
    ?device ssn:hasSubSystem ?entity .
    ?device ssn:onPlatform ?platform .
  }

  ?entity iot-lite:hasQuantityKind ?qk .
  ?entity iot-lite:hasUnit ?qu .
}
```

**Figure 5.** Example of minimal document validation for a resource.

This way, the content validation process for resource registration is done as a fully separate process, without interfering in the read and write operations on the TDB or the IoT registry's SPARQL execution engine. The same approach is applied for observations implementing the required adaptations in the process. Particularly, modifying the SPARQL query and testing that the associated resources are already registered.

Upon execution of the content validation procedure, and considering that the answer is not empty, we can assert that the information is valid and we can proceed to store it into the TDB. Otherwise, the full document is rejected, informing the user about the resources or observations that are not properly described.

It is worth mentioning that this process introduces a non-negligible computational overhead, as every request implies the execution of an internal SPARQL query. However, we consider that the benefit outweighs the expected registration delay as we have control over the data really stored and its usefulness. What is more, as described, it runs as a fully independent feature, not jeopardizing TDB performance.

### 5.3. TDB Organization Evaluation

IoT registry's triplestore stores semantic information that is grouped in two realms, resources and their observations, and as such, the TDB is organized. The dataset was divided in two named graphs keeping the IoT devices and the observations that they generate in two independent graphs that were only logically linked through the IoT device identifier.

It is clear that the difference in size between these two graphs can become large. The Resources graph is almost static since underlying platforms only have to register their assets once, at the very beginning, with small one-time updates when incorporating new resources. However, as long as sensors keep pushing data, the Observations graph will never stop growing, becoming difficult to handle.

Using one large graph for all the observations would make that even the simplest SPARQL queries had scalability problems. Thus, in order to mitigate the effect of Observations graph growth and to improve system performance, we came up with a solution that consists on splitting the graph into a number of subgraphs. Even though technically speaking, the IoT registry keeps saving all the observations into the TDB, the usage of various and independent named graphs can hold the information isolated when it comes to process SPARQL queries. This way, the system will have to seek into a portion of the dataset, limited to only selected graphs.

Figure 6 depicts the proposed solution. We graphically represent, first in Figure 6a, how observations (represented as bubbles) are stored into either a standalone graph or, on the other hand, in Figure 6b, sliced into various subgraphs, whose name can include indexing information that helps on accessing the stored information.

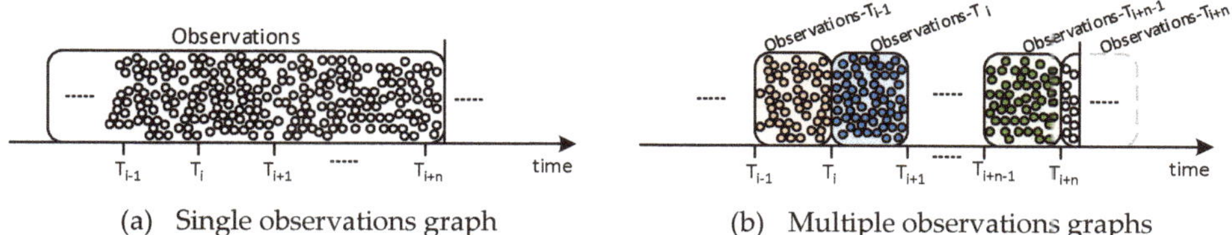


(a) Single observations graph (b) Multiple observations graphs

**Figure 6.** Single- vs. multi-observation graphs.

In our case, we consider time as the basis for the generation of new subgraphs. Another option would have been grouping the observations by resource, but this would only have helped when fetching information from one specific sensor. However, typically information is requested per location and period. The subgraph's creation time is appended to graph name to create a unique subgraph identifier. These subgraphs are created periodically and the interval between two consecutive graphs is fixed and preconfigured. For instance, a name like *observations:201803151316* corresponds to the subgraph created at 2018/03/15 13:16 UTC. This graph will store all observations posted to IoT Registry from that time on during the fixed interval.

From the end user standpoint, the existence of multiple subgraphs is mostly hidden. The REST API that gives access to RM and Semantic data query engine includes two query parameters (*from* and *to*) to set the time constraints of the underlying query, that is, to define the FROM statements to be included to the SPARQL queries. By fixing these parameters, the consumer of context information stored at the IoT Registry can directly specify the time interval they want to focus on, instead of having to perform an exhaustive search onto the whole TDB. If none of these parameters is present, then the query is only solved against the observations stored in the latest subgraph.

Figure 7 presents the reduction in the total number of RDF statements per graph of the proposed solution. Following the previous example, where a SPARQL query to retrieve observations in a specific time interval would search into the whole graph (see Figure 7a), with the implemented configuration, the IoT registry only attach graphs *observations-*$T_{i-1}$, *observations-*$T_i$, and *observations-*$T_{i+1}$ to the process (Figure 7b), significantly reducing the time and complexity of the search.

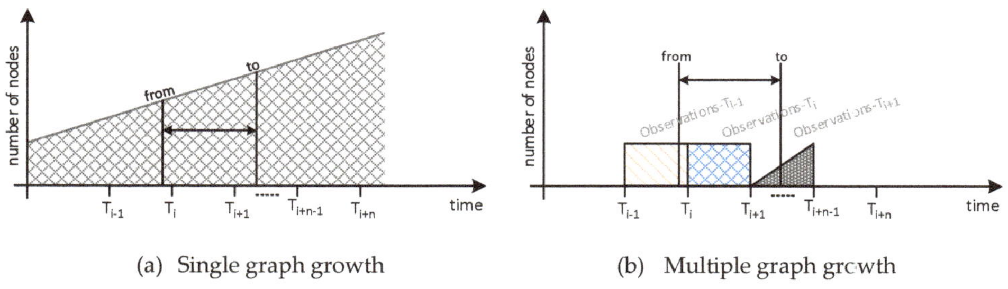


(a) Single graph growth (b) Multiple graph growth

**Figure 7.** Single- vs. multi-observation graph(s) operation.

It is important to note that it is not possible to guarantee that all observations stored in a time-based subgraph have been taken in the corresponding period, as the observations are stored as per time of arrival not as per generation time. This latter approach would imply that the timestamp included in its semantic description would be checked and it would introduce some non-negligible delay in the observation processing time. Hence, the correspondence between the observation's time and subgraphs' name is highly dependent on underlying platforms good practices. Data consumers have to take into account this potential lack of synchronization between the interval included in the request and actual timestamps of the measurements.

Summarizing, thanks to the configuration described, the system experiences various improvements that, altogether, lead to a better overall performance:

1. Since most of the queries address recent data, (i.e., typically only last values are of interest due to the stream nature of IoT) the reduction in size of the target graph enables a much more agile response. Thus, with respect to an end user, the quality of service, and experience is significantly enhanced.
2. Extrapolating to a general point of view, operations on reduced datasets mean less computational load. As a number of users will be interacting with the platform at the same time, the shorter the time dedicated per request, the higher the number of requests that can be processed without saturating the system.
3. Access to historical data is still possible for those data consumers that require information beyond the most recent context data. For those users, the capacity to sequentially query against consecutive time intervals allows them avoiding queries that would result excessively heavy if made over the compounded period.

## 6. Conclusions and Open Issues

Internet of Things testbed deployments have been growing at an incredible pace over the last few years, although ontologies and access tools are not yet fully standardized for IoT resources. Therefore, each testbed provides its own solution to access them. In this paper, we have presented the IoT Registry—an enabler of the Semantic Web of Things—which address data access problematic across heterogeneous IoT testbeds, providing a common semantic-aware homogenous access to their resources.

The IoT registry is a fully-fledged warehouse that stores all the data injected by different testbeds that belong to the FIESTA-IoT federation [15]. On top of this repository-like behavior, it provides several means of access for experimenters to collect testbed's data, where all resource descriptions and observations are provided in a testbed agnostic approach. Upon the scalability issues brought about by the huge amount of data pushed by these IoT testbeds (tens of thousands of observations per day), we have introduced a solution that with a simple-yet-effective time-slicing approach, the overall system performance is maintained below acceptable margins [42].

Among the future steps envisioned, we plan to carry out a thorough assessment on all IoT Registry's operations in order to characterize its behavior. Furthermore, we would also like to compare its performance with that of other mainstream platforms, such as FIWARE and OneM2M, which are called to stand out as the future references for IoT and M2M. Additionally, some policy has to be settled for long-term historical datasets and at some point, this data should not be available through the IoT Registry but on another kind of platforms (more oriented to Big Data than to WoT). Even when the IoT Registry implementation described in this paper has been running from beginning of 2017 till the end of 2018 and, during this period, it has been able to support several experiments (FIESTA-IoT Project Experiments (http://fiesta-iot.eu/index.php/fiesta-experiments/)) on different application domains, from scalability and usability point of view, it is not appropriate to store data at the IoT Registry forever. During the IoT Registry aforementioned operation time (almost two years), disk space was not a particularly challenging aspect, since storage space is quite cheap at the cloud, and, precisely, the solutions that we have implemented have allowed reasonably good behavior even under complex real scenario.

**Author Contributions:** Conceptualization, J.L. and L.S.; Methodology, J.L. and L.S.; Software, J.L.; Validation, D.G., and J.L.; investigation, J.L. and D.G. Resources, J.R.S. and P.S.; Writing—Original Draft Preparation, D.G., J.L. and L.S.; Writing—Review and Editing, L.S. and J.R.S.; Supervision, L.S.; Funding Acquisition, L.S.

**Funding:** This work was partially funded by the European project Federated Interoperable Semantic IoT/cloud Testbeds and Applications (FIESTA-IoT) from the European Union's Horizon 2020 Programme with the Grant Agreement No. CNECT-ICT-643943 and, in part, by the Spanish Government by means of the Project ADVICE "Dynamic Provisioning of Connectivity in High Density 5G Wireless Scenarios" under Grant TEC2015-71329-C2-1-R.

**Acknowledgments:** The authors would like to thank the FIESTA-IoT consortium for the fruitful discussions.

**Conflicts of Interest:** The authors declare no conflicts of interest.

## References

1. Niyato, D.; Maso, M.; Kim, D.I.; Xhafa, A.; Zorzi, M.; Dutta, A. Practical Perspectives on IoT in 5G Networks: From Theory to Industrial Challenges and Business Opportunities. *IEEE Commun. Mag.* **2017**, *55*, 68–69. [CrossRef]
2. Atzori, L.; Iera, A.; Morabito, G. The Internet of Things: A survey. *Comput. Netw.* **2010**, *54*, 2787–2805. [CrossRef]
3. Rachedi, A.; Rehmani, M.H.; Cherkaoui, S.; Rodrigues, J.J.P.C. IEEE Access Special Section Editorial: The Plethora of Research in Internet of Things (IoT). *IEEE Access* **2016**, *4*, 9575–9579. [CrossRef]
4. Ibarra-Esquer, J.; González-Navarro, F.; Flores-Rios, B.; Burtseva, L.; Astorga-Vargas, M.; Ibarra-Esquer, J.E.; González-Navarro, F.F.; Flores-Rios, B.L.; Burtseva, L.; Astorga-Vargas, M.A. Tracking the Evolution of the Internet of Things Concept Across Different Application Domains. *Sensors* **2017**, *17*, 1379. [CrossRef]
5. Neirotti, P.; De Marco, A.; Cagliano, A.C.; Mangano, G.; Scorrano, F. Current trends in Smart City initiatives: Some stylised facts. *Cities* **2014**, *38*, 25–36. [CrossRef]
6. Zanella, A.; Bui, N.; Castellani, A.; Vangelista, L.; Zorzi, M. Internet of Things for Smart Cities. *IEEE Internet Things J.* **2014**, *1*, 22–32. [CrossRef]
7. Mehmood, Y.; Ahmad, F.; Yaqoob, I.; Adnane, A.; Imran, M.; Guizani, S. Internet-of-Things-Based Smart Cities: Recent Advances and Challenges. *IEEE Commun. Mag.* **2017**, *55*, 16–24. [CrossRef]
8. Riazul Islam, S.M.; Kwak, D.; Humaun Kabir, M.D.; Hossain, M.; Kwak, K.-S. The Internet of Things for Health Care: A Comprehensive Survey. *IEEE Access* **2015**, *3*, 678–708. [CrossRef]
9. Elsts, A.; Fafoutis, X.; Woznowski, P.; Tonkin, E.; Oikonomou, G.; Piechocki, R.; Craddock, I. Enabling Healthcare in Smart Homes: The SPHERE IoT Network Infrastructure. *IEEE Commun. Mag.* **2018**, *56*, 164–170. [CrossRef]
10. Trasviña-Moreno, C.; Blasco, R.; Marco, Á.; Casas, R.; Trasviña-Castro, A.; Trasviña-Moreno, C.A.; Blasco, R.; Marco, Á.; Casas, R.; Trasviña-Castro, A. Unmanned Aerial Vehicle Based Wireless Sensor Network for Marine-Coastal Environment Monitoring. *Sensors* **2017**, *17*, 460. [CrossRef]
11. Brewster, C.; Roussaki, I.; Kalatzis, N.; Doolin, K.; Ellis, K. IoT in Agriculture: Designing a Europe-Wide Large-Scale Pilot. *IEEE Commun. Mag.* **2017**, *55*, 26–33. [CrossRef]
12. Majumder, S.; Aghayi, E.; Noferesti, M.; Memarzadeh-Tehran, H.; Mondal, T.; Pang, Z.; Deen, M.; Majumder, S.; Aghayi, E.; Noferesti, M.; et al. Smart Homes for Elderly Healthcare—Recent Advances and Research Challenges. *Sensors* **2017**, *17*, 2496. [CrossRef] [PubMed]
13. Chen, B.; Wan, J.; Shu, L.; Li, P.; Mukherjee, M.; Yin, B. Smart Factory of Industry 4.0: Key Technologies, Application Case, and Challenges. *IEEE Access* **2018**, *6*, 6505–6519. [CrossRef]
14. Sundmaeker, H.; Guillemin, P.; Friess, P.; Woelfflé, S. Vision and challenges for realising the Internet of Things. *Clust. Eur. Res. Proj. Internet Things Eur. Comm.* **2010**, *3*, 34–36.
15. Sánchez, L.; Lanza, J.; Santana, J.; Agarwal, R.; Raverdy, P.-G.; Elsaleh, T.; Fathy, Y.; Jeong, S.; Dadoukis, A.; Korakis, T.; et al. Federation of Internet of Things Testbeds for the Realization of a Semantically-Enabled Multi-Domain Data Marketplace. *Sensors* **2018**, *18*, 3375. [CrossRef] [PubMed]
16. Semantic Sensor Network Ontology. Available online: https://www.w3.org/TR/vocab-ssn/ (accessed on 10 May 2018).
17. Bermudez-Edo, M.; Elsaleh, T.; Barnaghi, P.; Taylor, K. IoT-Lite: A Lightweight Semantic Model for the Internet of Things. In Proceedings of the Intl IEEE Conferences on Ubiquitous Intelligence & Computing, Advanced and Trusted Computing, Scalable Computing and Communications, Cloud and Big Data Computing, Internet of People, and Smart World Congress, Toulouse, France, 18–21 July 2016; pp. 90–97.

18. Bassi, A.; Bauer, M.; Fiedler, M.; Kramp, T.; Van Kranenburg, R.; Lange, S.; Meissner, S. *Enabling Things to Talk*; Bassi, A., Bauer, M., Fiedler, M., Kramp, T., van Kranenburg, R., Lange, S., Meissner, S., Eds.; Springer: Berlin/Heidelberg, Germany, 2013; ISBN 978-3-642-40402-3.
19. Swetina, J.; Lu, G.; Jacobs, P.; Ennesser, F.; Song, J. Toward a standardized common M2M service layer platform: Introduction to oneM2M. *IEEE Wirel. Commun.* **2014**, *21*, 20–26. [CrossRef]
20. Alaya, M.B.; Medjiah, S.; Monteil, T.; Drira, K. Toward semantic interoperability in oneM2M architecture. *IEEE Commun. Mag.* **2015**, *53*, 35–41. [CrossRef]
21. Witt, K.J.; Stanley, J.; Smithbauer, D.; Mandl, D.; Ly, V.; Underbrink, A.; Metheny, M. Enabling Sensor Webs by Utilizing SWAMO for Autonomous Operations. In Proceedings of the8th annual NASA Earth Science Technology Conference (ESTC2008), College Park, MD, USA, 24–26 June 2008; pp. 263–270.
22. Hepp, M. GoodRelations: An Ontology for Describing Products and Services Offers on the Web. In *Knowledge Engineering: Practice and Patterns*; Springer: Berlin/Heidelberg, Germany, 2008; pp. 329–346.
23. Daniele, L.; den Hartog, F.; Roes, J. Created in Close Interaction with the Industry: The Smart Appliances REFerence (SAREF) Ontology. In *International Workshop Formal Ontologies Meet Industries*; Springer: Cham, Switzerland, 2015; pp. 100–112.
24. Prud'hommeaux, E.; Labra Gayo, J.E.; Solbrig, H. Shape expressions. In Proceedings of the 10th International Conference on Semantic Systems—SEM '14; ACM Press: New York, NY, USA, 2014; pp. 32–40.
25. Knublauch, H.; Kontokostas, D. Shapes Constraint Language (SHACL), W3C Recommendation 20 July 2017. Available online: https://www.w3.org/TR/shacl/ (accessed on 7 February 2019).
26. Web of Things at W3C. Available online: https://www.w3.org/WoT/ (accessed on 10 May 2018).
27. Paganelli, F.; Turchi, S.; Giuli, D. A Web of Things Framework for RESTful Applications and Its Experimentation in a Smart City. *IEEE Syst. J.* **2016**, *10*, 1412–1423. [CrossRef]
28. Karim, F.; Karim, F.; Frihida, A. Monitoring system using web of things in precision agriculture. *Procedia Comput. Sci.* **2017**, *110*, 402–409. [CrossRef]
29. Keppmann, F.L.; Maleshkova, M.; Harth, A. Semantic Technologies for Realising Decentralised Applications for the Web of Things. In Proceedings of the 2016 21st International Conference on Engineering of Complex Computer Systems (ICECCS), Dubai, UAE, 6–8 November 2016; pp. 71–80.
30. Nagib, A.M.; Hamza, H.S. SIGHTED: A Framework for Semantic Integration of Heterogeneous Sensor Data on the Internet of Things. *Procedia Comput. Sci.* **2016**, *83*, 529–536. [CrossRef]
31. Wu, Z.; Xu, Y.; Zhang, C.; Yang, Y.; Ji, Y. Towards Semantic Web of Things: From Manual to Semi-automatic Semantic Annotation on Web of Things. In Proceedings of the 2016 International Conference on Big Data Computing and Communications, Shenyang, China, 29–31 July 2016; pp. 295–308.
32. Dell'Aglio, D.; Calbimonte, J.-P.; Della Valle, E.; Corcho, O. Towards a Unified Language for RDF Stream Query Processing. In *International Semantic Web Conference*; Springer: Cham, Switzerland, 2015; pp. 353–363.
33. Alamri, A.; Ansari, W.S.; Hassan, M.M.; Hossain, M.S.; Alelaiwi, A.; Hossain, M.A. A Survey on Sensor-Cloud: Architecture, Applications, and Approaches. *Int. J. Distrib. Sens. Netw.* **2013**, *9*, 917923. [CrossRef]
34. Ullah, S.; Rodrigues, J.J.P.C.; Khan, F.A.; Verikoukis, C.; Zhu, Z. Protocols and Architectures for Next-Generation Wireless Sensor Networks. *Int. J. Distrib. Sens. Netw.* **2014**, *10*, 705470. [CrossRef]
35. Xively. Available online: https://xively.com (accessed on 10 May 2018).
36. ThingSpeak. Available online: https://thingspeak.com/ (accessed on 10 May 2018).
37. NGSI Context Management, Candidate Version 1.0. Available online: http://technical.openmobilealliance.org/Technical/release_program/docs/NGSI/V1_0-20101207-C/OMA-TS-NGSI_Context_Management-V1_0-20100803-C.pdf (accessed on 10 May 2018).
38. Hypercat 3.00 Specification. Available online: http://www.hypercat.io/uploads/1/2/4/4/12443814/hypercat_specification_3.00rc1-2016-02-23.pdf (accessed on 10 May 2018).
39. Project Haystack. Available online: https://project-haystack.org/ (accessed on 10 May 2018).
40. IoT Toolkit, Tools for the Open Source Internet of Things. Available online: http://iot-toolkit.com (accessed on 10 May 2018).
41. RDF Schema. Available online: https://www.w3.org/TR/2014/REC-rdf-schema-20140225/ (accessed on 10 May 2018).
42. Lanza, J.; Sanchez, L.; Santana, J.R.; Agarwal, R.; Kefalakis, N.; Grace, P.; Elsaleh, T.; Zhao, M.; Tragos, E.; Nguyen, H.; et al. Experimentation as a Service over Semantically Interoperable Internet of Things Testbeds. *IEEE Access* **2018**, *6*, 51607–51625. [CrossRef]

43. Agarwal, R.; Fernandez, D.G.; Elsaleh, T.; Gyrard, A.; Lanza, J.; Sanchez, L.; Georgantas, N.; Issarny, V. Unified IoT ontology to enable interoperability and federation of testbeds. In Proceedings of the IEEE 3rd World Forum on Internet of Things (WF-IoT), Reston, VA, USA, 12–14 December 2016; pp. 70–75.
44. SPARQL 1.1 Graph Store HTTP Protocol. Available online: https://www.w3.org/TR/sparql11-http-rdf-update/ (accessed on 10 May 2018).
45. Clark, K.G.; Grant, K.; Torres, E. SPARQL Protocol for RDF. Available online: http://www.w3.org/TR/rdf-sparql-protocol/ (accessed on 10 May 2018).
46. Zhao, M.; Kefalakis, N.; Grace, P.; Soldatos, J.; Le-Gall, F.; Cousin, P. Towards an Interoperability Certification Method for Semantic Federated Experimental IoT Testbeds. In *International Conference on Testbeds and Research Infrastructures*; Springer: Cham, Switzerland, 2017; pp. 103–113.

Article

# CDSP: A Solution for Privacy and Security of Multimedia Information Processing in Industrial Big Data and Internet of Things

**Xu Yang [1], Yumin Hou [2], Junping Ma [2] and Hu He [2,*]**

[1] School of Computer Science and Technology, Beijing Institute of Technology, Beijing 100081, China; yangxu@tsinghua.edu.cn

[2] Institute of Microelectronics, Tsinghua University, Beijing 100084, China; hou-ym12@mails.tsinghua.edu.cn (Y.H.); mjpyd41@126.com (J.M.)

* Correspondence: hehu@tsinghua.edu.cn; Tel.: +86-010-6279-5139; Fax: +86-010-6279-5104

Received: 10 January 2019; Accepted: 25 January 2019; Published: 29 January 2019

**Abstract:** With the widespread nature of wireless internet and internet of things, data have bloomed everywhere. Under the scenario of big data processing, privacy and security concerns become a very important consideration. This work focused on an approach to tackle the privacy and security issue of multimedia data/information in the internet of things domain. A solution based on Cryptographical Digital Signal Processor (CDSP), a Digital Signal Processor (DSP) based platform combined with dedicated instruction extension, has been proposed, to provide both programming flexibility and performance. We have evaluated CDSP, and the results show that the algorithms implemented on CDSP all have good performance. We have also taped out the platform designed for privacy and security concerns of multimedia transferring system based on CDSP. Using TSMC 55 nm technology, it could reach the speed of 360 MHz. Benefiting from its programmability, CDSP can be easily expanded to support more algorithms in this domain.

**Keywords:** privacy and security; internet of things; very long instruction word (VLIW); DSP; instruction set extension

## 1. Introduction

We have entered the era of Big Data. The widespread nature of internet of things and wireless network, is making multimedia communication systems, such as on-line chatting, video conference and surveillance systems, becoming more and more popular. Since the process of multimedia communication systems involves data generation, storage, sending, receiving, sharing and so on, various security issues should be concerned. Data encryption algorithms could be adopted in multimedia communication systems to guarantee the security of data. Many kinds of data encryption algorithms have been released. This provides people with more options to choose according to their own needs. However, it also imposes a challenge for hardware design to have a large level of flexibility to adapt to different kinds of data encryption algorithms with limited time.

In this work, we focused on solving the privacy and security issue of multimedia surveillance system in internet of things domain, where high quality video and audio recorded by cameras needs to be compacted, encrypted and transferred to the control center through network, and be replayed in real-time on the monitors.

When evaluating the feature of the data stream of multimedia surveillance system, one finds that:

1. Data in the stream always contain private information, so encryption of the raw data is required to protect the data stream from disclosure before transferring. Thus data encryption algorithms need to be supported.

2. Since surveillance audio and video can be used as legal evidence to justify the fact, it is necessary to ensure the truth and reliability of the data stream. Hash functions and authentication could be of use for this.
3. Sometimes it is required to verify whether the audio and video are from a specific user. Thus digital signature should also be supported.
4. Requirement of real-time. Time delay from the scene to the control center should be less than 170 ms, which means that the work of video/audio encoding, encryption, decryption and video/audio decoding should be finished in 170 ms. According to the practical application, the peak throughput of the data stream should reach 150 Mbps.

Thus, the middleware platform of this kind of surveillance system should support a lot of different kinds of data encryption algorithms. In this work, we present our middleware solution based on CDSP, combining the programmability of DSP and the high efficiency of dedicated designed special operations. Benefiting from the programmability of DSP, new kinds of algorithms can be easily implemented, providing a high level of flexibility. Furthermore, the special operations designed dedicated for some algorithms can significantly reduce the code size, and largely enhance the performance. This is a new attempt in cryptographic DSPs, and the results show that our approach is both feasible and efficient.

The remainder of this paper is organized as follows: Section 2 will introduce the proposed DSP architecture. The design of dedicated special operations is discussed in Section 3. Related works are presented in Section 4. Section 5 gives the result of the evaluation. Finally, we give a conclusion in Section 6.

## 2. The Proposed CDSP Architecture

Our middleware solution is built based on a DSP called CDSP, which is designed as a 6-issue VLIW DSP. It provides high instruction level parallelism, and can greatly improve the performance on cryptographic algorithms execution. In this section we will introduce it in more detail.

### 2.1. Design of CDSP Core

CDSP core [1] is composed of four main parts: Memory, Instruction Fetch Unit (IFU), Instruction Dispatch Unit (IDU), and Execution Unit (EU). The architecture of CDSP is shown in Figure 1. CDSP adopts Harvard architecture and has separate Program Memory (PMEM) and Data Memory (DMEM). PMEM is 24 KB SRAM, with a 256-bit width port. PMEM is used to store CDSP program, and can be initialized using DMA. DMEM is 16 KB dual-port SRAM, and each port is 64-bit width. DMEM is used to store data stream, and the data stream is transferred through AXI bus under the control of DMA.

IFU reads instruction packets from PMEM. Each instruction packet is 256-bit width, containing eight to 16 instructions.

IDU seeks available instructions in the package and hands them out to EU. At most, six instructions can be dispatched each cycle.

EU is clustered, and the two clusters are named as X cluster and Y cluster. EU includes six function units, which are named as XA, YA, XM, YM, XD and YD separately. XA, XM and XD belong to X cluster. YA, YM and YD belong to Y cluster. XA and YA are arithmetic units, executing arithmetic and logic instructions, such as ADD, SUB, AND, XOR, ASL and LSR. XM and YM are multiplication units, executing multiplication instructions, and some arithmetic logic instructions. XD and YD are load/store units, loading data from DMEM to register, or storing data from register to DMEM. Some arithmetic logic instructions and branch instructions can also be executed in XD and YD units.

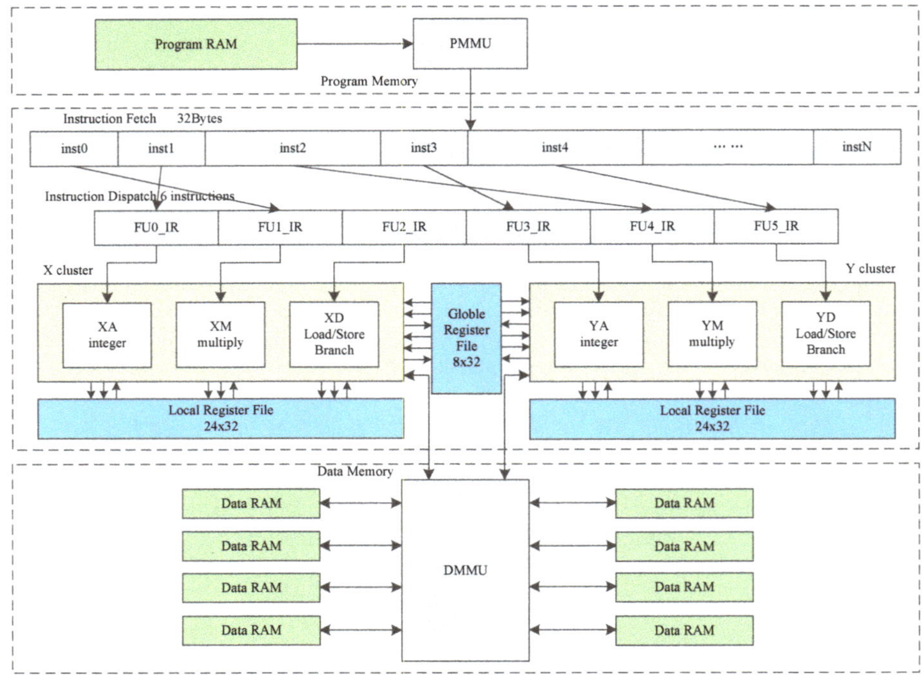


**Figure 1.** CDSP architecture.

### 2.2. Design of Pipeline

CDSP includes 11 pipeline stages, as shown in Figure 2.

PCG (PC generation): This stage generates the next PC, which is chosen from the interrupt PC, branch PC and PC + 4.

PCS (PC send): This stage passes the generated PC to the next stage.

PWT (PC wait): This stage checks whether the instruction is valid in the instruction cache or not. If the instruction is invalid, instructions should be fetched from the PMEM.

IR (instruction return): We get the instructions in this stage.

EXP (instruction expansion): As the instructions are either 16-bit or 32-bit, we expand the instructions to the same length in this stage.

IDP (instruction dispatch): Instruction dispatch is completed in this stage. Since the function unit information is encoded in the instructions, we can get this information and dispatch the instructions to the corresponding function units. This procedure can be interpreted as predecode and dispatch.

IDC (instruction decode): This stage decodes the instructions.

EX1∼EX4 (instruction execution): Instructions are executed in this stage. Different instructions need 2∼4 stages to complete execution.

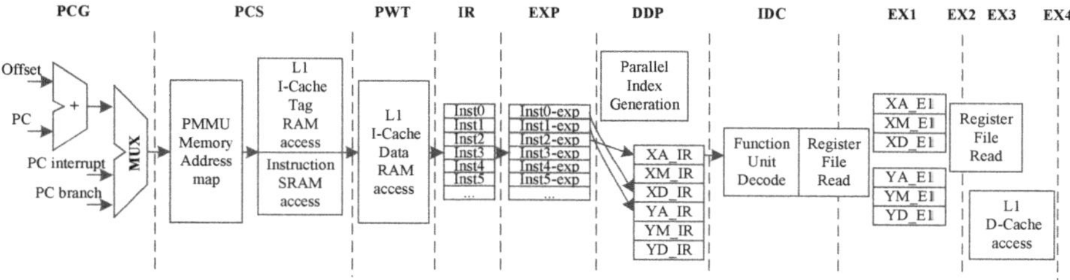


**Figure 2.** Design of CDSP pipeline.

## 2.3. Design of Register Files

CDSP has three register files, which are also clustered. Register files are also shown in Figure 1. X and Y cluster register files both contain 24 registers, which are named as X0~X23 and Y0~Y23 separately. There is a global register file G containing eight registers named as G0~G7. All the registers are 32-bit width. X unit instructions can use X cluster registers and G registers, and Y unit instructions can use Y cluster registers and G registers. Two adjacent registers can be used as a register pair, forming a bigger operand. For example, X1:X0 is a register pair and forms a 64-bit operand.

X and Y are two symmetrical clusters. Taking advantage of the symmetrical architecture, CDSP can access 56 registers using only 5 bits of the instruction. Usually, 5 bits can only address 32 registers. A large number of registers can help exploit Instruction Level Parallelism (ILP) and increase the performance. Some algorithms can also take advantage of the symmetrical architecture to run in parallel.

## 2.4. Design of General Instructions

CDSP is a RISC processor, and its Instruction Set consists of both general instructions and dedicated designed special instructions. General instructions include arithmetic and logic instructions, shift instructions, multiplication instructions, load instructions and branch instructions. Dedicated instructions are designed for specific cryptographic algorithms in order to improve the performance. General instructions will be described in this part.

Arithmetic instructions: Addition (ADD/ADC) and subtraction (SUB/SUBB). The operands could be 32-bit signed or unsigned numbers, and the operation could be addition with carry or subtraction with borrow.

Logic instructions: Bitwise AND, bitwise OR, bitwise XOR and bitwise Negate (NOT). Operands can be immediate data or come from registers.

Move instructions: Move a data from one register to another (MOV).

Shift instructions: Arithmetic Shift Right (ASR), Logic Shift Right (LSR), Arithmetic Shift Left (ASL), and Barrel Rotate Left (ROL).

Multiplication instructions: Multiplication (MUL) and Accumulation (ACC). Multiplication instructions occupy two pipeline stages.

Pack instructions: Pack instructions switch the byte sequence in a 32-bit operand, or select some bytes from two operands and build up a new word.

LOAD/STORE instructions: Load (LD) data from data memory to registers, or store (ST) data from register to data memory. A RISC processor always uses powerful load-store instructions to finish data transfer between registers and memory. CDSP LD instructions use base register and offset register to calculate the memory address, and the base register can increase or decrease automatically.

**Branch instructions:** Branch to the target PC and flush the instructions in pipeline (B); Branch to the target PC without flushing the instructions in pipeline (BD); Call subroutine (CALL); Return from subroutine (RET).

## 2.5. Design of Interfaces

The architecture of CDSP system is shown in Figure 3. There is an AXI-lite slave interface in CDSP, which is used to control work flow, configure CDSP and check the status of CDSP. AXI is a widely used bus standard, making CDSP easy to integrate.

An embedded Direct Memory Access (DMA) with an AXI master interface is designed in CDSP, which is used to read and write data stream though AXI bus, with high bandwidth. The DMA can be controlled either by CDSP or AXI-lite.

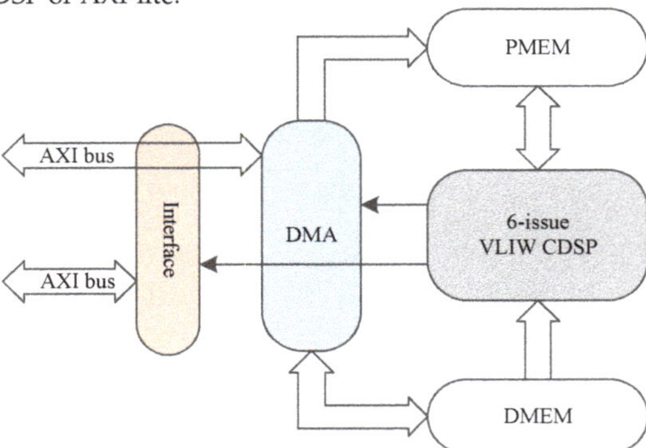


**Figure 3.** Architecture of CDSP system.

# 3. Design of Dedicated Instructions

The CDSP platform we presented combines both the programmability of DSP and the high efficiency of dedicated designed special instruction to: (1) provide flexibility for implementing of different kinds of encryption algorithms, and (2) achieve high performance. Dedicated instruction design for cryptographic algorithms is an important feature of CDSP, and is also an innovation in cryptographic DSP design.

Many kinds of data encryption algorithms have already been implemented on our platform:

1. Block cipher algorithms [2–5], including DES, 3DES, AES and IDEA. We have designed dedicated instruction for AES and DES, while IDEA is implemented using only general instructions.
2. Hash [6] functions and public key algorithms. The study of public key algorithms are mainly focused on the ASIC approach in the literature. It is an innovation to realize these algorithms on a DSP platform. These algorithms are implemented using general instructions.

In this section, the design of dedicated instructions for DES and AES algorithms will be introduced.

## 3.1. Design for DES Algorithm

### 3.1.1. Flow of DES Algorithm

DES encrypts 64-bit plain text, and generates a 64-bit cipher text, using a 64-bit key. DES algorithm could be divided into three function units as shown in Figure 4. The first part is an Initial Permutation (IP), which reorders the 64 bits of the plain text. The second part includes a Feistel (F) function and an operation of adding round key, which is iterated 16 times. The last part is a Final Permutation (FP), reordering the 64-bit intermediate data and outputting the final cipher text.

Figure 5 shows the process of DES round transformation. 64-bit intermediate text is divided into 2 parts, named as $L_i$ and $R_i$ separately, both of which are 32-bit width. DES round transformation mainly contains 4 steps, and The 4 steps are the Feistel function (F):

(1) $R_i$ is transformed using a Expansion (E) permutation;

(2) Result of step (1) xor with round key ($K_i$);
(3) Result of step (2) is substituted using lookup table S-box;
(4) Result of step (3) is rearranged with a new Permutation (P).

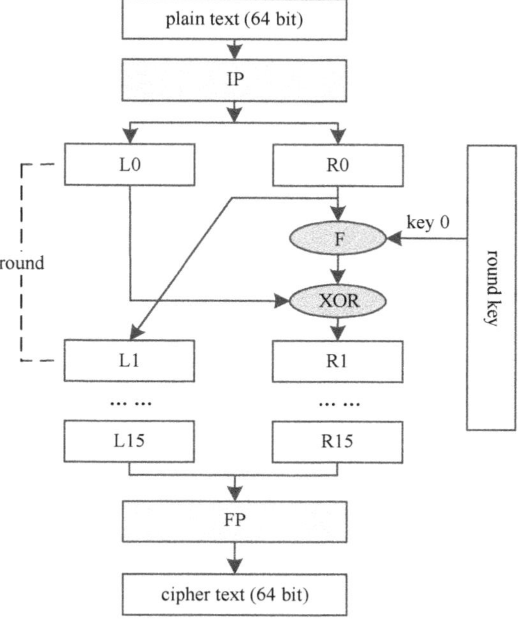


**Figure 4.** Architecture of DES.

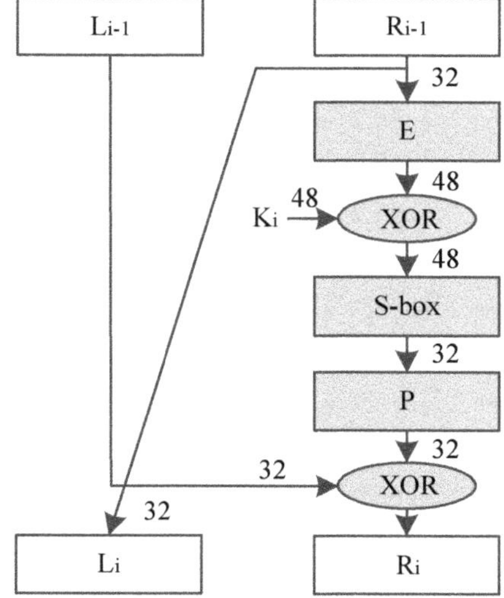


**Figure 5.** Data Encryption Standard (DES) round transformation.

#### 3.1.2. Dedicated Instruction Design for DES

If only general instructions are used to implement the DES algorithm, then a large number of instructions will be needed, considering the performing of the permutation, which would result in large code size. And also the performance might be low.

Three dedicated instructions are designed for the implementation of DES algorithm:

1. DESIP: Deigned to implement the IP operation. The source operand is a 64-bit data from a register pair, and the result is also 64-bit width and stored in a register pair.
2. DESFP: Deigned to implement the FP operation. Function of this instruction is the same as DSEIP.
3. DESRND: Designed to implement the DES round transformation. The source operand is a 64-bit intermediate result and a 48-bit round key. The output is a 64-bit intermediate result for the next round.

Table 1 shows the flow of DES algorithm with the help of dedicated designed special instructions. DES encryption can be finished in 20 cycles. Similarly, 3DES can be finished in 57 cycles. Since CDSP includes two clusters, two data blocks can be encrypted separately in X cluster and Y cluster in parallel, thus the performance will be doubled.

The comparison of code size is shown in Table 2. By the using of dedicated instructions, the code size can be more compacted, and the execution time also reduced.

**Table 1.** Assembly code for DES Algorithm.

| Cycle | Instruction |
|---|---|
| 1 | XD, G3:G2 = LD.D (X0++) |
| 2 | XD, G3:G2 = LD.D (X0++) |
| 3 | XA, G1:G0 = DESIP (G1:G0)<br>XD, G3:G2 = LD.D (X0++) |
| 4 | XA, G1:G0 = DESRND (G1:G0, G3:G2) |
| 5 | XA, G1:G0 = DESRND (G1:G0, G3:G2) |
| ... | ... ... |
| 20 | XA, G1:G0 = DESFP (G1:G0) |

**Table 2.** Code size comparison for DES (instruction number).

| Algorithm | Dedicated Inst. | General Inst. |
|---|---|---|
| DES_encryption | 56 | ~500 |
| DES_decryption | 57 | ~500 |
| 3DES_encryption | 155 | ~500 |
| 3DES_decryption | 158 | ~500 |

### *3.2. Design for AES Algorithm*

#### 3.2.1. Flow of AES Algorithm

AES encrypts 128-bit plain text, and generates a 128-bit cipher text, using a 128-bit key. AES consists of 10 round transformations. The intermediate result between two rounds is a 16-byte data called mid-state, which is usually arranged in a $4 \times 4$ matrix. AES round transformation is based on this matrix. AES could be divided into four main operations as shown in Figure 6:

1. SubBytes: Each byte is used as an address to look up a SubBytes table and output a new byte, substituting for the initial byte.

2. ShiftRows: The last three rows of the matrix shift left cyclically for 1 byte, 2 bytes and 3 bytes separately, as shown in Figure 6.
3. MixColumns: Combines 4 bytes in each column. It can be implemented by using a fix value matrix to multiply the mid-state, as shown in Figure 7.
4. AddRoundKey: Round key is a 128-bit key expanded from the initial key. AddRoundKey conducts bitwise XOR operation between mid-state and the round key.

ShiftRows can be moved to the front of SubBytes, and the sequence will become ShiftRows→SubBytes→MixColumns→AddRoundKey.

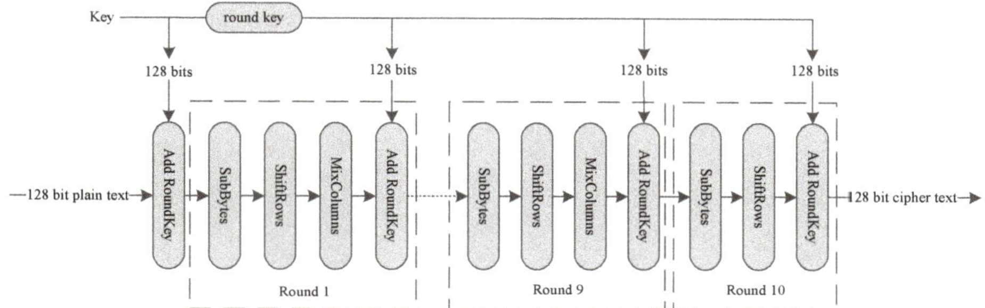


**Figure 6.** AES transformation.

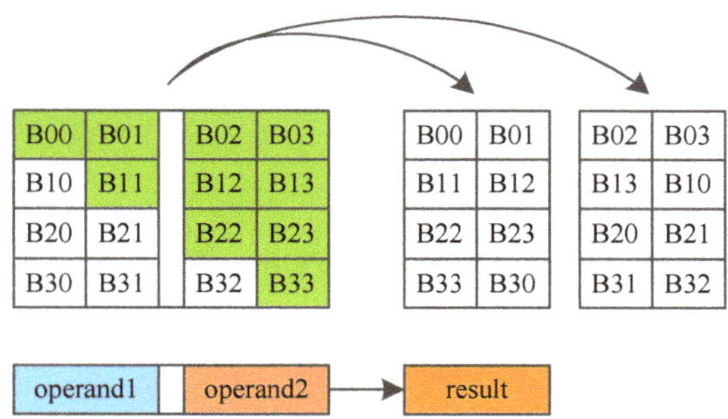


**Figure 7.** AES dedicated instructions AESSHF.

#### 3.2.2. Dedicated Instruction Design for AES

Using general instructions only to implement SubBytes, ShiftRows and MixColumns operations will result in low throughput and large code size. Dedicated instructions are designed to enhance the performance of AES.

Three dedicated instructions are designed for the implementation of DES algorithm:

1. AESSHF: AESSHF conducts ShiftRows, as shown in Figure 6. Each column of the matrix is stored in a register, and two registers make up a register pair, a register pair is used as an operand. AESSHF finishes the ShiftRows function using two source operands. The output is two new columns. AESSHF should run two times to finish the ShiftRows operation.
2. AESSUBMIX: AESSUBMIX conducts two operations, SubBytes and Mixcolumns, as shown in Figure 7. The instruction occupies two pipeline stages. At the first stage, 4 bytes are replaced according S-box table, and at the second stage, a fix-value matrix is multiplied by the S-box output, generating the result. Source operand is one column of the mid-state and the result is a column of the next mid-state.

3. AESSUB: AESSUB conducts SubBytes operation, as shown in Figure 8. AESSUB instruction outputs the S-box result and uses one pipeline stage. AESSUB instruction is used in the final round, because the final round dose not includes the MixColumns [7] operation.

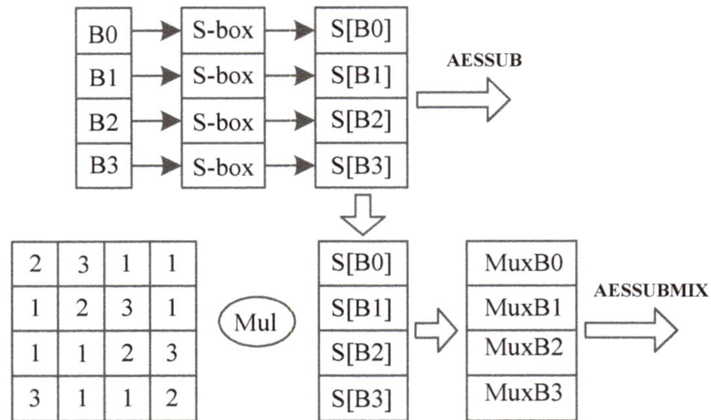


**Figure 8.** AES dedicated instructions AESSUB and AESSUBMIX.

Table 3 shows the flow of AES algorithm with the help of dedicated designed special instructions. One AES round can be finished in five clock cycles, shown as cycle 5 to 9. AES encryption can be finished in 59 cycles.

As shown in Table 4, code size can be noticeably reduced by using dedicated instructions.

**Table 3.** Assembly code for AES Algorithm.

| | |
|---|---|
| 1 | XD, X1:X0 = LD.D (G7)<br>YD, Y1:Y0 = LD.D (G7++[#2]) |
| 2 | NOP |
| 3 | NOP |
| 4 | XA, G0 = XOR (G0, X0)<br>XM, G1 = XOR (G1, X1)<br>XA, G2 = XOR (G2, Y0)<br>XM, G3 = XOR (G3, Y1) |
| 5<br>Loop | XA, G1:G0 = AESSHF (G1:G0, G3:G2)<br>YA, G3:G2 = AESSHF (G3:G2, G1:G0)<br>XD, X1:X0 = LD.D (G7)<br>YD, Y1:Y0 = LD.D (G7++[#2]) |
| 6 | XA, G0 = AESSUBMIX (G0)<br>YA, G1 = AESSUBMIX (G1) |
| 7 | XA, G2 = AESSUBMIX (G2)<br>YA, G3 = AESSUBMIX (G3) |
| 8 | XA, G0 = XOR (G0, X0)<br>XM, G1 = XOR (G1, X1) |
| 9<br>End | YA, G2 = XOR (G2, Y0)<br>YM, G3 = XOR (G3, Y1) |

**Table 4.** Code size comparison for AES (instruction number).

| Algorithm | Dedicated Inst. | General Inst. |
|---|---|---|
| AES_encryption | 65 | ∼300 |
| AES_decryption | 63 | ∼300 |

## 4. Related Works

There is not much research reported on the implementation of cryptographic algorithms on a DSP platform combined with dedicated instruction extension, while some are similar to ours. We divide them into two categories.

### 4.1. Cryptographic Algorithms on a DSP

T. Wollinger et al. [8] research how well-suited high-end DSPs are for the AES algorithms. Five AES candidates: Mars, RC6, Rijndael, Serpent and Twofish are investigated and realized on a TMS320C6201 DSP. They optimize the C code to speed up the algorithm. They provide single-block and multi-block processing to enable the data blocks to be executed in parallel; this method is limited to be used under certain confidentiality modes.

TMS320C6201 DSP has 32 32-bit registers and eight independent functional units. The architecture of TMS320C6201 DSP is also divided into two parts and each part includes four functional units and 16 registers.

They compare the result with Pentium-Pro processor, both working under 200 MHz. It shows that the performance of TMS320C6201 is better than Pentium-Pro processor by about 32.3% on average.

K. Itoh et al. [9] also implement public-key cryptographic algorithms on TMS320C6201 DSP, including Rivest-Shamir-Adleman (RSA) [10–12], DSA and ECDSA [13]. The performances of RSA1024, DSA1024 and ECDSA160 achieve 11.7 ms, 14.5 ms and 3.97 ms respectively. The result is achieved mainly be optimization of modular multiplication and elliptic doubling operations. D. Xu et al. [14] realize AES algorithm on a configurable VLIW DSP called Jazz DSP. The computation units can be configured by software. They implement AES on three configurations and the best performance reaches 10.56 cycles/byte, in which case, eight VLIW slots are configured and two functions SubByte and Mixcolomn are converted to the designer defined computation unit to improve the performance.

Y. S. Zhang et al. [15] have designed a low-cost and confidentiality-preserving data acquisition framework for IoMT. They first used chaotic convolution and random subsampling to capture multiple image signals, assembled these sampled images into a big master image, and then encrypted this master image based on Arnold transform and single value diffusion. The encrypted image is delivered to cloud servers for storage and decryption service.

### 4.2. Instruction Set Extension for Cryptographic Algorithms

Intel proposes Advanced Encryption Standard (AES) new instruction set in 2010 [16]. The new instruction set includes six instructions designed for AES, of which four instructions realize AES encryption and decryption, and the other two support AES key expansion. The new instruction set makes AES simple to implement with small code size. The performance of AES with 128-bit key achieves 4.44, 4.56 and 4.49 cycle/byte under ECB, CBC and CTR mode respectively. The result comes from a processor based on Intel microarchitecture running at 2.67 GHz.

The IBM Power8 [17] processor also improves the performance on data encryption. It adds 11 new instructions to improve the performance of cryptographic algorithms, including AES, Galois Counter Mode (GCM) of operation for AES, SHA-2, and CRC. Vector and Scalar Unit (VSU) is also added to enhance the performance. Under CBC mode, the throughput of AES128 reaches about 680,000 KB/s with the processor running at 3.59 GHz.

Our implementation combines the features of these two approaches, which realizes cryptographic algorithms on a DSP with an extended instruction set.

## 5. Experimental Results

### 5.1. Experimental Framework

Function simulation of CDSP is achieved by Synopsis VCS and Verdi joint simulation, working under Redhat system. FPGA verification is conducted on XC5VLX330T from Xilinx vertex 5 series. Verification is based on Synopsis VMM (Verification Methodology Manual), using system Verilog. Using VMM, we can generate random test cases or test cases with constraints.

C language is used to build a C-model for certain cryptographic algorithm. Furthermore, we use C-model to generate the correct result, called the golden result.

The programs implementing different cryptographic algorithms are written in CDSP assembly language. The assembler of CDSP is designed based on GNU Binutils binary tool set. We compiled the assembly programs, sent the firmware, together with the test cases, to the DSP, and got the simulation result. The result will be compared with the golden result, and whether the result is correct will be reported.

### 5.2. Results

#### 5.2.1. Results for Block Cipher Programs

Table 5 lists the performance of DES, 3DES, AES and IDEA in CDSP. The clock cycles and throughput for one block encryption are given, when CDSP is running at 360 MHz. The 3rd column gives the performance with no encryption mode. The following columns give the performance in different confidentiality modes, including ECB, CBC, CFB, OFB and CTR, which are defined in FIPS [18]. In CFB and OFB confidentiality modes, the block length is select as 64-bit for DES, 3DES, and IDEA, and 128-bit for AES. The programs running in CDSP can be easily modified to support more confidentiality modes.

**Table 5.** Block cipher performance (360 MHz).

| | | No Mode | ECB | CBC | CFB/OFB | CTR |
|---|---|---|---|---|---|---|
| DES | cycles | 21 | 41 | 43 | 43 | NA |
| | throughput (Mbps) | 1097 | 562.0 | 535.8 | 535.8 | NA |
| 3DES | cycles | 53 | 73 | 75 | 75 | NA |
| | throughput (Mbps) | 434.7 | 315.6 | 307.2 | 307.2 | NA |
| AES | cycles | 59 | 71 | 73 | 73 | 75 |
| | throughput (Mbps) | 781.0 | 649.0 | 631.2 | 631.2 | 614.4 |
| IDEA | cycles | 57 | 69 | 71 | 71 | NA |
| | throughput (Mbps) | 404.2 | 333.9 | 324.5 | 324.5 | NA |

Performance comparison of AES128 with general purpose processors is shown in Table 6. We compare the performance of our implementation with ARM7 and ARM9. The result shows that CDSP has dramatic advantage over the two ARM processors. In comparison, the architecture of our design provides higher level of parallelism and the dedicated instructions are useful to speed up CDSP.

We also compare the performance of CDSP with other approaches of ISA extension, as shown in Table 7. Power8 processor adds 11 instruction to enhance the efficiency of AES. Intel processor adopts the AES new instruction set. The result shows that our implementation is better than Power8 in the performance of CPU cycles per Byte. According to Ref. [17], the throughput of Power8 for 128 bit encryption is about 680,000 KB/s, and the processor run under 3.59 GHz. We calculate that the performance of CPU cycles per Byte is 5.53. The result of Intel is based on Intel microarchitecture codename Westmere running at 2.67 GHz [16]. Our implementation shows close performance with Intel under CBC mode. Power8 and Intel processors are both ASIC. ASIC always has higher frequency than DSP. The parameter of CPU cycles per Byte is not affected by frequency and it provides a more fair comparison. The difference in the design of instructions and hardware results in different performance.

Table 8 shows the performance comparison of AES128 between CDSP and other DSP approaches. The result shows that CDSP has far better performance than Ref. [8,14]. DSPs always provide high level of parallelism, but the dedicated instructions make our design outweigh other DSPs in cryptographic algorithms.

**Table 6.** AES128 performance comparison with general purpose processors (Cycles/Byte).

| | | |
|---|---|---|
| Ref. [19] | ARM7 | 104.69 |
| Ref. [19] | ARM9 | 86.5 |
| Ours | CDSP | 4.56 |

**Table 7.** AES128 performance comparison with other ISA extension methods (Cycles/Byte).

| | | ECB | CBC | CTR |
|---|---|---|---|---|
| Ref. [17] | Power8 | - | 5.53 | - |
| Ours | CDSP | 4.44 | 4.56 | 4.49 |
| Ref. [16] | Intel | 1.28 | 4.15 | 1.38 |

**Table 8.** AES128 performance comparison with other DSP implementations (Cycles/Byte).

| | | |
|---|---|---|
| Ref. [8] | TMS320C6201 | 14.25 |
| Ref. [14] | Jazz DSP | 10.56 |
| Ours | CDSP | 4.56 |

5.2.2. Results for Hash Function

Table 9 lists the performance of MD5 [20,21] and SHA-1 in CDSP. The second column lists the clock cycles consumed in one data block compression, which is 512-bit width. The third column shows the time consumed compressing one data block. The forth column gives the throughput when CDSP is running under 360 MHz. Since CDSP owns 6 function units, the arithmetic and logic operations in SHA-1 and MD5 can run in parallel. The VLIW architecture with 6-issue is good at exploiting instruction level parallelism and achieving better performance. CDSP can also implement other Hash functions through software development, such as SHA-256 and SHA-512.

**Table 9.** Hash function performance (360 MHz).

| | Cycles | Time (ns) | Throughput (Mbps) |
|---|---|---|---|
| MD5 | 325 | 902.8 | 567 |
| SHA-1 | 316 | 877.8 | 583 |

5.2.3. Results for Public Key Algorithms

Table 10 lists RSA and Elliptic Curve Cryptography (ECC) [22,23] performance when CDSP is running under 360 MHz. The first column shows the algorithms, including RSA using a 1024-bit key and a 2048-bit key, with and without applying CRT, and ECC using a 192-bit key and a 256-bit key. The second column lists the clock cycles used in Montgomery modular multiplication. The third column shows the number of multiplication operations used in these algorithms. The forth and the fifth columns list the clock cycles and total time consumed. The last column gives the number of executions of the algorithm per second.

**Table 10.** RSA and ECC performance.

| | Cycles (MUL) | MUL Operation | Cycles | Time (ms) | Tran./s |
|---|---|---|---|---|---|
| RSA1024 | 2840 | 1536 | 4,367,000 | 12.13 | 82.0 |
| RSA1024(CRT) | 815 | 1536 | 1,223,000 | 3.397 | 294.4 |
| RSA2048 | 10,512 | 3072 | 32,313,800 | 89.76 | 11.1 |
| RSA1024(CRT) | 2840 | 3072 | 8,751,000 | 24.31 | 41.1 |
| ECC192 | 163 | 1920M + 1536S | 742,200 | 2.062 | 485.0 |
| ECC256 | 301 | 2560M + 2048S | 1,651,000 | 4.586 | 218.1 |

In the third column, the number of multiplication operations are evaluated based on the assumption that half of the binary bits in a big number are 1, which is a worst case. Choice of the key and window width can affect the result significantly. In this paper, the window width is 2, and the performance can be raised by 8.7% on average compared with the performance when the window width is 1.

Table 11 shows that CDSP has equal RSA performance with Ref. [9]. Since both designs are DSPs and there are not dedicated instructions for RSA in our design, the result is understandable. In Ref. [9], the DSP works under 200 MHz, while our design works under higher frequency of 360 MHz.

**Table 11.** RSA performance comparison with other DSP implementations (ms).

| | | RSA1024 | RSA2048 |
|---|---|---|---|
| Ref. [9] | TMS320C6201 | 11.7 | 84.6 |
| Ours | CDSP | 12.13 | 89.76 |

In conclusion, CDSP shows satisfactory performance in cryptographic algorithms. Compared with other DSPs, CDSP shows far better performance for algorithms with dedicated instructions and close performance for algorithms without dedicated instructions. Compared with processor with dedicated cryptographic instructions, CDSP shows better performance than Power8 and close performance with Intel under CBC mode.

### 5.3. Silicon Implementation

We have implemented the multimedia surveillance system based on CDSP platform. The chip of the surveillance system is taped out and mass-produced. Using TSMC 55 nm technology, the synthesized frequency of CDSP achieves 360 MHz. The critical path is in the instruction dispatch stage. The dedicated instruction extension does not reduce the working frequency. The area is 186,000 gates and 40 KB SRAM (16 KB DMEM and 24 KB PMEM). Area consumption caused by ISA extension are 7862 gates, which is about 4% of the total area. The power consumption is 58 mW. Figure 9 is the layout of multimedia surveillance chip. At the bottom are 2 DSP clusters, each DSP cluster consisting of a CDSP. The chip can output at least 1080P or 4 channel D1 video format in real-time. Figure 10 is the photo of the chip.

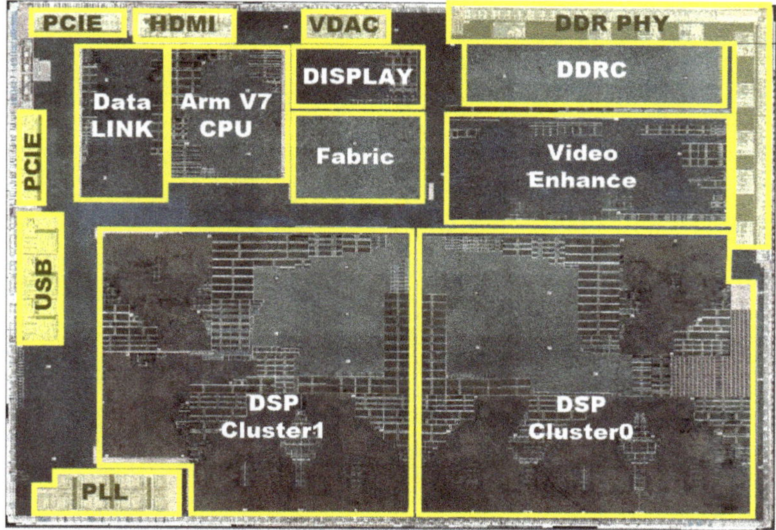


**Figure 9.** Layout of the multimedia surveillance system.

**Figure 10.** Photo of the chip.

## 6. Conclusions

This paper proposes our approach and experiences for designing a platform based on CDSP, a clustered VLIW DSP with ISA extension for cryptographic algorithms. CDSP is designed to target the solving of privacy and security issues in multimedia surveillance system. CDSP has 11 pipeline stages, making it achieve high frequency of 360 MHz. CDSP has six function units, and can run up to six instructions in one cycle, largely enhancing the calculation density. The architecture of CDSP is advantageous in exploiting instruction level parallelism and achieving better performance.

ISA of CDSP consists of both general instructions and dedicated instructions. In this work, we presented our experience for design dedicated instructions for DES and AES algorithms. The result shows that those dedicated designed instructions can significantly improve the performance, and reduce software code size. Since cryptographic algorithms usually consist of special complex computation-intensive operations, making software solution yield poor throughput, according to our results, adding dedicated instructions is a good choice to improve performance, and is more convenient compared with the co-processor scheme.

Many common cryptographic algorithms are already implemented in CDSP, including block cipher algorithms, hash functions and public key cryptographic algorithms. Using our approach, new data encryption algorithms could be easily implemented on CDSP platform, making CDSP a practical solution for building, establishing a complete network security system.

**Author Contributions:** Y.H. conceived the design of the VLIW architecture. Y.H. was responsible for hardware realization and simulation of CDSP. H.H. and X.Y. advised for the dedicated instruction design. J.M. revised the paper. All authors have read and approved the final manuscript.

**Funding:** This work is supported by the National Natural Science Foundation of China under Grant No. 91846303, the National Natural Science Foundation of China under Grant No. 61502032, and the Core Electronic Devices, High-End General Purpose Processor, and Fundamental System Software of China under Grant No. 2012ZX01034-001-002.

**Conflicts of Interest:** The authors declare that there is no conflict of interest.

## References

1. Liu, Y.; He, H.; Xu, T. Architecture design of variable lengths instructions expansion for VLIW. In Proceedings of the 8th IEEE International Conference on ASIC, Changsha, China, 20–23 October 2009; pp. 29–32.
2. National Institute of Standards and Technology (NIST). FIPS PUB 197: Advanced Encryption Standard (AES). Available online: http://csrc.nist.gov/publications/fips/fips197/fips-197.pdf (accessed on 28 January 2019).
3. National Institute of Standards and Technology (NIST). FIPS PUB 46-3: Data Encryption Standard (DES). Available online: http://csrc.nist.gov/publications/fips/fips46-3/fips46-3.pdf (accessed on 28 January 2019).
4. National Institute of Standards and Technology (NIST). FIPS PUB 180-1: Secure Hash Standar. Available online: http://www.itl.nist.gov/fipspubs/fip180-1.htm (accessed on 28 January 2019).
5. Gladman, B. implementation experience with the AES candidate algorithms. In Proceedings of the 2nd AES Candidate Conference, Rome, Italy, 22–23 March 1999; pp. 7–14.
6. Kang, Y.K.; Kim, D.W.; Kwon, T.W.; Choi, J.R. An efficient implementation of hash function processor for IPSEC. In Proceedings of the 3rd IEEE Asia-Pacific Conference on ASIC, Taipei, Taiwan, 6–8 August 2002; pp. 93–96.
7. Tillich, S.; Großschädl, J. Instruction set extensions for efficient AES implementation on 32-bit processors. In Proceedings of the Cryptographic Hardware and Embedded Systems (CHES 2006), Yokohama, Japan, 10–13 October 2006; Springer: Berlin/Heidelberg, Germany, 2006; pp. 270–284.
8. Wollinger, T.; Wang, M.; Guajardo, J.; Paar, C. How well are high-end DSPs suited for the AES algorithms. In Proceedings of the Third Advanced Encryption Standard Candidate Conference, New York, NY, USA, 13–14 April 2000; pp. 94–105.
9. Itoh, K.; Takenaka, M.; Torii, N.; Temma, S.; Kurihara, Y. Fast implementation of public-key cryptography on a DSP TMS320C6201. In *Cryptographic Hardware and Embedded Systems*; Springer: Berlin/Heidelberg, Germany, 1999; pp. 61–72.
10. Mclvor, C.; McLoone, M. Fast Montgomery Modular Multiplication and RSA Cryptographic Processor Architectures. In Proceedings of the Thirty-Seventh Asilomar Conference on Signals, Systems, and Computers, Pacific Grove, CA, USA, 9–12 November 2003; pp. 379–384.
11. Phillips, B.J. Implementing 1024-bit RSA exponentiation on a 32-bit processor core. In Proceedings of the IEEE International Conference on Application-Specific Systems, Architectures, and Processors, Boston, MA, USA, 10–12 July 2000; pp. 127–137.

12. Quisquater, J.; Couvreur, C. Fast Decipherment Algorithm for RSA Public-key Cryptosystem. *Electron. Lett.* **1982**, *18*, 905–907. [CrossRef]
13. Johnson, D.; Menezes, A.; Vanstone, S. The elliptic curve digital signature algorithm (ECDSA). *Int. J. Inf. Secur.* **2001**, *1*, 36–63. [CrossRef]
14. Xu, D.; Ussery, C.; Chen, H. AES implementation based on a configurable VLIW DSP. In Proceedings of the 2010 10th IEEE International Conference on Solid-State and Integrated Circuit Technology (ICSICT), Shanghai, China, 1–4 November 2010; pp. 536–538.
15. Zhang, Y.; He, Q.; Xiang, Y.; Zhang, L.Y.; Liu, B.; Chen, J.; Xie, Y. Low-Cost and Confidentiality-Preserving Data Acquisition for Internet of Multimedia Things. *IEEE Internet Things J.* **2018**, *5*, 3442–3451. [CrossRef]
16. Gueron, S. *Intel® Advanced Encryption Standard (AES) New Instructions Set*; Intel Corporation: Santa Clara, CA, USA, 2010.
17. Mericas, A.; Peleg, N.; Pesantez, L.; Purushotham, S.B.; Oehler, P.; Anderson, C.A.; King-Smith, B.A.; Anand, M.; Arnold, J.A.; Rogers, B.; et al. IBM POWER8 performance features and evaluation. *IBM J. Res. Dev.* **2015**, *59*. [CrossRef]
18. National Institute of Standards and Technology (NIST). FIPS PUB 81 Des Modes of Operation. Available online: http://csrc.nist.gov/publications/fips/fips81/fips81.htm (accessed on 28 January 2019).
19. Bertoni, G.; Breveglieri, L.; Fragneto, P.; Macchetti, M.; Marchesin, S. Efficient software implementation of AES on 32-bit platforms. In *Lecture Notes in Computer Science*; Springer: Berlin/Heidelberg, Germany, 2003; Volume 2523/2003, pp. 129–142.
20. Cao, D.; Han, J.; Zeng X. A Reconfigurable and Ultra Low-Cost VLSI Implementation of SHA-1 and MD5 functions. In Proceedings of the 7th International Conference on ASIC, Guilin, China, 22–25 October 2007; pp. 862–865.
21. Rivest R . The MD5 Message-Digest Algorithm. Available online: https://www.rfc-editor.org/rfc/pdfrfc/rfc1321.txt.pdf (accessed on 28 January 2019).
22. Aydos, M.; Yamk, T.; Koc, C.K. High-Speed Implementation of an ECC-based Wireless Authentication Protocol on an ARM Microprocessor. *IEE Proc. Commun.* **2010**, *148*, 273–279. [CrossRef]
23. Malik, M.Y. Efficient Implementation of Elliptic Curve Cryptography Using Low-power Digital Signal Processor. In Proceedings of the 12th International Conference on Advanced Communication Technology (ICACT), Phoenix Park, Korea, 7–10 February 2010; pp. 1464–1468.

*Article*

# Ontology-Defined Middleware for Internet of Things Architectures

**Víctor Caballero *, Sergi Valbuena, David Vernet and Agustín Zaballos**

Engineering Department, Universitat Ramon Llull (URL), La Salle, 08022 Barcelona, Spain; sergi.valbuena@students.salle.url.edu (S.V.); david.vernet@salle.url.edu (D.V.); agustin.zaballos@salle.url.edu (A.Z.)

* Correspondence: victor.caballero@salle.url.edu; Tel.: +34-93-290-2436

Received: 15 January 2019; Accepted: 26 February 2019; Published: 7 March 2019

**Abstract:** The Internet of Things scenario is composed of an amalgamation of physical devices. Those physical devices are heterogeneous in their nature both in terms of communication protocols and in data exchange formats. The Web of Things emerged as a homogenization layer that uses well-established web technologies and semantic web technologies to exchange data. Therefore, the Web of Things enables such physical devices to the web, they become Web Things. Given such a massive number of services and processes that the Internet of Things/Web of Things enables, it has become almost mandatory to describe their properties and characteristics. Several web ontologies and description frameworks are devoted to that purpose. Ontologies such as SOSA/SSN or OWL-S describe the Web Things and their procedures to sense or actuate. For example, OWL-S complements SOSA/SSN in describing the procedures used for sensing/actuating. It is, however, not its scope to be specific enough to enable a computer program to interpret and execute the defined flow of control. In this work, it is our goal to investigate how we can model those procedures using web ontologies in a manner that allows us to directly deploy the procedure implementation. A prototype implementation of the results of our research is implemented along with an analysis of several use cases to show the generality of our proposal.

**Keywords:** heterogeneity; middleware; semantic; ontology; behaviour; web-of-things

---

## 1. Introduction

The recent growth of the Internet has fostered the interaction between many heterogeneous technologies. The term Internet of Things (IoT) [1] has been coined as the umbrella term that refers to this new reality. The heterogeneity of the IoT encompasses both hardware and software. At the hardware level, an amalgamation of devices from multiple vendors take part in the IoT. At the software level, those devices use multiple Internet protocols and data exchange formats. Device manufacturers choose what seems to fit their needs in describing and operating their devices. Hence, the current IoT scenario is filled with devices/Things that are not fully interoperable. Even stronger lack of interoperability can be found in vertical domain specific silos of IoT. In order to facilitate the communication between devices and the integration of them into systems and systems of systems, researchers have conducted work that aims at providing a common communication interface. The Web of Things (WoT) [2] provides a common interface by translating the heterogeneity into the homogeneity of the Web 3.0. To achieve a common language, the WoT refers to Hyper-Text Transfer Protocol/Secure (HTTP/S) as the standard communication protocol and to Semantic Web [3] as the common information exchange structure.

An example that takes advantage of both IoT and WoT technologies is as follows. In the context of a Smart Home, there are several electrical appliances which consume or produce energy. On the one hand, home appliances that consume energy can be an air conditioner, vacuum cleaner, washing machine, clothes dryer, dishwasher, oven, stove, mobile charger, lights and so on. On the other hand, devices that produce energy usually take advantage of Renewable Energy Sources (RESs) like photo-voltaic panels. In order to take advantage of both production and consumption, those devices need to exchange information between themselves. A course of action would be, for example, to autonomously turn on the clothes dryer during the day when there is enough sunlight to produce enough energy for it to function. It is very unlikely, though, that they all use the same communication protocol and data exchange format. Some will use wireless protocols like Bluetooth, ZigBee, Z-Wave or 6LoWPAN, or wired ones like Ethernet, RS-232, or USB, to name a few. The WoT integration pattern proposes to use gateways as translators between those protocols (some are proprietary) and HTTP/S, along with data exchange format translation, which occurs from a variety of formats such as EXI, JSONB or unstructured data to a semantically annotated Web 3.0 standards, such as JSON or XML. Scenarios like the one presented are very likely to occur. Each scenario, however, will comprehend different types of devices that use different protocols and different data exchange formats. The simple scenario of pairing a light bulb from brand A with a switch from brand B can be different in each household. In such a simple use case, the task of describing each specific scenario can be tedious. Therefore, there is a need to describe the execution flow or behaviour of those devices in an abstract manner, thus the description can be used for multiple and diverse devices with the same capabilities.

### 1.1. Motivation

A large number of heterogeneous devices share the same capabilities and concrete services, with concrete, shared behaviours can be requested from them. Normally, the description of the services and the execution process that they follow are specified and implemented in multiple manners, which pose an obstacle for the shareability and interpretation of their execution flow. A great effort in providing a common abstraction and interoperability layer is being done in the framework of the IoT/WoT. The World Wide Web Consortium (W3C), for example, is leading an effort in order to provide a common communication and abstraction layer. There is also a growing tendency to describe services using Semantic Web standards such as OWL-S [4], W3C Web of Things (WoT) Thing Description (TD) [5] or the oneM2M Base Ontology [6]. Our vision is that we can share and reuse a concrete flow of execution among multiple and heterogeneous IoT deployments using existing standards. We can leverage Semantic Web Ontologies to model, and execute by means of a specialized engine, the behaviour of the devices that lay in the lower layer of the IoT, by providing a common template (abstraction) that defines such behaviour. This template can be reused by as many Things and services as needed in a variety of heterogeneous deployments.

### 1.2. Contributions

This work aims to address the challenge of describing and executing the same behaviour in multiple and heterogeneous IoT scenarios. In heterogeneous IoT scenarios, where devices from different manufacturers are aimed at providing the same service, it is often found that their heterogeneity prevents developers from reusing the same solution for each device. We address that challenge by using Finite-State Machines (FSMs) to model their behaviour, translating the FSM into in abstract description framework, namely Resource Description Framework (RDF) [7], and developing an engine capable of interpreting and executing the RDF model. The main contributions of this work are:

1. Abstraction of flow of control or behaviour over the heterogeneity of the IoT using existing standards such as RDF, SPARQL Protocol and RDF Query Language (SPARQL), FSM and HTTP/S. This abstraction brings reusability.
2. A prototype implementation of an engine and web interface that allows direct deployment, interpretation and execution of a concrete instance obtained by materializing the abstraction.
3. Analysis of several use cases that demonstrates the generality of the abstraction.

This article is organized as follows. Section 2 gives a brief overview of IoT/WoT standards relevant to our goal and analyses related work on control of flow/behaviour modelling using Semantic Web technologies. Section 3 gives an overview of the architecture that supports our approach. Section 4 introduces and models the use case that is used to explain our solution and Section 5 enhances the model using ontologies. Section 6 explains how we use the template (abstraction) and execute the behaviour described in it. Section 7 analyses several use cases in relation to our approach and finally Section 8 concludes the paper and presents future work.

## 2. State of the Art

Given the massive number of services and processes that IoT enables, it has become almost mandatory to describe such services and processes to enhance interoperability, allowing them to be automatically discovered and called by multiple, heterogeneous systems [8]. Ontologies such as OWL-S [4] and low-level specifications such as the TD [5] or the oneM2M Base Ontology [6] can be used together to describe IoT/WoT systems, fostering interoperability. OWL-S helps software agents to discover the web service that will fulfil a specific need. Once discovered, OWL-S provides the necessary language constructs to describe how to invoke the service. It allows describing inputs and outputs. Thanks to the high-level description of the service, it is possible to compose multiple services to perform more complex tasks. In OWL-S, service description is organized into four areas: the *process model*, the *profile*, the *grounding* and the *service*. Specifically, the process model describes how a service performs its tasks. It includes information about inputs, outputs, preconditions and results. Similarly, the oneM2M Global Initiative [9] defines a standard for machine-to-machine communication interoperability at the semantic level, the one M2M Base Ontology, which is a high-level ontology designed to foster interoperability among other ontologies using equivalences and alignments. The TD "is a central building block in the W3C Web of Things (WoT) and can be considered as the entry point of a Thing [...]. The TD consists of semantic metadata for the Thing itself, an interaction model based on WoT's Properties, Actions, and Events paradigm, a semantic schema to make data models machine-understandable, and features for Web Linking to express relations among Things". Both oneM2M Base Ontology and the ontology defined in TD strive for interoperability among multiple IoT applications and platforms, each one covering a large set of use cases, so there is also a work in progress to align the oneM2M ontology with the TD ontology [10]. Orthogonal to these ontologies, the SOSA/SSN ontology is an ontology for describing sensors and their observations. Among other concepts, it defines the class sosa:Procedure, which "is a reusable workflow, protocol, plan, algorithm, or computational method that can be used, among others, to specify how an observation activity has been made [...], (the sosa:Procedure) can be either a record of how the actuation has been performed or a description of how to interact with an actuator (i.e., the recipe for performing actuations)" [11]. How much detail is provided to define such procedures is beyond the scope of SOSA/SSN. It is our vision that detailed and deployable procedure descriptions could fit into and be orthogonal to the models just presented. This work aims at modelling the flow of control for a given service. Therefore, our focus is on modelling the procedure/behaviour of Things using ontologies in a manner that allows us to directly deploy the behaviour implementation.

Work aimed at modelling the behaviour of Things using FSMs and Web Ontology Language (OWL) exists in the literature. On [12], they aim to represent Unified Modeling Language (UML) FSMs using OWL, performing an almost one-to-one translation between UML concepts and OWL classes. Although their mapping from UML to OWL allows for a more machine-readable information structure, its complexity makes it unpractical to use for our objective. The work done in [13] presents a simpler FSM model. UML is used to specify platform independent navigation guides in web applications. They use OWL to describe a model for FSMs which serves as a meta-model for semantic web descriptions of the navigation guides on the Semantic Web. There also exists some scientific literature devoted not only to create a model to express the behaviour of a service but also to interpret and execute the behaviour. In [14], the aim is to develop an FSM that a special server can read and translate to executable entities. These executable entities are executed later by robots. They successfully satisfy their objective, their OWL FSM is domain specific and it includes properties that solve the complexity of their use case but make the FSM too complex for our goal. The goal in [15] is to model an FSM that can be easily translated into Programmable Logic Controller (PLC) code. This model is domain-agnostic as it is not directly merged with the PLC domain. Works [14,15] show how a semantic FSM can be translated and executed by machines, although their application field is industrial and not on the WoT. In our approach, we do not want to express or translate the model in languages other than OWL so it maintains the benefits that OWL gives. Our aim is to build a model that is directly interpreted by different machines.

Our solution aims to be as middleware-agnostic as possible, being a building block for middleware that implement other technical challenges. Therefore, we rely on standard and well-known architectures to contextualize our proposal. We have considered two big standardization and recommendation bodies devoted to promoting interoperability among IoT verticals. An international partnership project for IoT standards, the oneM2M Global Initiative [9] aims to minimize M2M service layer standards by consolidating standardization activities among several member organizations. The W3C Web of Things Working Group [5] aims to develop standards for the Web of Things; their goal is also to reduce fragmentation between IoT silos. For this work, we are interested in the architecture standards and recommendations that oneM2M and W3C WoT Working Group propose. As our objective is to contextualize and provide a prototype implementation for our solution, the W3C WoT Architecture recommendation seems more suitable for that purpose for its simplicity. The architecture is explained in Section 3.

## 3. Architecture

The objective of this paper to provide an abstraction mechanism to describe and execute the behaviour of Things in any WoT architecture. This section explains the architecture used in our proposal. There is a common trend on IoT architectures to represent physical devices with virtual representations or agents that enhance their capabilities. Such virtual representations are called Virtual Object (VO) [16]. Virtual Objects (VOs) are usually placed in fog or cloud middleware [16–18]. The behaviour templates developed in this work are interpreted and executed where those VO reside, thus more resources are available to execute the behaviour that the templates describe. The first layer in the seminal work of the WoT [2] defines an Accessibility Layer that enables each physical device to the web by translating each IoT protocol to HTTP/S and providing a common RESTful (REpresentational State Transfer) Application Programming Interface (API) to interact with them. As seen in Figure 1, we assume that physical devices are exposed via HTTP/S in order to interact with the VOs. A basic division between layers can be made. The first layer is composed of physical devices, which is the Perception layer. The second layer is responsible for enabling physical devices to the web, the Accessibility Layer. The third layer, the Virtual Object layer, is composed of VOs that interact with the Perception layer via the Accessibility layer. As explained in [2] and in the W3C WoT Architecture [5], Things can become Web Things (web-enabled Things) by three mechanisms. If they have enough computing resources to implement the HTTP/S protocol, they can be directly connected to the web. Otherwise, they can use Gateways (GWs) or cloud/fog infrastructures that translate their proprietary protocols or non-standard data exchange formats into web standards. The template developed in this paper is embodied by a VO. Each VO is given the capability to interact with physical devices through the Accessibility layer. Therefore, actions encoded by an HTTP/S call the Accessibility layer, which in turn informs the physical device to actuate in the real world. Thanks to this action abstraction, relationships can be made between one VO and multiple physical devices. It is only necessary to change the Uniform Resource Identifier (URI) actions refer to; this is explained in more detail in Section 5.3. The architecture is shown in Figure 1 where different settings can be appreciated. The following lists the settings, from left to right:

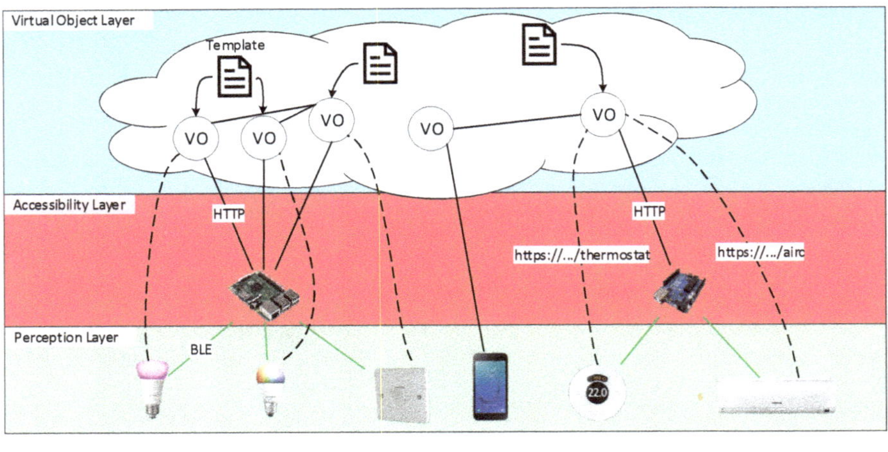


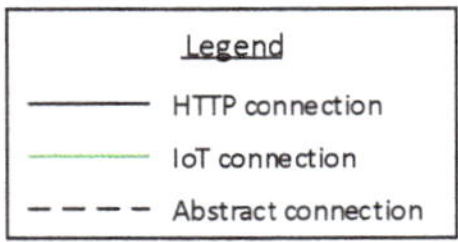


**Figure 1.** Network architecture used in our proposal. Three layers are depicted: (1) Perception layer, where physical devices reside; (2) the translation between Internet of Things (IoT) protocols and Hyper-Text Transfer Protocol/Secure (HTTP/S) is made at the Accessibility layer and; (3) the Virtual Object layer is comprised of virtual representations of physical devices, each Virtual Object (VO) embodies a template that describes the behaviour for physical devices. Straight lines represent direct connections between nodes, which means that no intermediate protocol or data exchange format is needed (nodes at the Accessibility layer are responsible for performing such actions). Dashed lines represent that the VO at one end of the line handles the physical device at the other end of the line. Finally, black straight lines represent that nodes are connected through HTTP/S and that there exists an HTTP/S Application Programming Interface (API) compatible with nodes at both ends of the line. Green, straight lines represent that nodes are connected through IoT protocols, such as Bluetooth Low Energy (BLE).

1. The first setting is composed of two smart light bulbs from different brands (they use different communication protocols and/or data exchange formats) and a light switch at the Perception layer. There is a Raspberry Pi at the Accessibility layer that translates IoT protocols from the physical devices to an HTTP/S API. Finally, three VOs dwell in the VO layer and each one embodies a template. Note that the same template is used for each one of the VOs that communicate with the smart light bulbs, which means that the same template is used to control both of them. The template for the light switch describes the behaviour for controlling (turn on/off) both smart light bulbs. The control is made at the VO layer. More precisely, the switch VO sends commands and receives messages from the two VOs that represent the smart light bulbs.
2. The second setting is composed of a smartphone at the Perception layer that directly connects to a VO at the VO layer through HTTP/S. To lighten the image, the template the VO embodies is not drawn. The interaction is made possible via an app installed at the smartphone.
3. The third setting comprehends a thermostat and an air conditioner machine at the Perception layer. They both connect to an Arduino board at the Accessibility layer, which exposes their functionality as an HTTP/S API. In this scenario, a single VO is responsible for managing

the thermostat and the air conditioner. This can be achieved by simply using both Uniform Resource Locators (URLs), *https://.../thermostat* and *https://.../airc*, as the target for the template actions. Finally, this VO communicates with the VO that represents the smartphone, enabling for controlling the temperature of the room.

The model architecture depicted in Figure 1 shows the interaction patterns between physical devices and our proposal (templates) in a WoT architecture. One of the features of interest in our proposal is that it is middleware-agnostic. Therefore, critical challenges such as full-fledged security and privacy are implemented by concrete middleware. For example, the VICINITY project [19] defines a Virtual Neighbourhood and is a platform that links various ecosystems providing interoperability as a service for infrastructures in the IoT. It tackles privacy by allowing the owner of Things to define sharing access rules of devices and services. The communication between VICINITY Nodes occurs between a Peer-to-Peer (P2P) network based on those rules. This network supports VICINITY Nodes with security services such as end-to-end encryption and data integrity [20]. The IoT ecosystems are connected to VICINITY through the VICINITY Node. Hence, security and privacy outside VICINITY is handled by the manager of the IoT ecosystem. Security in that ecosystem could be an exchange of identifiers and encryption keys between VOs and physical devices or gateways, representing the weakest phase when security could be compromised.

Integration with other middleware involves three main efforts: (1) to implement an engine to interpret the flow of execution defined by the FSM template (which can be reused by middleware developed with the same programming language); (2) to adapt the input and output channels of the FSM engine with an HTTP/S interface provided by the middleware; and to provide an RDF and SPARQL API to the FSM engine.

As shown in Figure 2, our approach complements third-party middleware by providing several desired properties. By combining RDF and the model of execution flow that FSMs provide, we enable a programming abstraction. The programming abstraction can be shared among multiple engine implementations in different middleware. The ease of deployment is facilitated by the fact that behaviour and executor are decoupled. Behaviours for web-enabled physical devices can be created ad hoc when needed and deployed, as the necessary software to execute the behaviour already exists. A Triplestore Database (TDB) or RDF API and SPARQL endpoint is also needed for the FSM engine in order to store the template and save the data received during runtime. Guards, conditions and actions use these data (see Section 5). As seen in Figure 2, our approach relies on HTTP/S in order to enable communication with the Perception Layer. Therefore, it is mandatory for third-party middleware to have an HTTP/S interface. Two pathways are shown in Figure 2. The first pathway assumes a sensor (thermostat/thermometer) with an HTTP/S client that pushes data to the middleware. The HTTP/S server in the middleware receives this request and sends it to the input interface of the FSM engine and triggers the FSM logic, which is read from the template instance. The FSM does not return any response with data as it may not be available at that time. However, the middleware is still able to respond to the HTTP/S request. The second pathway is the actuator pathway. There, the actuator is directly addressable as it is an HTTP/S server. Following the process triggered by the sensor, the FSM may generate an action to turn on the actuator. For example, if the temperature is higher than 27 °C, the course of action would be to turn on the air conditioner. This logic and action are written in the FSM template instance. The FSM generates an action which is sent to the HTTP/S client provided by the middleware. An HTTP/S request is sent to the actuator and the actuator may respond with valuable data, which is then used as the input of the FSM.

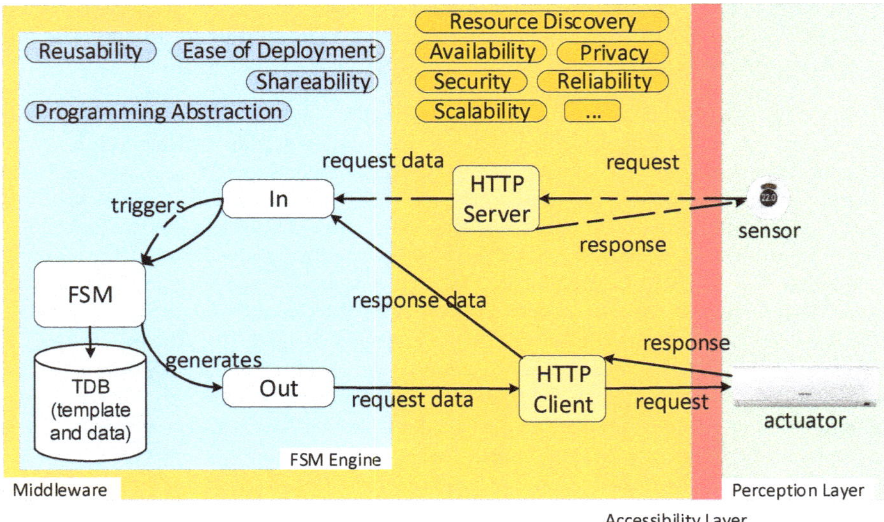


**Figure 2.** Functional blocks for third-party middleware integration. The layers are: Finite-State Machine (FSM) Engine and Third-party middleware (Middleware), Accessibility Layer and Perception Layer. The FSM Engine complements the middleware. Only two components are mandatory in the middleware layer to enable our approach. Arrows indicate the flow of information between components in each layer. Dashed lines represent the sensor pathway and straight lines represent the actuator pathway.

## 4. An Example to Model

This section introduces the use case that is modelled throughout the paper. Our objective is to model a template for a VO that allows users to exchange their energy profile information and check if they are compatible. For the sake of simplicity, we define "compatibility" of energy profiles as "similarity" of energy profiles. The use case is framed in the context of the Social Internet of Things (SIoT), which endows devices with the ability to belong to social networks of devices. The Lysis [21] platform is an implementation that enables the SIoT. There, VOs are extended to become social and create relationships; they become Social Virtual Objects (SVOs). Several types of relationships can be created. We are interested in Ownership Object Relationship (OOR) and Social Object Relationship (SOR) relationships. OORs are established among heterogeneous objects that belong to the same user. SORs are defined when objects come into contact (close range) because their owners come in touch with each other during their lives. The overall context of the use case we want to model is as follows:

1. A user interested in being part of a prosumer community to benefit from the community and achieve certain goals installs a discovery app on his/her smartphone. The smartphone becomes an SVO [21].
2. The potential Prosumer Community Group (PCG) user installs apps or allocates apps in the cloud that enable retrieving data from his/her energy-related devices (devices that produce or consume energy), enabling social capabilities for each one (SVO).
3. Devices and Distributed Energy Resources (DERs) that consume and/or produce energy owned by an owner establish an OOR relationship (Figure 3, 1). This relationship is then leveraged to

obtain energy-related data from DERs that belong to the same owner. These data can be both aggregated and stored or stored individually for each device.

4. The data obtained is used to build an energy user profile during a time period. The profile is then stored as part of the discovery app, and thus enabling the smartphone and other social devices to retrieve such information. A social device is a device that the user (prosumer) usually carries with himself/herself.
5. When the user establishes a social relationship with another potential user (Figure 3, 2), both social devices (i.e., smartphones) also establish social relationships (Figure 3, 3) and exchange energy profiles and goals. Then, the SVOs determine if they are compatible or not.

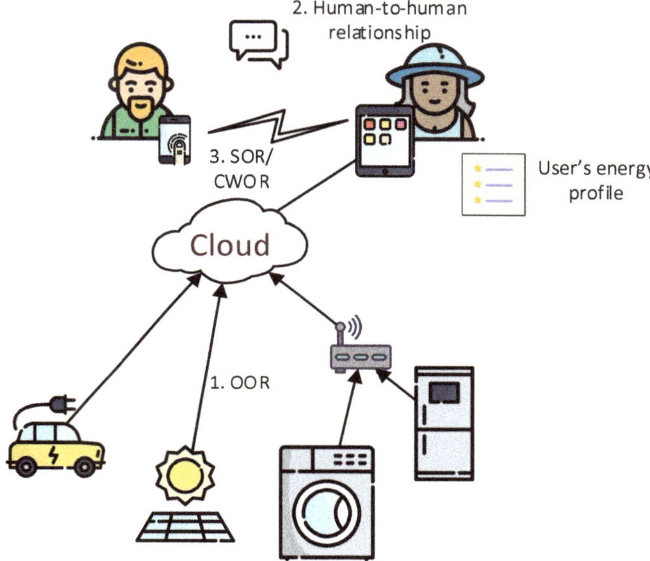


**Figure 3.** Prosumer Community Group formation leveraging Social Internet of Things. Firstly, prosumers and appliances establish an Ownership Object Relationship (OOR) via the augmented Social Virtual Object (SVO) (1). When prosumers come close together and establish and interact (2), their devices establish Social Object Relationship (SOR) or Co-work Object Relationship (CWOR) relationships and exchange the energy profiles of the prosumers (3).

Our objective is to model the behaviour expressed in Item 5. The model starts describing the steps that take place when one SVO meets another SVO and ends when the compatibility results between them have been exchanged. The initial peering process is assumed to be done by a service implemented in Lysis [21]. In addition, the energy profile is supposed to exist in the device. The process is described as:

1. Wait for a peer to be available.
2. Send the energy profile to the peer.
3. Wait until the peer's energy profile is received.
4. Check the compatibility between the device's and peer's profiles.
5. Inform of the compatibility result to the peer. The results are exchanged because the compatibility check of each device could be different.
6. Wait until the peer's compatibility result has been received.

To model the behaviour, we have chosen an FSM because it enables us to describe a sequence of actions to be executed and the use of conditions to modify the execution's flow. In order to visualize the FSM, we first model the behaviour using StarUML [22] following UML standards. Figure 4 shows the execution flow using states and conditions. The aim of UML is to provide a standard manner to model the design and behaviour of a system; it is made to be comprehended by humans and not by machines. However, our interest is to enable machines to interpret and execute this behaviour. OWL provides explicit meaning, making it easier for machines to automatically process and integrate information. From the options discussed in Section 2, we have chosen the FSM ontology developed in [13]. Some properties and classes have been added to integrate it with the web, others have been renamed to follow W3C naming conventions on ontologies. The modified ontology is used to describe the system's behaviour. Note that the FSM is designed in a manner that both VOs (each one interacting with one of the physical devices shown in Figure 3) share the same behaviour.

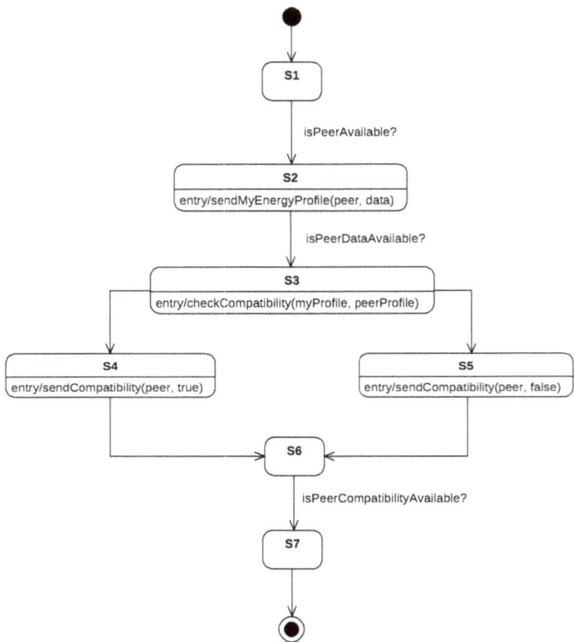


**Figure 4.** Finite-State Machine representation using Unified Modeling Language.

## 5. Behaviour Modelling Using Ontologies

Throughout the rest of the paper, the prefix fsm: is used to shorten its URI. The siot: prefix represents some classes and properties specific to the domain of the use case. Finally, the: prefix is used to reference the file or ontology where the own model is stored.

### 5.1. *Skeleton of the FSM*

The first step is to define the skeleton of the FSM, it contains the instance of the machine, its states and its transitions.

As shown in Listing 1, an FSM is defined by the class fsm:StateMachine and the states as fsm:State. The initial state is also defined as fsm:InitialState, the final state as fsm:FinalState and intermediate states as fsm:SimpleState, these three classes extend from fsm:State. An fsm:StateMachine instance should have their states associated with the property fsm:hasStateMachineElement. Each state can have some exit or entry actions; they are associated with the property fsm:hasEntryAction or fsm:hasExitAction.

The transitions between states are described by the class fsm:Transition. As seen in Listing 2, a transition has a source state with the property fsm:hasSourceState and a target state with fsm:hasTargetState. A transition can also contain none or more guards associated with the property fsm:hasGuard.

```
### The FSM
:siot_fsm rdf:type owl:NamedIndividual ,
                   fsm:StateMachine ;
          fsm:hasStateMachineElement :S1_WatingForPeerState ,
                                     :S2_InformationExchangeState ,
                                     :S3_CheckCompatibilityState ,
                                     :S4_IsCompatibleState ,
                                     :S5_NotCompatibleState ,
                                     :S6_WaitingForPeerCompatibilityState ,
                                     :S7_CompareCompatibilitiesState .


### States
:S1_WatingForPeerState rdf:type owl:NamedIndividual ,
                                fsm:InitialState ,
                                fsm:State .

:S2_InformationExchangeState rdf:type owl:NamedIndividual ,
                                      fsm:SimpleState ,
                                      fsm:State ;
                             fsm:hasEntryAction :sendMyData .

:S3_CheckCompatibilityState rdf:type owl:NamedIndividual ,
                                     fsm:SimpleState ,
                                     fsm:State ;
                            fsm:hasEntryAction :checkCompatibility .

:S4_IsCompatibleState rdf:type owl:NamedIndividual ,
                               fsm:SimpleState ,
                               fsm:State ;
                      fsm:hasEntryAction :sendImCompatible .

### ...

:S7_CompareCompatibilitiesState rdf:type owl:NamedIndividual ,
                                         fsm:FinalState ,
                                         fsm:State .
```

**Listing 1.** Finite-State Machine and states.

### *5.2. Guards and Conditions*

The purpose of a guard, defined by the class fsm:Guard, is to provide a method to evaluate if the transition has to be travelled. The guard can contain none or more conditions to express what has to be evaluated. It can also have none or more actions to be executed if the guard evaluates true.

A condition is defined with the class fsm:Condition; it contains a body object marked as fsm:Body that represents the condition's content. This body has a string associated with fsm:hasContent that contains the actual condition represented as a SPARQL query. The body should also include its type associated with fsm:hasBodyType to indicate the type of the body's content. Even though it is possible to use different types, an ASK SPARQL query is needed when used on conditions, as the main focus is to work directly with RDF data and ontologies. The reason to use the type property is that the body object is also used as the body and the value for the header "Content-Type" to be sent on HTTP/S requests, where it can take other formats such as RDF or JavaScript Object Notation (JSON). Some conditions are modelled on Listing 4. The procedure to evaluate the transitions is:

- Retrieve all the state's exit transitions. Note that only one of these transitions should be true at the same time, otherwise different executions of the same FSM may lead to different results if the order of the evaluation of transitions changes.
- For each transition, retrieve its guards (Listing 3). If it has no guards, then that transition is considered feasible. If any guard is true, then the transition is also feasible, which means that an OR condition is applied between all the guards. The utility of having multiple guards is to have different actions that are executed only under certain cases and not each time the transition is travelled.
- For each guard, retrieve its conditions. Another OR operation is applied between these conditions, and if any of them is true, the guard also evaluates to true. It is important to evaluate all the guards and to not stop when one of them is true, as all the guards' actions must be executed if their guard is true.
- The body of each condition should be an ASK query. An ASK query is a type of query that contains a pattern of triples, it searches for a set of triples that match the pattern. If any matching set is found, the query returns true, and it is false otherwise. The query is evaluated against the RDF database. The condition evaluates the same as the ASK query.

```
:S1_to_S2_Transition rdf:type owl:NamedIndividual ,
                              fsm:Transition ;
                     fsm:hasSourceState :S1_WatingForPeerState ;
                     fsm:hasTargetState :S2_InformationExchangeState ;
                     fsm:hasTransitionGuard :S1_to_S2_T_Guard .

:S3_to_S4_Transition rdf:type owl:NamedIndividual ,
                              fsm:Transition ;
                     fsm:hasSourceState :S3_CheckCompatibilityState ;
                     fsm:hasTargetState :S4_IsCompatibleState ;
                     fsm:hasTransitionGuard :S3_to_S4_T_Guard .
```

**Listing 2.** Transitions.

```
:S1_to_S2_T_Guard rdf:type owl:NamedIndividual ,
                           fsm:Guard ;
                  fsm:hasGuardCondition :IsPeerAvailable .

:S3_to_S4_T_Guard rdf:type owl:NamedIndividual ,
                           fsm:Guard ;
                  fsm:hasGuardCondition :IsPeerCompatible .
```

**Listing 3.** Guards.

```
:IsPeerCompatible rdf:type owl:NamedIndividual ,
                           fsm:Condition ;
                  fsm:hasContent
        """
            PREFIX siot: <file:///D:/projects/ontologies/siot/siot.owl#>
            PREFIX owl: <http://www.w3.org/2002/07/owl#>
            PREFIX fsm: <file:///D:/projects/ontologies/fsm/fsm.owl#>

            ASK {
                self: siot:hasPeer ?peer .
                self: siot:hasCompatibility ?compatibility .
                ?compatibility siot:withPeer ?peer .
                ?compatibility fsm:hasContent \"true\" .
            }
        """ .
```

**Listing 4.** Conditions.

### 5.3. Actions

An action requests or sends data via HTTP/S to another node in the architecture. The HTTP Vocabulary in RDF 1.0 [23] ontology by W3C is used to define actions. An action is defined with the class fsm:Action and http:Request at the same time. An action has a body, associated with fsm:hasBody, which is sent with the request. The body follows the same structure used in the conditions; it can have different types like RDF (Listing 5), JSON and others. Moreover, a timeout can be specified to limit the amount of time that an action waits for the response, it is specified in milliseconds with the property fsm:hasTimeoutInMs. A request method like GET or POST is indicated with the property http:mthd. This property references an object and not a literal, W3C has already defined the standard method objects in [23], these objects are the preferred ones to use. Therefore, as templates are not bound to perform actions using the same URI, a template can perform actions on multiple nodes. This, in conjunction with the FSM model, enables the templates or, more precisely, their instantiation, the VOs, to interact with multiple physical devices and services, such as other VOs. This is the case of the third scenario described in Figure 1, where a single VO is able to orchestrate two different physical devices. The VO also communicates with the smartphone described in the second scenario via its VO.

The action has a URI to express where the request is done. This URI can be expressed in different ways; our two approaches are:

- Use an absolute URI with the property *http:absoluteURI* to express the final address as in Listing 6. This is used when we know in advance where the request is done and that the address will not change.
- Use a prototype URI with the property fsm:hasPrototypeURI, shown in Listing 7. A prototype is an object of the class fsm:Prototype that defines the structure of a URI with one or more placeholders that are replaced during runtime. A schema of the classes and properties involved is presented in Figure 5. This is used when we do not know in advance the final address or when it changes frequently. An absolute URI is computed from the prototype each time the action needs to be executed. The replacements of parameters are not done on the ontology; this newly computed URI is built separately in program memory. A prototype defines the URI's structure with a string linked with fsm:hasStructure, for example *http://localhost:9000/[fsm_id]/send_data*, where *[fsm_id]* is the placeholder. The prototype has an fsm:Parameter for each placeholder, linked via the property fsm:hasParameter. The purpose of the parameter is to indicate what placeholder needs to be replaced and to provide a function that generates the value to fill the placeholder. Each parameter has the placeholder itself as a string (like *[fsm_id]*) linked with fsm:hasPlaceholder and a SPARQL query as a string linked with fsm:hasQuery. The query must be a SELECT query that returns only

one value. The value returned by the query is used to replace on the URI the placeholder with the same parameter key.

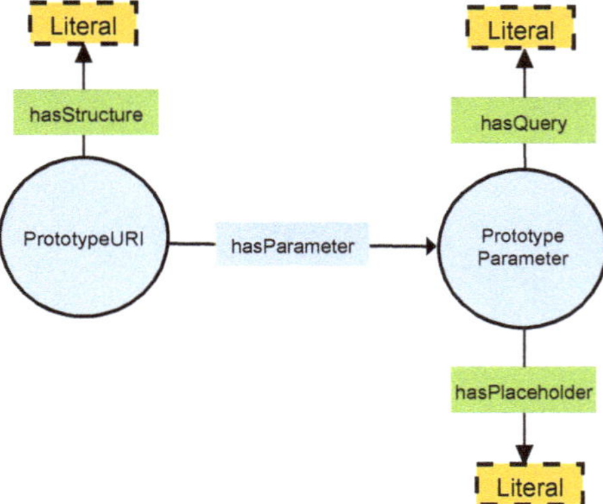


**Figure 5.** Schema of PrototypeURI.

```
:sendImNotCompatibleBody rdf:type owl:NamedIndividual ,
                                  fsm:Body ;
                         fsm:hasBodyType fsm:rdf ;
                         fsm:hasContent
        """
            @prefix siot: <file:///D:/projects/ontologies/siot/siot.owl#> .
            @prefix owl: <http://www.w3.org/2002/07/owl#> .
            @prefix fsm: <file:///D:/projects/ontologies/fsm/fsm.owl#> .

            self: siot:hasCompatibility ?compatibility .
            ?compatibility siot:withPeer self:myPeer .
            ?compatibility fsm:hasContent \"false\" .
        """ .
```

**Listing 5.** Bodies.

```
:checkCompatibility rdf:type owl:NamedIndividual ,
                             fsm:Action ,
                             http:Request ;
                    fsm:hasBody :checkCompatibilityBody ;
                    http:absoluteURI <http://localhost:9000/check_compatibility> ;
                    http:mthd <http://www.w3.org/2011/http-methods#POST> .
```

**Listing 6.** Action with absolute Uniform Resource Identifier.

```
:sendImNotCompatible rdf:type owl:NamedIndividual ,
                              fsm:Action ,
                              http:Request ;
                     fsm:hasBody :sendImNotCompatibleBody ;
                     fsm:hasPrototypeURI :sendDataPrototype ;
                     http:mthd <http://www.w3.org/2011/http-methods#POST> .

:sendDataPrototype rdf:type owl:NamedIndividual ,
                            fsm:PrototypeURI ;
           fsm:hasParameter :peerUriParameter ;
           fsm:hasStructure "[peer_uri]/send_data" .

:peerUriParameter rdf:type owl:NamedIndividual ,
                           fsm:PrototypeParameter ;
           fsm:hasPlaceholder "[peer_uri]" ;
           fsm:hasQuery
        """
            PREFIX siot: <file:///D:/projects/ontologies/siot/siot.owl#>

            SELECT (str(?peer) as ?peer_uri)
            WHERE
            {
                self: siot:hasPeer ?peer
            }
        """ .
```

**Listing 7.** Action with prototype Uniform Resource Identifier.

## 6. Execution

This section explains how the proposed ontology-defined behaviours can be executed. We have implemented the template interpreter and embedded it in a web server. We have used Java as the programming language to implement the template interpreter and executor. Apache Jena [24] has been used as the TDB and SPARQL engine. Play Framework 2 [25] has been used to implement the web service. Akka [26], an Actor Model toolset, has facilitated the task of deploying the two VOs used in the example and enabled us to endow the engine with a reactive programming model. This means that the engine only "wakes up" when there is a new request for that VO, that is when new data is available. Both Play Framework 2 and Akka ease the task of implementing asynchronous actions. As explained in Section 5.3, the amount of time an action waits for a response (HTTP/S response) can be specified. Nevertheless, actions are performed sequentially and asynchronously. If actions $A_1$ and $A_2$ with fsm:hasTimeoutInMs 1 s and 5 s respectively are to be performed when entering a state, both actions will execute at the same time, without $A_2$ waiting for $A_1$ response. No other FSM operation like checking guards will be performed until the actions finish. An action finishes if it receives a response within its response timeout or it has not received any response during that time. We have made this implementation decision to prevent the VO from being blocked while waiting for the response.

As shown in Figure 6, users are provided with a graphic interface where they can upload their models. Users are asked to upload the file with the FSM ontology and the Internationalized Resource Identifier (IRI) that identifies the instance of fsm:StateMachine to be executed. When accessing the web the user is given an ID that identifies the FSM instance. This ID is also used to identify the VO's unique endpoint that it will be listening HTTP/S requests from. The VO is also able to send HTTP/S requests to both web interfaces of physical Things and other VOs. Each VO is provided with a RDF database that stores all the data received and generated during the execution of the FSM. The RDF database uses a base prefix ":" that uniquely identifies the data. The IRI of this prefix is built from the server URL (e.g., *http://localhost:9000/*) and the previous auto-generated ID. The IRI has a # appended at the end to refer to objects inside this domain. The final IRI looks like *http://localhost:9000/[fsm_id]#*.

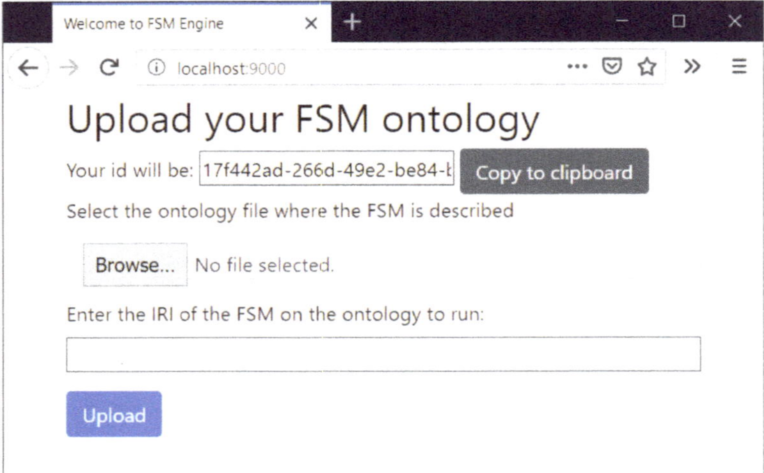


**Figure 6.** Graphic interface where users can upload their models of Finite-State Machines. Users are requested to load their Finite-State Machine (FSM) ontology and the Internationalized Resource Identifier (IRI) that identifies the concrete instance of the FSM. Users are also provided with an ID that uniquely identifies the Virtual Object (VO) that will use the FSM.

In order to ease how conditions and actions are built, the VO automatically adds the RDF database's ":" prefix to all RDF data and SPARQL queries that are written on the ontology or sent to the server. The VO also adds a prefix called "self:" that refers to the same IRI that ":" but without the final #, so the FSM instance is able to reference itself. The IRI of self: is then *http://localhost:9000/[fsm_id]*.

The endpoint of the VO provides an API with three generic actions that enable communication with the FSM instance:

- Get data: it returns the result of a SPARQL SELECT query. The query is sent to the server on the request's body. It is a GET request and the URI is *http://localhost:9000/[fsm_id]/get_data*.
- Send data: it saves RDF data on the RDF database. The RDF data to be stored is sent on the request's body. It is a POST request and the URI is *http://localhost:9000/[fsm_id]/send_data*.
- Execute operation: it executes a SPARQL query like INSERT on the RDF database. The query to execute is sent on the request's body. It is a POST request and the URI is *http://localhost:9000/[fsm_id]/execute_operation*.

After the user sends the data via the initial form, a VO is deployed. During the deployment process, the FSM ontology is read and pre-loaded. The process is as follows:

1. Load the FSM ontology into Apache Jena.
2. Search on the model and read the FSM instance of type fsm:StateMachine that has the IRI specified by the user.
3. Read all the states and states' actions. All entry and exit actions are also read.
4. For each state, read the transitions that have the state as the source are searched. For each transition, retrieve their guards, conditions and guards' actions.

In the initialization process, the VO applies default values to some properties if they are not specified on the ontology:

- Action timeout: if no timeout is specified for an action, the VO sets a default timeout of 1 second for that action.
- Action method: if no method is specified for an action, the VO sets the method to GET.
- Action body: if the body is not specified, it is set to an empty string.
- Body type: if no type is specified for a body, it is treated as plain text.

After the initialization process, the VO starts to execute the FSM. The model explained in Section 4 and modelled through the paper has been executed using the prototype implementation. Two VOs are deployed and instantiated using the same template. Conditions *isPeerAvailable?*, *isPeerDataAvailable?* and *isPeerCompatibilityAvailable?* are fulfilled by external services. Those services are mocked as the main purpose of the prototype implementation is to validate the feasibility and consistency of our approach. They return true or the necessary data so the FSM can transition to the next state. In order to visualize the execution flow of the FSM in real time via a web browser, we have endowed the VO with a WebSocket interface.

## 7. Use Case Analysis

This section analyses several use cases to assess the suitability and to identify the strengths and weaknesses of our solution.

### 7.1. Use Cases

We have considered seven use cases with different qualitative characteristics. Those use cases are gathered or related to European projects where the authors participate. Three of them relate to the Healthcare IoT for equipment tracking, personal monitoring and medicine dispense, they are related to the Advanced Training in Health Innovation Knowledge Alliance (ATHIKA) project. One of them considers logistics in Industry 4.0, gathered from the project Strategic Programme for Innovation and Technology Transfer (SPRINT). The aim of both projects is to transfer knowledge, and thus a learning enhancement system is analyzed. The Smart City use case involving intelligent trash bins is related to the ENVISERA project. Finally, an analysis is performed over the use case explained in Section 4 which is related to the Smart Grid [17,27], it is referred to as the "Prosumers" use case.

We also provide a qualitative analysis of the effort needed to integrate our approach in different types of middleware. In order to do that, we classify middleware in use cases according to the types described in [28,29]. We assume that they support the HTTP/S protocol both as server and client. Healthcare middleware is considered service-based. Service-based middleware is more secure than cloud-based middleware in that cloud security and privacy are managed by the cloud provider. However, end-to-end encryption is still challenging in IoT middleware. Actor-based middleware provide the capability of embedding lightweight actors in resource-constrained, thus enabling security capabilities to end devices. Actor-based middleware is not considered for healthcare use cases as many challenges are still to be solved [28]. An event-based middleware is considered for the use cases of logistics in Industry 4.0, intelligent trash bins and prosumers. Concretely, an event-based and actor-based (no embedded actors in the physical device) middleware is assumed [17]. Event-based architectures are recurrent in Smart City, Industry 4.0 and Smart Grid systems as they allow massive data ingestion. Finally, the prototype implementation presented in Section 6 is considered to be used between web-enabled devices and a Learning Management System (LMS) system.

#### 7.1.1. Hospital Equipment Tracking

Hospital equipment, like stretchers and wheelchairs, can be tracked to obtain its position and whether they are being used or not. The tracking is used to optimize the utilization of resources, to keep track of the inventory and to reserve the equipment. Our VO can be used to represent each piece of equipment. The VO will be asking the sensors for the current position and for other values like the weight they are carrying. For instance, when the VO detects a significant change on the weight it can send a message to a centralized service.

#### 7.1.2. Hospital Personal Monitoring

On a hospital, patients identity, status or position can be monitored through IoT to increase their safety. The medical staff can also be tracked to optimize their workflow. The VO used for the patient and the staff are different and there is a one-to-one relationship between any user (patient or staff) and a VO. The VO can be used to retrieve the current position and patients' vital signs. If the patient exits their room or some value of a vital sign gets out of a healthy range, the VO can send a notification to the medical team. The staff position is also useful, for example, to alert the professional closer to the patient.

#### 7.1.3. Intelligent Medication Dispenser

An intelligent medication dispenser is a machine that gives patients the correct medication at the right time and also checks for incompatibilities between medicines. This is a perfect tool to avoid problems like the ones explained by the authors in [30] such as a lethal combination of medicines, recalled medicines, not taking medicines on the right time or taking the same dose twice. The VO can implement the logic of deciding which medicines to dispense. First, it will wait for the identification of the patient (for instance, with an Radio-Frequency IDentification (RFID) card). Second, the VO will retrieve the information about the medication assigned to the patient by a doctor (this information has been previously added). Then, it will select the medication that needs to be taken at that time and verify that it had not been taken already. The VO also checks the compatibility with other medicines. It uses an external database that describes medicine compatibility in RDF format. Then, the VO executes a SPARQL query with the necessary reasoning logic. If any incompatibility is found, the patient is notified. The VO can be also programmed to alert patients at the time they have to take the medication through a smart-alarm or a smartphone.

#### 7.1.4. Logistics on Industry 4.0

Logistics on warehouses can be made more efficient through automatic processing of the incoming items. The author in [31] introduces a use case where all pallets of items have an RFID tag. The RFID tags are read on their arrival by RFID readers. Multiple tags can be read at the same time as their position on the pallet is not important as long as they are in the range of the reader. The tag can provide information like the product identifier, the number of items in the pallet or additional information like fragile content or notes written by the sender or the driver. The stock of the warehouse is updated through an Enterprise Resource Planning (ERP) application. A VO can implement this process. For each RFID tag processed, the VO sends the information of the tag to the ERP system. The VO can ask the ERP for the best place to store the incoming item or it can even implement the logic of deciding the best place to store it via queries and reasoning. Finally, the VO will notify a worker or robot to carry the item to its storage place.

#### 7.1.5. Learning Enhancement

The learning performance of students can be improved by monitoring the environment of the classroom and the status of the student. The work done in [32] explains a way to achieve this; they use the technology of Wireless Sensor Networks (WSNs) to integrate temperature, humidity, illumination, the concentration of $CO_2$ sensors and human emotions detection cameras on wireless networks. Some parameters like temperature are used to automatically adjust the room temperature, but others like the students' mood may need the attention of the professor. A VO can be used to actively monitor all the parameters and adjust the conditions of the room to optimal values; it can also alert the professor about the students that are losing focus. We assume that the VO has a one-to-one relationship with the classroom (temperature, humidity, $CO_2$, ...) and one-to-many with the students in a classroom, but it is a relationship with a reduced and limited set of entities (students).

#### 7.1.6. Intelligent Trash Bin

The intelligent trash bin is an already used technology on Smart Cities that monitors the state of trash bins and enables optimization on the waste collection. Authors in [33] propose an architecture that implements this technology; the bin contains an Arduino and attached sensors that collect information about waste level, temperature, humidity, motion and weight. The information of the sensors is sent to the cloud where patterns to optimize the collection of waste are extracted. A previous step can be achieved using edge computing to analyze the physical state of the bin to prevent heavy damage, as the analysis on the cloud may not be real time. The one-to-one or one-to-many relation between a VO and an intelligent trash bin or set of trash bins (if they are placed close to one another) can be established. The VO is placed on the edge cloud. Each time, the VO receives the report from the Arduino, before sending it to the cloud, it will analyze the values of humidity, toxicity (with additional sensors), motion or weight to detect potential problems like fire, vandalism/bad use or a possible toxic hazard. If any of these problems are detected, the VO sends an alert to the trash management system.

### *7.2. Analysis and Discussion*

This section provides a qualitative analysis of the described use cases. A summary of the analysis is provided in Table 1. The following metrics are considered to provide the analysis:

- Time reliability. This metric considers the limitations of the technologies that have been used to develop our solution, namely RDF, SPARQL and HTTP/S. They allow interoperability and shareability, the main focus of our proposal. However, they may be computationally heavy, especially SPARQL. This solution may not be suitable for time-critical applications. Specialized systems are better suitable to run time-critical applications. The speed of the network due to congestion can be mitigated with a dedicated Local Area Network (LAN). A Low score means that the VO cannot keep up with the timing requirements of the use case. A High score means that the use case does not need accurate timing and that some delay is acceptable.
- Scalability. Scalability represents how the VO performance scales with the number of Things that it manages. The total performance of the system degrades when the number of Things controlled increases. Usually, the performance degradation is linear and the biggest impact will be at time reliability because the available resources will be shared between more Things. Having more Things to control usually means that there will be more states and actions to perform, and that will translate in more delays. A Low score means that the manner in which the system is designed provokes a performance degradation as more Things are integrated into the system.
- Reusability. We consider reusability as the degree of how easy it is to reuse the generated template or templates for other similar use cases. For example, a template used by a VC that senses the room temperature and actuates accordingly to increase the comfort of the inhabitants can be easily used on other rooms or buildings. It may be only necessary to change the temperature thresholds. The temperature is represented as the value to be queried and not hardcoded in the execution flow or template. This example has a High reusability. On the other hand, a template that describes the behaviour of a robot that assembles car pieces has Low reusability. There will be substantial differences in the assembly logic of different models of cars and between companies.
- Suitability. Suitability is the degree of fitness between the behaviour design that the current FSM ontology allows and the desired functionalities/behaviour of the use case. If none or few modifications are needed, the suitability is considered High. If major modifications are needed, then suitability is Low.
- Incompatibilities (with domain-specific ontologies). The degree of incompatibilities between the base ontologies used by the proposed solution and the ontologies of the use case domain.

A score of High means that it is hard to integrate our solution with domain-specific ontologies. It is possible to have a score of None if there are no incompatibilities.

- Outsourcing. Outsourcing refers to the need to use external services that provide operations that the FSM cannot do. If the VO is independent or uses few external services, then the outsourcing is considered Low. VOs that act as service aggregators have high outsourcing. The value of this metric does not have an explicit positive or negative meaning, its purpose is to describe a characteristic of the VO behaviour.
- Query complexity. Query complexity is an estimation of how heavy the queries of the use case are. Heavy means that the number of returned triples is very high or/and the query execution time is expected to be long (more than half a second). A High complexity means that heavyweight queries will be performed during the execution flow.
- Knowledge complexity. Knowledge complexity refers to how much knowledge about ontologies, SPARQL and behaviour representation with FSMs the user requires to implement the system for a concrete use case. A High score means that the user necessitates a High degree of knowledge to represent the behaviour of the system. A Low score denotes that few experiences are needed as it is easy to represent the behaviour.
- Integration Effort. Integration effort considers the required effort to integrate our system in the middleware/architectures assumed for each use case. If the middleware supports actors or VOs, the effort is considered Low or Medium.

**Table 1.** Use cases benchmark.

| | Hospital Equipment Tracking | Hospital Personal Monitoring | Intelligent Medication Dispenser | Logistics on Industry | Learning Enhancement | Intelligent Trash Bin | Prosumers |
|---|---|---|---|---|---|---|---|
| Time reliability | High | Medium | High | Medium | High | High | High |
| Scalability | High | Medium | High | High | High | High | High |
| Reusability | High | Medium | High | Medium | High | High | High |
| Suitability | High | High | High | High | High | High | High |
| Incompatibilities | None | None | None | None | None | None | None |
| Outsourcing | Low | Low | Medium | Medium | Low | Low | Medium |
| Query complexity | Low | Low | Medium | Low | Low | Low | Low |
| Knowledge complexity | Low | Low | Medium | Low/Med. | Low | Low | Low |
| Integration Effort | Medium | Medium | Medium | Medium | Low | Medium | Medium |

Time reliability is High in the cases where it is not important if the system has delays greater than one second, for instance, it is not critical if the VO of the medicine dispenser has a delay of two seconds. It is Medium where a second of difference is important like monitoring the health of a patient (a fast alert could save the patient's life). In the case of Logistics, it also critical if, for each item to be unloaded, there is a delay of some seconds because a lot of items are expected to be processed fast and that can sum up to a big delay.

Scalability gets a Medium score on hospital personal monitoring because adding more Things to the VO may cause bigger delays that are not desired. This is because if more constants are to be monitored on the patient, the tasks and logic that the VO performs increase.

Reusability is Medium on the logistics use case as the desired behaviour about how the items are stored and processed may change for each company or warehouse. The idea behind templates is to define a behaviour abstracted from the physical implementation to provide reusability. It is also Medium in the use case of hospital personal monitoring as each hospital may have different workflows to optimize the performance of the staff. Reusability is High in other use cases as we consider that the behaviour/execution flow can be defined generically.

Suitability is High in all use cases as the VO is able to execute the desired behaviour. There are no incompatibilities between our proposed solution and domain-specific ontologies. Our solution relies on ontologies in order to be as compatible as possible with other ontologies. Domain-specific ontologies are expected to extend higher level ontologies such as SOSA/SSN in order to promote template reusability.

Outsourcing is Medium where the system has some operations that rely on external processes. The medication dispenser relies on an external catalogue in order to check medicine compatibility. The logistics use case scores Medium at outsourcing because it relies on the ERP to perform its tasks; some operations are too complex to be implemented by a FSM or need specialized software.

Finally, query complexity is Low in almost all cases as the queries are usually with local data and are simple. It is Medium on the medication dispenser because there may be some heavy queries that check the compatibility between all the medications of the patient. Knowledge complexity is Medium on the dispenser as the queries for medicine compatibility check can be complex to write. On the logistics, the queries can be also complex if the ERP does not provide the location to store the items; queries of searching the optimal place should be written in this case.

Our approach is easily applicable in learning enhancement, intelligent trash bin and prosumers use cases. We consider that they have a generic structure (High reusability) and that the modelling of their behaviour is straightforward with Low query and knowledge complexity They are not time-critical use cases so they score High in time reliability. Their score is High on scalability, as each VO has a relation one-to-one or one to a reduced and limited set of entities to control. Prosumers use case has a Medium score on Outsourcing as it uses external services to get informed if a peer is available or for the complex operation of checking the energy profile compatibility.

The integration effort with any concrete middleware comprehends (1) implementing the FSM engine in the programming language used by the middleware; (2) implementing a software adapter to connect the FSM inputs and outputs to the HTTP/S server and client interfaces of the middleware; and (3) providing an RDF and SPARQL API to the FSM engine. If the middleware provides a TDB, it is only necessary to connect the FSM engine to the database. Otherwise, it needs to be implemented. For that reason, the integration effort is considered Medium in almost all use cases and respective type of middleware. Once this initial effort is done, the effort remains to create the template of the desired behaviour using the ontologies and considerations described in this manuscript. Regarding the learning enhancement use case, a simple web service like the one presented in Section 6 can be used as the orchestrator between web-enabled devices and the LMS. Note that the purpose of the prototype is only to provide a web execution environment to the FSM engine. It lacks common web security and privacy mechanisms such as authentication and authorization. If the integration between our approach and the middleware is not possible or requires too much effort (e.g., in middleware that does not have the concept of VO/(Web-)Thing), our approach can be used at the application layer as a microservice that composes with the middleware's HTTP/S API.

## 8. Conclusions and Future Work

Given the heterogeneity that characterizes the IoT, a novel middleware-agnostic approach that allows for describing and executing the behaviour of devices has been proposed and implemented. The objective is to allow the reusability and shareability of the execution flow among multiple and heterogeneous IoT deployments. The approach relies heavily on existing standards to promote interoperability and reusability. FSMs are used as the model to create the execution flow using ontologies. Our work is contextualized in a reference architecture recommended by W3C, the W3C WoT Architecture. We have used the concept of VO, which are virtual counterparts of physical devices as the computational entities that run the concrete instances of the templates. Several use cases have been analyzed to asses the viability and suitability of our solution. Relying on standards such as RDF, SPARQL and HTTP/S has some drawbacks. They tend to be more heavyweight than an ad hoc solution. For example, HTTP/S can be replaced by Constrained Application Protocol (CoAP) (a lightweight, IoT version of HTTP/S) and the data model can be replaced by a SQL or NoSQL database. For that reason, our approach is not well suited for time-critical applications such as monitoring and reasoning over patients' vital signs in a hospital. However, our approach is well-suited for IoT scenarios that are non-time-critical and with a low level of variability between each deployment. It allows for reusing behaviours for heterogeneous physical devices with the same set or subset of capabilities.

In future works, we expect to fully apply a TD interface to the VOs. Our aim is to extract and generate part of the TD definition from the FSM instance. This will also enable the alignment with other ontologies such as the oneM2M Base Ontology and enable service composition. Given that each state has its own inputs and outputs, research is needed to identify which Properties, Actions and Events (according to the TD model) should be exposed. SPARQL performance is computationally heavy and our solution only allows to send SPARQL queries to external services. We plan to add the capability of sending non-SPARQL requests to external services described using ontologies such as TD ontology or oneM2M Base Ontology. Our goal was to facilitate deployments in similar IoT scenarios. Nevertheless, the task of creating the template ontology can be tedious, especially if the template has multiple states and actions with prototype URIs. Our approach has some restrictions imposed by the FSM model and the WoT architecture (we assume that physical devices expose an HTTP/S API). Therefore, we plan to create a visual tool to hasten the creation of FSM templates using visual building blocks like in StarUML. The IDE will be used to create the template in a visual manner and to automatically translate the visual representation to the ontological one.

**Author Contributions:** V.C. and D.V. conceived the idea and decided the technologies involved in the prototype; V.C. and S.V. implemented the prototype; V.C. contributed with insightful knowledge about the technologies involved; A.Z. and D.V. supervised the execution of the work and contributed with insightful discussions and ideas to improve the overall quality of the work; A.Z. is also the link with the European projects ATHIKA and SPRINT. All authors contributed equally to writing the paper.

**Funding:** This work received funding from the "Agència de Gestió d'Ajuts Universitaris i de Recerca (AGAUR)" of "Generalitat de Catalunya" (grant identification "2017 SGR 977") and ENVISERA project, funded by Spanish Ministerio de Economía y Competitividad (Plan Estatal de Investigación Científica y Técnica y de Innovación 2013–2016, grant reference CTM2015-68902-R). Part of this research has also been funded by the "SUR" of the "DEC" of the "Generalitat de Catalunya" and by the European Social Funds (grant reference 2018FI_B1_00171). In addition, we would like to thank La Salle—Universitat Ramon Llull for their support.

**Acknowledgments:** Icons used in Figure 3 made by Freepick and Gregor Cresnar at www.flaticon.com

**Conflicts of Interest:** The authors declare no conflict of interest.

## References

1. Atzori, L.; Iera, A.; Morabito, G. Understanding the Internet of Things: Definition, potentials, and societal role of a fast evolving paradigm. *Ad Hoc Netw.* **2017**, *56*, 122–140, doi:10.1016/j.adhoc.2016.12.004. [CrossRef]
2. Guinard, D. A Web of Things Application Architecture—Integrating the Real-World into the Web. Ph.D. Thesis, University of Fribourg, Fribourg, Switzerland, 2011.
3. Shadbolt, N.; Hall, W.; Berners-Lee, T.; Hall, W. The Semantic Web Revisited. *IEEE Intell. Syst.* **2006**, *21*, 96–101, doi:10.1109/MIS.2006.62. [CrossRef]
4. Martin, D.; Burstein, M.; Hobbs, J.; Lassila, O.; McDermott, D.; McIlraith, S.; Narayanan, S.; Paolucci, M.; Parsia, B.; Payne, T. OWL-S: Semantic Markup for Web Services. Available online: https://www.w3.org/Submission/OWL-S/ (accessed on 17 February 2019).
5. World Wide Web Consortium. W3C Web of Things Working Group. Available online: https://www.w3.org/WoT/WG/ (accessed on 17 February 2019).
6. OneM2M. oneM2M Base Ontology. Available online: www.onem2m.org/technical/published-drafts (accessed on 13 February 2019).
7. World Wide Web Consortium. RDF 1.1 Concepts and Abstract Syntax. Available online: https://www.w3.org/TR/rdf11-concepts/ (accessed on 18 January 2019).
8. Charpenay, V.; Kabisch, S.; Kosch, H. Semantic data integration on the web of things. In Proceedings of the 8th International Conference on the Internet of Things—IOT '18, Santa Barbara, CA, USA, 15–18 October 2018; pp. 1–8.
9. OneM2M. oneM2M Global Initiative. Available online: http://www.onem2m.org/ (accessed on 18 February 2019).
10. OneM2M. WoT Interworking. Technical Report. 2018. Available online: http://www.onem2m.org/ (accessed on 18 February 2019).

11. Haller, A.; Janowicz, K.; Cox, S.J.; Lefrançois, M.; Taylor, K.; Le Phuoc, D.; Lieberman, J.; García-Castro, R.; Atkinson, R.; Stadler, C. The modular SSN ontology: A joint W3C and OGC standard specifying the semantics of sensors, observations, sampling, and actuation. *Semant. Web* **2018**, 1–24, doi:10 3233/SW-180320. [CrossRef]
12. Belgueliel, Y.; Bourahla, M.; Brik, M. Towards an Ontology for UML State Machines. *Lect. Notes Softw. Eng.* **2014**, *2*, 116–120, doi:10.7763/LNSE.2014.V2.106. [CrossRef]
13. Dolog, P. Model-Driven Navigation Design for Semantic Web Applications with the UML—Guide. In Proceedings of the Workshops in Connection with the 4th International Conference on Web Engineering, ICWE 2004, Mubich, Germany, 28–30 July 2004; pp. 1–12.
14. Haage, M.; Malec, J.; Nilsson, A.; Stenmark, M.; Topp, E.A. Semantic Modelling of Hybrid Controllers for Robotic Cells. *Procedia Manuf.* **2017**, *11*, 292–299, doi:10.1016/j.promfg.2017.07.108. [CrossRef]
15. Pessemier, W.; Deconinck, G.; Raskin, G.; Saey, P.; van Winckel, H. Developing a PLC-friendly state machine model: Lessons learned. *Proc. SPIE* **2014**, *9152*, 915208, doi:10.1117/12.2054881. [CrossRef]
16. Nitti, M.; Pilloni, V.; Colistra, G.; Atzori, L. The Virtual Object as a Major Element of the Internet of Things: A Survey. *IEEE Commun. Surv. Tutor.* **2016**, *18*, 1228–1240, doi:10.1109/COMST.2015.2498304. [CrossRef]
17. Caballero, V.; Vernet, D.; Zaballos, A.; Corral, G. Prototyping a Web-of-Energy Architecture for Smart Integration of Sensor Networks in Smart Grids Domain. *Sensors* **2018**, *18*, 400, doi:10.3390/s18020400. [CrossRef] [PubMed]
18. Negash, B.; Westerlund, T.; Tenhunen, H. Towards an interoperable Internet of Things through a web of virtual things at the Fog layer. *Future Gener. Comput. Syst.* **2019**, *91*, 96–107, doi:10.1016/j.future.2018.07.053. [CrossRef]
19. Guan, Y.; Vasquez, J.C.; Guerrero, J.M.; Samovich, N.; Vanya, S.; Oravec, V.; Garcia-Castro, R.; Serena, F.; Poveda-Villalon, M.; Radojicic, C.; et al. An open virtual neighbourhood network to connect IoT infrastructures and smart objects—Vicinity: IoT enables interoperability as a service. In Proceedings of the 2017 Global Internet of Things Summit (GIoTS), Geneva, Switzerland, 6–9 June 2017; pp. 1–6.
20. The Vicinity Consortium. 2nd Open Call. Technical Details. Technical Report; Vicinity Consortium. 2019. Available online: https://vicinity2020.eu/vicinity/ (accessed on 18 February 2019).
21. Girau, R.; Martis, S.; Atzori, L. Lysis: A platform for iot distributed applications over socially connected objects. *IEEE Internet Things J.* **2017**, *4*, 40–51, doi:10.1109/JIOT.2016.2616022. [CrossRef]
22. MKLab. StarUML. Available online: http://staruml.io (accessed on 18 January 2019).
23. World Wide Web Consortium. HTTP Vocabulary in RDF 1.0. Available online: https://www.w3.org/TR/HTTP-in-RDF10/ (accessed on 10 February 2019).
24. Apache. Apache Jena. Available online: https://jena.apache.org/ (accessed on 16 February 2019).
25. Lightbend. Play Framework 2. Available online: https://www.playframework.com/ (accessed on 16 February 2019).
26. Lightbend. Akka. Available online: https://akka.io/ (accessed on 16 February 2019).
27. Martín de Pozuelo, R.; Zaballos, A.; Navarro, J.; Corral, G. Prototyping a Software Defined Utility. *Energies* **2017**, *10*, 818, doi:10.3390/en10060818. [CrossRef]
28. Ngu, A.H.H.; Gutierrez, M.; Metsis, V.; Nepal, S.; Sheng, M.Z. IoT Middleware: A Survey on Issues and Enabling technologies. *IEEE Internet Things J.* **2016**, *4*, 1–20, doi:10.1109/JIOT.2016.2615180. [CrossRef]
29. Razzaque, M.A.; Milojevic-Jevric, M.; Palade, A.; Cla, S. Middleware for Internet of Things: A Survey. *IEEE Internet Things J.* **2015**, *3*, 70–95, doi:10.1109/JIOT.2015.2498900. [CrossRef]
30. Gomes, C.E.M.; Lucena, V.F.; Yazdi, F.; Gohner, P. An intelligent medicine cabinet proposed to increase medication adherence. In Proceedings of the 2013 IEEE 15th International Conference on e-Health Networking, Applications and Services, Healthcom 2013, Lisbon, Portugal, 9–12 October 2013; pp. 737–739.
31. Gilchrist, A. *Industry 4.0—The Industrial Internet of Things*; Apress: Berkeley, CA, USA, 2016.

32. Chiou, C.K.; Tseng, J.C. An Intelligent Classroom Management System based on Wireless Sensor Networks. In Proceedings of the 2015 8th International Conference on Ubi-Media Computing, UMEDIA 2015, Colombo, Sri Lanka, 24–26 August 2015; pp. 44–48.
33. Gutierrez, J.M.; Jensen, M.; Henius, M.; Riaz, T. *Smart Waste Collection System Based on Location Intelligence*; Procedia Computer Science; Elsevier: Amsterdam, The Netherlands, 2015; Volume 61, pp. 120–127.

Article

# Microservice-Oriented Platform for Internet of Big Data Analytics: A Proof of Concept

**Zheng Li *, Diego Seco † and Alexis Eloy Sánchez Rodríguez**

Department of Computer Science, University of Concepción, Concepción 4070409, Chile; dseco@udec.cl (D.S.); alexisanchez@udec.cl (A.E.S.R.)

* Correspondence: imlizheng@gmail.com or zli@udec.cl; Tel.: +56-41-220-3686

† Millennium Institute for Foundational Research on Data, Chile.

Received: 15 January 2019; Accepted: 28 February 2019; Published: 6 March 2019

**Abstract:** The ubiquitous Internet of Things (IoT) devices nowadays are generating various and numerous data from everywhere at any time. Since it is not always necessary to centralize and analyze IoT data cumulatively (e.g., the Monte Carlo analytics and Convergence analytics demonstrated in this article), the traditional implementations of big data analytics (BDA) will suffer from unnecessary and expensive data transmissions as a result of the tight coupling between computing resource management and data processing logics. Inspired by software-defined infrastructure (SDI), we propose the "microservice-oriented platform" to break the environmental monolith and further decouple data processing logics from their underlying resource management in order to facilitate BDA implementations in the IoT environment (which we name "IoBDA"). Given predesigned standard microservices with respect to specific data processing logics, the proposed platform is expected to largely reduce the complexity in and relieve inexperienced practices of IoBDA implementations. The potential contributions to the relevant communities include (1) new theories of a microservice-oriented platform on top of SDI and (2) a functional microservice-oriented platform for IoBDA with a group of predesigned microservices.

**Keywords:** big data analytics; Internet of Things; microservices architecture; microservice-oriented platform; software defined infrastructure

---

## 1. Introduction

The emerging age of big data is leading us to an innovative way of understanding our world and making decisions. In particular, it is the data analytics that eventually reveals the potential values of datasets and completes the value chain of big data. When it comes to big data analytics (BDA), in addition to theories, mathematics, and algorithms, suitable infrastructures and platforms are also prerequisites to efficient BDA implementations. In practice, it is the managed and scheduled resources that deliver computing power to data processing logics to fulfill BDA jobs at runtime. In this article, we clarify the physical resources to be infrastructure, while treating the intermediate supporting mechanisms as layered platforms between the runtime jobs and their infrastructure.

Driven by increasing BDA demands, various and numerous frameworks and tools have emerged as BDA platforms [1]. According to their functional specifications, these BDA platforms generally combine data processing logics together with computing resource management. For instance, to perform Hadoop MapReduce jobs, a Hadoop system including its distributed filesystem needs to be installed and configured on a dedicated cluster managed by YARN [2]. Consequently, the current BDA implementations still require significant effort on environmental configurations and platform manipulations.

Furthermore, based on the de facto platforms, BDA applications tend to be environmentally monolithic because they are essentially stuck to predefined computing resources no matter how redundant the resources are. This environmental monolith is generally acceptable for traditional BDA problems with centralized workloads. When it comes to the booming ecosystem of Internet of Things (IoT), there is no doubt that such a mechanism will require tremendously extra overhead for big data collection from largely distributed sensors/devices for later analytics [3]. Nevertheless, IoT-oriented data analytics could eventually become inefficient and expensive if we always transfer and process data cumulatively in a central repository, because many IoT-oriented BDA problems (which we name "IoBDA") can be addressed without combining the originally distributed data [4,5].

Inspired by software-defined infrastructure (SDI) [6,7], we propose the use of a standard microservice-oriented platform to relieve the complexity in IoBDA implementations and to break the environmental monolith of IoBDA applications. In fact, it has been identified that deploying physical infrastructures for BDA applications can be costly and challenging [8], not to mention addressing the possibly tremendous heterogeneity and diversity in hardware. To reduce the infrastructural cost, as well as obtain deployment agility and automation, there has arisen a software-driven trend in making the computing environment programmable and software-defined [9]. Typical examples of SDI include software-defined network (SDN) and software-defined storage (SDS). SDN decouples the data transmission control from networking devices, such as switches and routers [10], SDS separates the data store management from storage systems [2], and both of them leverage heterogeneous hardware to facilitate support of workload demands via open-interface programming. In other words, SDI is mainly focused on the programmable management and control of computing resources. In contrast, our proposed microservice-oriented platform is expected to further isolate domain-specific business logics from the resource control and management.

This article mainly portrays the proof of concept of the microservice-oriented platform for IoBDA, and it demonstrates this idea by using Monte Carlo analytics and Convergence analytics that both fit in the characteristics of the IoT environment. We believe such an idea can theoretically be justified by the philosophy of osmotic computing [11] and technically be supported by the existing underlying frameworks [12]. We are still investigating IoT-friendly data processing logics and gradually migrating them to the microservices architecture (MSA) in order to further solidify our proposed platform.

On the basis of our ongoing efforts and current outcomes, we expect this article to make a twofold contribution to both the IoT and BDA communities:

- A new theory of a microservice-oriented platform on SDI. The theory essentially evolves the architecture of BDA implementations in general, i.e., further decoupling data processing logic from computing resource management. Such an architecturally loose coupling will generate more research opportunities in academia and will better guide BDA practices, particularly in the dynamic IoT environment.
- A functional microservice-oriented platform with predesigned microservices. In addition to facilitating IoBDA implementations for practitioners, a functional platform will, in turn, drive its own evolution along two directions. Vertically, our use cases can help validate and improve the underlying technologies (deeper-level platforms or frameworks) [12] while strengthening their compatibility with our platform. Horizontally, the initial functionalities can inspire more research efforts that aim to enrich microservice-oriented data processing logics and expand the applicability of this platform.

The remainder of this article is organized as follows. Several studies related to our work are summarized in Section 2. Section 3 elaborates on the definition of the microservice-oriented platform for IoBDA and briefly justifies its feasibility. Sections 4 and 5 demonstrate two IoT-friendly data processing logics to facilitate implementing Monte Carlo analytics and Convergence analytics, respectively, which will, in turn, be used to prototype our microservice-oriented platform. Section 6 draws conclusions and outlines two directions for our future work.

## 2. Related Work

By offering many benefits, including better scalability, reliability, and maintainability, microservice technologies have been employed successfully in practical projects (e.g., data organization and analysis in a Solar Nexus project [13]). However, these projects are still based on a dedicated computing environment, and their application scenarios have nothing to do with distributed data sources in the dynamic IoT environment.

On the basis of observations that recent technological advances are moving Cloud's centralized computing capacity close to users, osmotic computing has been proposed to represent the new paradigm of dynamic microservice-oriented computing that exploits both Cloud and Edge resources [11]. Unlike our proposed platform in this article that aims at technical solutions to IoBDA, the authors of [11] essentially supply a philosophy of MSA in smart environments. On the other hand, this philosophy can be used to justify the novelty and feasibility of our proposed microservice-oriented platform.

As a matter of fact, given its advantages of agility and scalability, MSA has been considered promising middleware architecture for IoT in general [14]. When it comes to the BDA domain, the closest work to ours is the distributed BDA framework for IoT [12]. It also emphasizes handling different data streams from various sources. Nevertheless, the layered framework in this study is mainly a microservice management mechanism including management policies, a messaging infrastructure, etc. Although this mechanism is supposed to facilitate the development of MSA-based applications, it does not care about the design and implementation of individual microservices, not to mention their boundary identification. On the contrary, our work is intended to support specific data processing logics with predesigned standard microservices (and/or templates), which will largely help inexperienced developers facing immature IoBDA implementations. Considering the microservice management at runtime, in particular, this framework [12] can act as a supporting layer beneath our proposed platform.

## 3. Conceptual Explanation and Justification

Emerging from the agile practitioner communities, MSA emphasizes addressing the shortcomings of monolithic architecture by decomposing the monolith into multiple small-scale, loosely coupled, and independently deployable microservices and making them communicable with each other through lightweight mechanisms (e.g., REST APIs) [15]. In fact, the booming container technology has enabled virtualization for many types of edge devices [16], and this eventually supports building container-based microservices within those devices. Note that it is possible to introduce agent microservices (similar to the concept of "infrastructure mediate systems" [17]) to represent or coordinate non-virtualizable binary sensors.

Unfortunately, despite the various benefits of MSA, there is still a lack of systematic guidelines for microservice boundary identification in general [18]. Thus, it would be more feasible and practical to narrow our focus down to specific problems and reach consensus on concrete microservices in the same context. In turn, the predefined microservices can be shared and employed to facilitate constructing MSA-based applications in the specific problem domain. In other words, given a particular problem context, it will be worthwhile and beneficial to build a scaffold-like platform in advance by designing a family of domain-specific microservices (or microservice templates) that predefine different pieces of business logics potentially involved in the problem.

In this work, we dedicate ourselves to developing a microservice-oriented platform for IoBDA: i.e., the problem context here is IoBDA. As mentioned previously, in contrast to different instances of SDI, our platform emphasizes the microservice-level abstraction of IoT-friendly BDA logics. The so-called microservices contained in the platform essentially cater loosely coupled and distributable tasks of those BDA logics, as demonstrated in Sections 4 and 5. Ideally, this context-specific platform will be able to cover various BDA logics, and each logic should have been well-analyzed into suitable tasks to match the predesigned microservices (and/or microservice templates).

Since there is no one-size-fits-all approach to BDA implementations, we are gradually investigating different data processing logics that are applicable to the IoT environment, such as Monte Carlo analytics, which can obtain approximate results through random observations or sampling, and Convergence analytics, which can rely on cascade MapReduce processes to gather increasingly dense results. It is noteworthy that both scenarios are compatible with the analytics characteristics in IoT, i.e., to divide a problem into micro-portions and to reduce unnecessary data transfer without sacrificing meaningful insights gained from the whole data [4,5].

In practice, the microservice-oriented platform will enable the implementation of data processing logics in the IoT environment via the orchestration of a set of standard microservices (and/or microservice templates). Benefiting from the nature of loose coupling, early-task microservices can be deployed close to where the data are generated so as both to avoid unnecessary data accumulation and to conduct data processing jobs more efficiently. In particular, programming a data processing logic will be realized as functional service calls without necessarily being concerned with or even being aware of the backend resource management.

More importantly, the microservice approach can conveniently address the challenges in the integration of heterogeneous sensor technologies [19,20] and in the communication between IoT sensors/devices. Similar to the early computer industry, today's sensor industry has not yet employed or reached any global agreement on standard protocols and universal interfaces [21]. Given the various niche markets, different sensor vendors even tend to continue the existing incompatible protocols and keep their own proprietary standards in order to secure market shares against their competitors. By wrapping different data processing tasks into microservices, sensors will be able to expose their interfaces as unified and language-neutral APIs (e.g., typical REST APIs look like "cloudpracticeaesr.appspot.com/mci/1000000", as demonstrated in Section 4), while diverse sensor communications will then become standard microservice communications that mainly rely on the predominant TCP/UDP-based Internet protocols [22].

## 4. Case I: Microservice-Oriented Logic for Monte Carlo Analytics

### *4.1. Architectural Design with Microservice Identification*

By stressing the independence and distribution of IoT sensors/devices, we came up with a microservice-oriented star-topology logic (cf. Figure 1) for implementing IoT-based Monte Carlo solutions for random sampling problems that can represent a broad class of easy-to-parallel applications. To address a random sampling problem is to statistically perform a set of observations from a tremendous range of (or even infinite) possibilities to approximate the answer under a particular condition. Following the Monte Carlo method, suppose the exact probability of a situation is $P$, and we randomly perform observations; then, the percentage of observations that match the situation will approximately be equal to $P$ if the observation number is large enough. In this case, the workload of performing observations (e.g., population estimation through traffic monitoring in a city) can conveniently be divided into any size of pieces and distributed widely.

We are mainly concerned with three roles in the generic logic of Monte Carlo analytics, as illustrated in Figure 1. Correspondingly, we predefine microservices with respect to these roles and briefly explain them as follows.

- **Observer** is a microservice template to be instantiated for specific observation tasks. Multiple observation tasks can be accomplished either by a group of observer instances or by multiple observation activities of a single observer instance. As the name suggests, observer instances are supposed to be deployed or migrated to virtualization-friendly IoT sensors/devices or their governors.
- **Central Processor** splits a whole job into independent pieces as observation tasks, assigns individual tasks to available observer instances, and correspondingly receives observation results. In addition to passing the observation results to **Observation Aggregator**, the central processor

can also incrementally retrieve and manipulate the aggregated observation results into continuous and value-added outputs if needed.

- **Observation Aggregator** collects, stores, and can directly output observation results if they are immediately ready to use. Note that the observation results here should not be a simple transition of original data from the observer side. Taking a sampling task as an example, the observation will send back a statistical result instead of the information about detailed samples (see the demonstration in Section 4.2). In other words, the observation aggregation here does not conflict with the principle of collectionless analytics in the IoT environment.

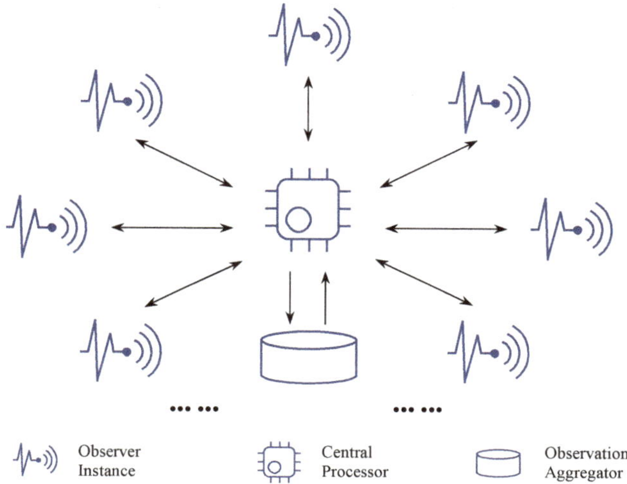


**Figure 1.** Star topology of the microservice-oriented logic for Monte Carlo analytics.

Given these predefined microservices and the pre-settled Monte Carlo processing logic, an IoT-friendly random sampling problem would quickly be addressed on such a platform. Note that the underlying resource management mechanisms (e.g., Google App Engine mentioned in Section 4.2) are completely transparent to the platform users.

### 4.2. Conceptual Validation Using Double Integral Estimation

To demonstrate and initially validate the effectiveness and efficiency of such a platform for Monte Carlo analytics, we selected the approximation of a double integral as the random sampling problem. A double integral of a function can be interpreted as the volume between the surface given by the function and its x–y region. In particular, the area of any x–y region that is represented as a single integral can also be estimated through the Monte Carlo method, as described by Equation (1).

$$\int_a^b f(x)dx \approx \frac{b-a}{N}\sum_{i=1}^{N} f(x_i) \quad (1)$$

where $f(x_i)$ is the $i$th random evaluation of the function within the interval $[a,b]$, and $N$ indicates the total number of trials in the Monte Carlo estimation.

Although the generic description in Equation (1) aims at a single integral, we can conveniently extend it to a double integral by iterating two similar steps, i.e., (1) area approximation and (2) volume approximation according to the estimated area, as concisely outlined in Algorithm 1.

**Algorithm 1** Monte Carlo Approximation of Double Integral

**Input:** N: the total number of trials in Monte Carlo estimation.

**Output:** The approximation result of the predefined double integral $\int_a^b \int_{g(x)}^{h(x)} f(x,y)dydx$.

```
valid_points ← 0                                  ▷ for counting the number of points inside the valid x–y region.
function_evaluations ← 0                          ▷ for summing up individual evaluations of f(x,y).
X ← a random number between a and b
Y ← a random number between MIN{g(x)} and MAX{h(x)}, x ∈ [a,b]
for i ← 1, N do
    if g(X) <= Y and Y <= h(X) then               ▷ If the random point (X,Y) is located inside the valid x–y region.
        valid_points ← valid_points + 1
        function_evaluations ← function_evaluations + f(X,Y)
    end if
end for
area ← (b − a) · (MAX{h(x)} − MIN{g(x)}) · valid_points/N, x ∈ [a,b]   ▷ Estimating the area of the valid x–y region.
volume ← area · function_evaluations/valid_points   ▷ Note that the function was evaluated valid_points times.
return volume                                     ▷ The estimated volume as the double integral result.
```

We explain the complete process by using a concrete example specified in Equation (2). The projection of the corresponding function $f(x,y) = (4xy - y^3)$ onto the x–y plane is illustrated in Figure 2. It is noteworthy that the x–y plane covering the projection happens to be a unit square in this case, and it is constrained by $x \in [0,1]$, $y \in [x^3, \sqrt{x}]$.

$$\int_0^1 \int_{x^3}^{\sqrt{x}} (4xy - y^3)dydx \tag{2}$$

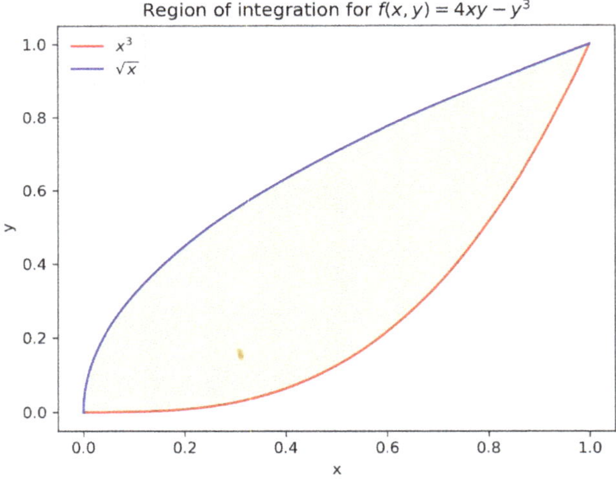


**Figure 2.** Projection of the function $f(x,y) = (4xy - y^3)$ onto the x–y plane, $x \in [0,1]$, $y \in [x^3, \sqrt{x}]$.

When it comes to the area approximation, imagine that we blindly draw points inside the unit square; there must be some points drawn inside the valid x–y region of the projection and others that are not. After spreading numerous points randomly over the square, its area can be replaced with the number of points, as can the x–y region's area. Then, we will be able to use the ratio of point numbers to satisfy the Monte Carlo approximation of the projection area, i.e., $Amount_{projection}/Amount_{all} = Area_{projection}/Area_{all}$.

While drawing points, the evaluations of $f(x,y)$ also accumulate at the points located inside the valid x–y region. Once the projection area is estimated, we can use the accumulated evaluations to calculate and sum up the corresponding solid volumes and eventually use the average volume to approximate the double integral result.

Recall that, in general, the larger the number of random samples, the more accurate the Monte Carlo approximation. Here, we decided to split this unlimited size of sampling workload into individual tasks and define each task as drawing 1 million points in a unit square, i.e., $x, y \in [0, 1]$. Then, we implemented the aforementioned observer role to fulfill such a task and deployed a number of observer instances on multiple Python runtimes of Google App Engine (one of the observer instances is at https://cloudpracticeaesr.appspot.com/mci/1000000, which returns (1) the number of points that fall inside the projection area from 1 million random-point generations and (2) the accumulated evaluations of the function $f(x, y) = (4xy - y^3)$ at those points). The experimental result of this Monte Carlo analytics experiment is visualized in Figure 3. Compared with the double integral value 55/156 or 0.3525641, it is clear that as the number of observer instances (or observation activities) grows, we can expect progressively more precise approximations of the double integral. Meanwhile, the local computing workload of the central processor will increase trivially, as it only deals with the random sampling results.

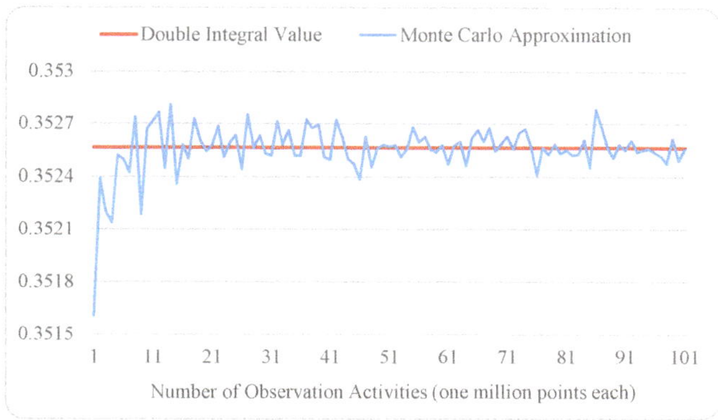


**Figure 3.** Monte Carlo approximation of the double integral $\int_{0}^{1}\int_{x^3}^{\sqrt{x}} (4xy - y^3)dydx$ as the sampling size (number of observation activities) grows.

To help validate the potential efficiency of Monte Carlo analytics in the distributed IoT environment, we further conducted experiments to compare the performance of finishing the same integral approximation job (on the basis of 10 million random points generated) between the local centralized calculation and the application of different numbers of observer instances. In particular, our local environment is a laptop with an Intel Core i3-6100U @2.3 GHz processor and 4 GB of RAM. To simulate the distributed locations of sensors, we randomly deployed observer instances to different geographical regions of Google App Engine [23], as shown in Table 1.

**Table 1.** Multiple observer instances randomly deployed to different regions of Google App Engine.

| Number of Deployed Observer Instances | Region | Number of Deployed Observer Instances | Region |
|---|---|---|---|
| Two | us-central (Iowa) | One | us-east1 (South Carolina) |
| One | us-west2 (Los Angeles) | One | us-east4 (Northern Virginia) |
| One | northamerica-northeast1 (Montréal) | One | southamerica-east1 (São Paulo) |
| One | europe-west2 (London) | One | europe-west3 (Frankfurt) |

By manually balancing the workload among observer instances and triggering multiple observation activities if needed, we obtained experimental the results illustrated in Figure 4. Note that all the experiments were repeated 10 times so that we could expect relatively more accurate results by using the average values. We also calculated the standard deviation of the 10-trial results from every experimental scenario, as indicated via the error bars in Figure 4. It seems that fewer observer instances incur higher performance variations. This phenomenon is due to fewer observer instances undertaking more observation activities to accomplish the same-sized job and, consequently, exaggerating the uncertainty in data transmission latency across the Internet.

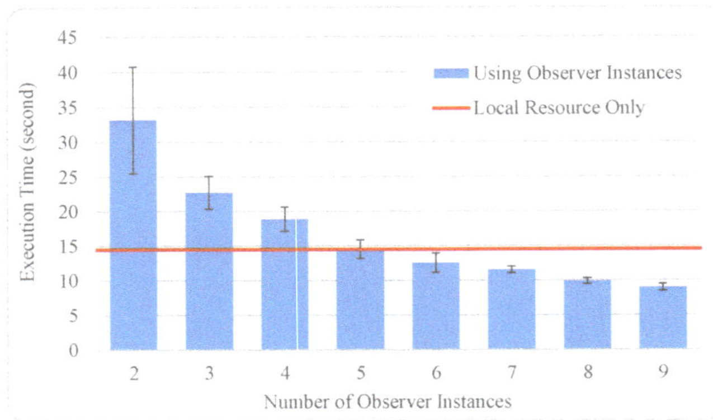


**Figure 4.** Monte Carlo approximation of the double integral $\int_{0}^{1}\int_{x^3}^{\sqrt{x}}(4xy-y^3)dydx$, with the fixed workload (10 million sampling points) as the number of observer instances growing. The error bars indicate the performance variations in the Monte Carlo approximation job.

Acting as a baseline for comparison, the average execution time of locally approximating the double integral $\int_{0}^{1}\int_{x^3}^{\sqrt{x}}(4xy-y^3)dydx$ (without considering any data transmission) is about 14.453 s. Since it does not make practical sense to compare the baseline with the case of a single-observer instance, we intentionally removed the single-observer performance from Figure 4. Overall, it can be seen that the Star-topology Monte Carlo analytics can beat its local version after involving more than five observer instances in this case, even though the distributed "sensors" require extra Internet communication overheads. In fact, there could be an even higher networking cost if collecting all the raw data together before analytics (cf. Section 5.2). Thus, the scenario of IoT-based Monte Carlo analytics would particularly be suitable for the applications when assuming "the more sensors, the better performance", which essentially takes advantage of horizontally scaling the whole system. Such an advantage is also one of the reasons that we chose MSA to fit IoBDA, i.e., the microservice technology naturally matches the characteristics of IoT, particularly in terms of better scalability and maintainability [24].

### 4.3. Prospect of Practical Application

On the basis of this proof of concept for Monte Carlo analytics, we expect to prototype the platform by aligning with practical use cases. For example, a possible use case would be a traffic monitoring system that can facilitate urban design or help adjust local traffic policies. In detail, a single task to be fulfilled by the Observer will be counting vehicles and pedestrians (analogous to the points inside and outside the projection area, respectively, in Figure 2). The distributed observer instances can be deployed together with sensors at different crossroads. Then, our platform prototype with suitable central processing functionalities will be able to monitor and analyze the traffic information of a city.

By emphasizing the Star topology of data processing in IoT, the Monte Carlo logic can be conveniently extended to broader application types as long as the applications can take advantage of the aforementioned horizontal scalability. We take sorting as a generic example application other than Monte Carlo analytics. Imagine a job is to sort a large total of random numbers. The facilities of our platform (e.g., the possible Docker files and deployment routines) for supporting Monte Carlo analytics will remain the same, while developers only need to adapt the job's implementation to the Star-topology architecture and instantialize the corresponding microservice templates. For example, a single task deployed in an observer instance can be (1) generating a small group of random numbers and (2) using the Quick Sort algorithm to sort the generated random numbers; the central processor can implement the Merge Sort algorithm to gradually receive and sort the numbers from all the observer instances.

## 5. Case II: Microservice-Oriented Logic for Convergence Analytics

### 5.1. Architectural Design with Microservice Identification

At this current stage, we mainly use the well-known MapReduce logic to represent Convergence analytics, which can significantly reduce the data size/amount during analytical processing, e.g., by merging original and intermediate data. Furthermore, to better fit the characteristics of IoT, we consider a Tree topology of Convergence analytics with cascade MapReduce logic in practice, as illustrated in Figure 5. From the perspective of topology, Convergence analytics can be viewed as an extension of Monte Carlo analytics. However, we distinguish between Monte Carlo analytics and Convergence analytics by emphasizing different concerns of collectionless BDA. Specifically, Monte Carlo analytics focuses on spreading observations, while Convergence analytics focuses on reducing the size of data transmission.

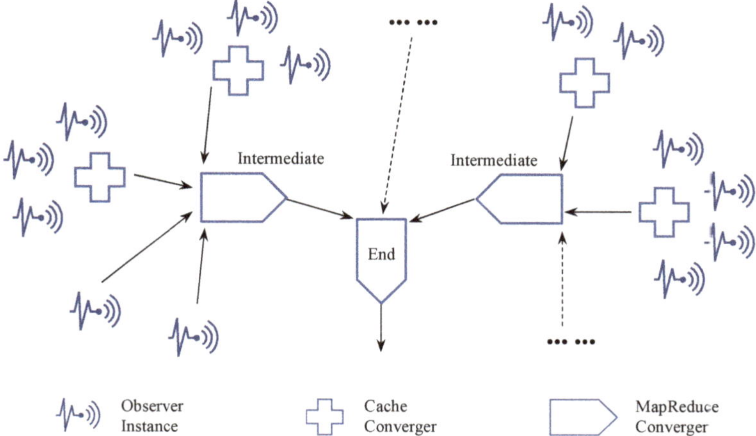


**Figure 5.** Tree topology of the microservice-oriented logic for Convergence analytics.

Similarly, we identify three main roles in Convergence analytics, as shown in the legends of Figure 5. For the purpose of conciseness, although the MapReduce logic can further be broken down into more specific roles (i.e., mapper and reducer) to be implemented as microservices, we do not elaborate on those well-known component roles here. In addition, to avoid duplication, there is no need to respecify the reusable role Observer (cf. Section 4.1). Thus, we only focus on and explain different convergers as follows.

- **Cache Converger** prepares data blocks by merging small pieces of data from a limited range/cluster of IoT sensors/devices, whereas it does not reduce the overall data size. Cache convergers could particularly be helpful for passing a large number of discrete data records to the subsequent MapReduce logic, as dealing with tiny-data transactions would be inefficient in terms of both execution time and energy expense [25]. In fact, caching data before transmission has become an energy optimization strategy, especially for mobile devices [26]. Note that cache convergers should be located at (or at least close to) the Internet edge in order to take advantage of the relatively trivial overhead of edge communication for receiving small data pieces.
- **Intermediate MapReduce Converger** either receives preprocessed data blocks from observer instances and cache convergers or receives pre-converged data from antecedent (also intermediate) MapReduce convergers and then uses the MapReduce mechanism to further converge the received data. Since we do not expect cache convergers to reduce data size/amount tremendously, the outermost MapReduce convergers should also be located close to the edge of the Internet.
- **End MapReduce Converger** receives final-stage intermediate convergence results and still uses the MapReduce mechanism to complete the whole analytics job. In contrast, the end MapReduce converger can be located remotely from the Internet edge. There is no doubt that the region-wide and cross-region communications will incur increasingly higher overhead; however, here we can expect to transfer less data as compensation, because the intermediate convergence results should have been much smaller than the sum of their raw inputs.

Note that, in practice, the Tree topology of a cascade convergence logic can be more flexible and comprehensive than that illustrated in Figure 5. As mentioned above, it is possible to plug and play more intermediate MapReduce convergers in series and/or parallel connections to conduct iterative Convergence analytics if needed for various workload distributions.

### 5.2. Conceptual Validation Using Word Count with Cascade Convergence

Here, we employ Word Count, which is the most popular MapReduce demo, to conceptually validate the microservice-oriented cascade convergence logic. Imagine that the requirement is to count words in a distributed text retrieval scenario, including numerous voice-to-text recognition and optical character recognition (OCR) sensors. Instead of transferring all the recognized texts to a central repository, the word counting job can be done on our platform with cascade convergers (cf. Figure 5). In particular, each cache converger can be implemented as a microservice which joins distributed "words" into a single "file" (i.e., a text document) and meanwhile performs simple data preprocessing or initial convergence tasks (e.g., restructuring data into predefined formats like JSON or XML). As for MapReduce convergers, the predesigned mapper/reducer microservice templates can be instantiated with the word-count-related functionalities and then be deployed/migrated to different and proper processing locations. The whole process of counting words through cascade convergence is specified in Algorithm 2. Note that the individual <word, 1> pairs are supposed to be manipulated when caching the input data (cf. the function Pairify($S$) in Algorithm 2), which enables the unification of the Map procedures of all the MapReduce convergers, including the outermost one.

To facilitate monitoring of the changes in data size during the cascade convergence process, we conducted several rounds of word count experiments in our local environment. In particular, we replaced data caching with preparing a set of text files (ranging from around 2 MB to around 20 MB) by copying, pasting, and intentionally duplicating large amounts of random Wikipedia contents. As mentioned previously, the texts in each file were further restructured into <word, 1> pairs before going through the MapReduce convergers. When it comes to the MapReduce convergers, we employed three intermediate ones to imitate the outermost MapReduce processes within three different sensor regions so as to make the tree topology here consistent with the demonstration in Figure 5.

**Algorithm 2** Cascade-Convergence-based Word Count

**Input:** S: the continuous string-format "sensor" data.
**Output:** The word count result.

```
function PAIRIFY(S)
    P ← ∅
    for each word w ∈ S do                 ▷ Restructuring data into <key, value> pairs and storing them.
        P ← P+ < w,1 >
    end for
    return P
end function

function MAPREDUCE(F list)
    procedure MAP(F)
        for each line l ∈ F do             ▷ Splitting the file into <key, value> pairs.
            Parse l into < w,v >
            EmitIntermediate(< w,v >)
        end for
    end procedure
    procedure REDUCE(key, value_array)
        value_new ← 0
        for each value v ∈ value_array do  ▷ Counting the number of a particular word key.
            value_new ← value_new + v
        end for
        Emit(< key, value_new >)
    end procedure
    return < word, count > list
end function

while receiving S do
    repeat
        F ← ∅                              ▷ F is the data block to be cached.
        repeat
            F ← F+PAIRIFY(S)
        until reaching the threshold size of data block
    until having a F list
    for each non-end MapReduce convergers do
                                           ▷ F list is for the outermost MapReduce convergers,
                                           ▷ while intermediateResult list is for the other intermediate MapReduce convergers.
        intermediateResult ← MAPREDUCE(F list or intermediateResult list)
    end for
end while
result ← MAPREDUCE(final intermediateResult list)   ▷ Delivering final result by the end MapReduce converger.
return result
```

Since different experimental results vary significantly because of different sizes of professional vocabularies in the input files (e.g., same-field texts vs. multi-field texts), we only show the rough data size changes as an average during the MapReduce convergence process, as portrayed in Figure 6.

It is clear that by tremendously reducing the data size through the outermost (as well as intermediate) MapReduce convergers, the rest of the convergence process enjoys much less overhead for data transmission. We further define the data transmission saving rate as a metric to quantitatively investigate the efficiency of Convergence analytics, as specified in Equation (3).

$$R = \frac{D_{in} - D_{out}}{D_{in}} \times 100\% \quad (3)$$

where $R$ represents the data transmission saving rate of a particular convergence process (by any type of converger). $D_{in}$ and $D_{out}$ respectively indicate the input data size before the convergence process and the output data size for the subsequent transmission. After applying this metric to our Word Count experiment, we illustrate the quantitative measurement in Figure 7. Note that, for the purpose of conciseness, we used 10 MB as the representative size of input data (i.e., the prepared text files ranging from around 2 MB to around 20 MB) for the intermediate convergence process to calculate its data transmission saving rate.

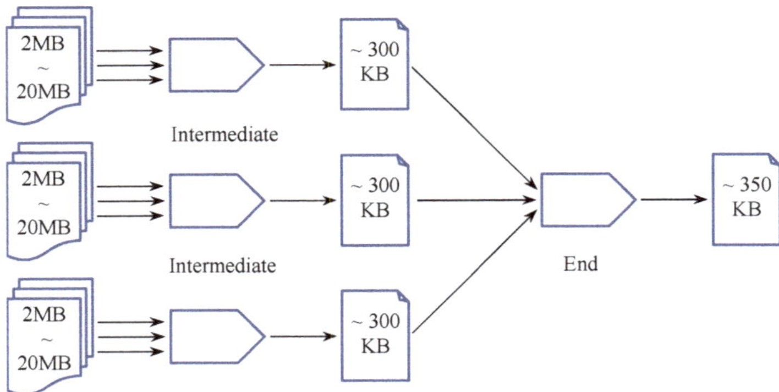


**Figure 6.** Data size changes during the cascade convergence process of word count experiments.

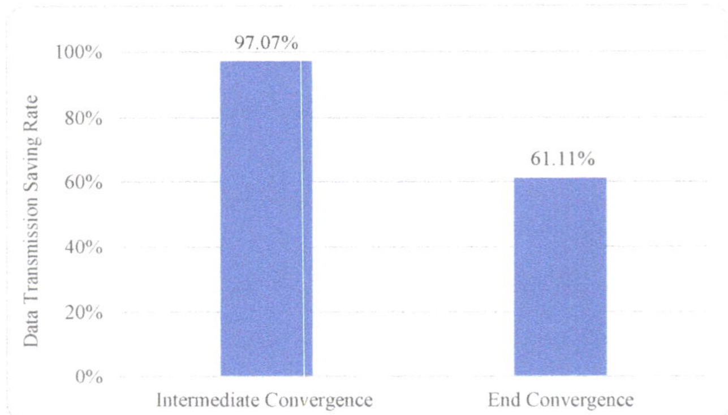


**Figure 7.** Data transmission saving rates of the intermediate convergence process and the end convergence process in the Word Count experiment.

Specifically, the data transmission saving rates of intermediate convergence process and end convergence process within our experiment are roughly 97.07% and 61.11%, respectively. We believe that such high rates are the result of the extreme case of counting words, because the vocabulary size of human languages is fairly small, especially at the conversational level [27]. We reckon that other applications of Convergence analytics could have more moderate rates of data transmission saving.

### *5.3. Prospect of Practical Application*

On the basis of the same logic of the cascade word count, we expect to solidify our microservice-oriented platform with respect to Convergence analytics through a real project, i.e., visualizing large spatial datasets in a Web-based map viewer [28], which requires handling tons of position events generated by GPS devices (vehicles) and quickly visualizing integration results at the client side, especially when switching zoom levels.

Similarly, without changing the Tree topology of data processing in IoT, our proposed platform can naturally support more application types beyond Convergence analytics. Still taking sorting as an example, we can reuse the same architecture (cf. Figure 5) and microservices (and templates) to quickly satisfy such a different job. In detail, the cache convergers can directly be reused to collect numbers from nearby sensors and pass data blocks to the subsequent MapReduce convergers, and the MapReduce convergers should be equipped with suitable sorting functionalities [29]. Note that, benefiting from the cascade convergence logic, users will not have to rely on a single MapReduce converger with heavyweight implementations (e.g., what Google did includes thousands of nodes [29]).

## 6. Conclusions and Future Work

Different types of SDI, such as SDN and SDS, have widely been accepted to make physical computing environments programmable. In essence, the idea of SDI is still from the computing resource's perspective. In contrast, we propose a microservice-oriented platform from the application's perspective in order to help make domain-specific business logics microservice-able and further decouple business logic implementations from resource control and management. Considering the tremendous distribution of IoT data, such a microservice-oriented platform will particularly be valuable and useful in the IoBDA domain by relieving the complexity in and breaking the environmental monolith of traditional BDA mechanisms.

In summary, we use this article to initialize an ambitious proposal at its proof-of-concept stage. Although it is impossible (and unnecessary) to address all kinds of BDA problems, we have shown that suitable scenarios such as Monte Carlo analytics and Convergence analytics in IoT can benefit from the proposed microservice-oriented platform. We plan to follow a reinforcement process to develop this platform for IoBDA. The platform prototype will stick to limited data processing logics and applications and then use new applications and include new BDA logics to gradually validate, improve, and enrich the platform features.

Thus, our future work will unfold along two directions. Firstly, we will try to extend the two conceptual validation demos described in this article to more practical case studies. Considering the relatively high failure ratio and long latency from IoT sensors/devices, we will particularly focus on investigating fault-tolerant mechanisms within the practical IoT environment. In fact, it has been identified that MSA can naturally help isolate failures [30] and increase application resilience [31], i.e., one failed microservice does not necessarily affect the other microservices, at least when there is no chaining between them. Secondly, we will keep enriching this proposed platform, e.g., including MSA-friendly machine learning logics [32]. By integrating more types of data processing logics, the eventually developed platform will provide a catalog of IoBDA scenarios together with demo applications, which can guide users to select suitable microservices and instantialize relevant microservice templates for their specific IoBDA implementations.

**Author Contributions:** Conceptualization, Z.L. and D.S.; Methodology, Z.L. and D.S.; Validation, Z.L., D.S. and A.E.S.R.; Investigation, Z.L. and D.S.; Data curation, Z.L. and A.E.S.R.; Writing—original draft preparation, Z.L.; Writing—review and editing, D.S.; Visualization, A.E.S.R.; Supervision, Z.L. and D.S.; Funding acquisition, Z.L.

**Funding:** This work was supported in part by Comisión Nacional de Investigación Científica y Tecnológica (CONICYT) under Grant FONDECYT Iniciación 11180905, in part by the University of Concepción under Grant VRID INICIACION 218.093.017-1.0 IN, and in part by Millennium Institute for Foundational Research on Data (IMFD).

**Conflicts of Interest:** The authors declare no conflict of interest.

## Abbreviations

The following abbreviations are used in this manuscript:

| | |
|---|---|
| API | Application Program Interface |
| BDA | Big Data Analytics |
| IoBDA | Internet of Big Data Analytics |
| IoT | Internet of Things |
| JSON | JavaScript Object Notation |
| MSA | Microservices Architecture |
| OCR | Optical Character Recognition |
| REST | Representational State Transfer |
| SDI | Software-Defined Infrastructure |
| SDN | Software-Defined Network |
| SDS | Software-Defined Storage |
| XML | Extensible Markup Language |

## References

1. PAT Research. Top 50 Bigdata Platforms and Bigdata Analytics Software. 2018. Available online: https://www.predictiveanalyticstoday.com/bigdata-platforms-bigdata-analytics-software/ (accessed on 7 January 2019).
2. Darabseh, A.; Al-Ayyoub, M.; Jararweh, Y.; Benkhelifa, E.; Vouk, M.; Rindos, A. SDStorage: A Software Defined Storage Experimental Framework. In Proceedings of the 3rd International Conference on Cloud Engineering (IC2E 2015), Tempe, AZ, USA, 9–13 March 2015; IEEE Computer Society: Tempe, AZ, USA, 2015; pp. 341–346.
3. Djedouboum, A.C.; Ari, A.A.A.; Gueroui, A.M.; Mohamadou, A.; Aliouat, Z. Big Data Collection in Large-Scale Wireless Sensor Networks. *Sensors* **2018**, *18*. 4474. [CrossRef] [PubMed]
4. Froehlich, A. How Edge Computing Compares with Cloud Computing. Available online: https://www.networkcomputing.com/networking/how-edge-computing-compares-cloud-computing/1264320109 (accessed on 7 January 2019).
5. IEEE. Cloud-Link: IoT and Cloud. Available online: https://cloudcomputing.ieee.org/publications/cloud-link/march-2018 (accessed on 7 January 2019).
6. Kang, J.M.; Bannazadeh, H.; Rahimi, H.; Lin, T.; Faraji, M.; Leon-Garcia, A. Software-Defined Infrastructure and the Future Central Office. In Proceedings of the 2nd Workshop on Clouds Networks and Data Centers, Budapest, Hungary, 9–13 June 2013; IEEE Press: Budapest, Hungary, 2013; pp. 225–229.
7. Kang, J.M.; Lin, T.; Bannazadeh, H.; Leon-Garcia, A. Software-Defined Infrastructure and the SAVI Testbed. In *TridentCom 2014: Testbeds and Research Infrastructure: Development of Networks and Communities*; Leung, V.C., Chen, M., Wan, J., Zhang, Y., Eds.; Springer: Cham, Switzerland, 2014; Volume 137, pp. 3–13.
8. Harris, R. Myriad Use Cases. Available online: https://cwiki.apache.org/confluence/display/MYRIAD/Myriad+Use+Cases (accessed on 7 January 2019).
9. Li, C.S.; Brech, B.L.; Crowder, S.; Dias, D.M.; Franke, H.; Hogstrom, M.; Lindquist, D.; Pacifici, G.; Pappe, S.; Rajaraman, B.; et al. Software defined environments: An introduction. *IBM J. Res. Dev.* **2014**, *58*, 1–11. [CrossRef]
10. Nunes, B.A.A.; Mendonca, M.; Nguyen, X.N.; Obraczka, K.; Turletti, T. A Survey of Software-Defined Networking: Past, Present, and Future of Programmable Networks. *IEEE Commun. Surv. Tutor.* **2014**, *16*, 1617–1634. [CrossRef]
11. Villari, M.; Fazio, M.; Dustdar, S.; Rana, O.; Ranjan, R. Osmotic Computing: A New Paradigm for Edge/Cloud Integration. *IEEE Cloud Comput.* **2016**, *3*, 76–83. [CrossRef]
12. Vögler, M.; Schleicher, J.M.; Inzinger, C.; Dustdar, S. Ahab: A Cloud-based Distributed Big Data Analytics Framework for the Internet of Things. *Softw. Pract. Exp.* **2017**, *47*, 443–454. [CrossRef]

13. Le, V.D.; Neff, M.M.; Stewart, R.V.; Kelley, R.; Fritzinger, E.; Dascalu, S.M.; Harris, F.C. Microservice-based Architecture for the NRDC. In Proceedings of the 13th IEEE International Conference on Industrial Informatics (INDIN 2015), Cambridge, UK, 22–24 July 2015; IEEE Press: Cambridge, UK, 2015; pp. 1659–1664.
14. Kang, R.; Zhou, Z.; Liu, J.; Zhou, Z.; Xu, S. Distributed Monitoring System for Microservices-Based IoT Middleware System. In *ICCCS 2018: Cloud Computing and Security*; Sun, X., Pan, Z., Bertino, E., Eds.; Springer: Cham, Switzerland, 2018; Volume 11063, pp. 467–477.
15. Newman, S. *Building Microservices: Designing Fine-Grained Systems*; O'Reilly Media: Sebastopol, CA, USA, 2015.
16. Morabito, R.; Cozzolino, V.; Ding, A.Y.; Beijar, N.; Ott, J. Consolidate IoT Edge Computing with Lightweight Virtualization. *IEEE Netw.* **2018**, *32*, 102–111. [CrossRef]
17. Ding, D.; Cooper, R.A.; Pasquina, P.F.; Fici-Pasquina, L. Sensor technology for smart homes. *Maturitas* **2011**, *69*, 131–136. [CrossRef] [PubMed]
18. Chen, R.; Li, S.; Li, Z. From Monolith to Microservices: A Dataflow-Driven Approach. In Proceedings of the 24th Asia-Pacific Software Engineering Conference (APSEC 2017), Nanjing, China, 4–8 December 2017; IEEE Computer Society: Nanjing, China, 2017; pp. 466–475.
19. Augusto, J.C. Past, Present and Future of Ambient Intelligence and Smart Environments. In *Agents and Artificial Intelligence*; Filipe, J., Fred, A., Sharp, B., Eds.; Springer International Publishing: Berlin, Germany, 2010; Volume 67, pp. 3–15.
20. Viani, F.; Robol, F.; Polo, A.; Rocca, P.; Oliveri, G.; Massa, A. Wireless Architectures for Heterogeneous Sensing in Smart Home Applications: Concepts and Real Implementation. *Proc. IEEE* **2013**, *101*, 2381–2396. [CrossRef]
21. Gorman, B.L.; Resseguie, D.; Tomkins-Tinch, C. Sensorpedia: Information sharing across incompatible sensor systems. In Proceedings of the 2009 International Symposium on Collaborative Technologies and Systems (CTS 2009), Baltimore, MD, USA, 18–22 May 2009; IEEE Press: Baltimore, MD, USA, 2009; pp. 448–454.
22. Vresk, T.; Čavrak, I. Architecture of an Interoperable IoT Platform Based on Microservices. In Proceedings of the 39th International Convention on Information and Communication Technology, Electronics and Microelectronics (MIPRO 2016), Opatija, Croatia, 30 May–3 June 2016; IEEE Computer Society: Opatija, Croatia, 2016; pp. 1196–1201.
23. Google. App Engine Locations. 2019. Available online: https://cloud.google.com/appengine/docs/locations (accessed on 11 February 2019).
24. Butzin, B.; Golatowski, F.; Timmermann, D. Microservices approach for the Internet of Things. In Proceedings of the 21st International Conference on Emerging Technologies and Factory Automation (ETFA 2016), Berlin, Germany, 6–9 September 2016; IEEE Press: Berlin, Germany, 2016; pp. 1–6.
25. Chen, F.; Grundy, J.; Yang, Y.; Schneider, J.G.; He, Q. Experimental Analysis of Task-based Energy Consumption in Cloud Computing Systems. In Proceedings of the 4th ACM/SPEC International Conference on Performance Engineering (ICPE 2013), Prague, Czech Republic, 21–24 April 2013; ACM Press: Prague, Czech Republic, 2013; pp. 295–306.
26. Balasubramanian, N.; Balasubramanian, A.; Venkataramani, A. Energy Consumption in Mobile Phones: A Measurement Study and Implications for Network Applications. In Proceedings of the 9th ACM SIGCOMM conference on Internet measurement (IMC 2009), Chicago, IL, USA, 4–6 November 2009; ACM Press: Chicago, IL, USA, 2009; pp. 280–293.
27. Gibbons, J. The Numbers Game: How Many Words Do I Need to Know to Be Fluent in a Foreign Language? 2018. Available online: https://www.fluentu.com/blog/how-many-words-do-i-need-to-know/ (accessed on 7 February 2019).
28. Cortiñas, A.; Luaces, M.R.; Rodeiro, T.V. A Case Study on Visualizing Large Spatial Datasets in a Web-Based Map Viewer. In *ICWE 2018: Web Engineering*; Mikkonen, T., Klamma, R., Hernández, J., Eds.; Springer: Cham, Switzerland, 2018; Volume 10845, pp. 296–303.
29. Hamilton, J. Google MapReduce Wins TeraSort. 2008. Available online: https://perspectives.mvdirona.com/2008/11/google-mapreduce-wins-terasort/ (accessed on 21 February 2019).
30. Taibi, D.; Lenarduzzi, V.; Pahl, C. Architectural Patterns for Microservices: a Systematic Mapping Study. In Proceedings of the 8th International Conference on Cloud Computing and Services Science (CLOSER 2018), Madeira, Portugal, 19–21 March 2018; Science and Technology Press: Madeira, Portugal, 2018; pp. 221–232.

31. Bogner, J.; Zimmermann, A. Towards Integrating Microservices with Adaptable Enterprise Architecture. In Proceedings of the 20th International Enterprise Distributed Object Computing Workshop (EDOCW 2016), Vienna, Austria, 5–9 September 2016; IEEE Computer Society: Vienna, Austria, 2016; pp. 1–6.
32. Slepicka, J.; Semeniuk, M. *Deploying Machine Learning Models as Microservices Using Docker*; O'Reilly Media: Sebastopol, CA, USA, 2017.

*Article*

# Wireless Middleware Solutions for Smart Water Metering

**Stefano Alvisi [1], Francesco Casellato [1], Marco Franchini [1], Marco Govoni [1], Chiara Luciani [1], Filippo Poltronieri [1,*], Giulio Riberto [1], Cesare Stefanelli [1] and Mauro Tortonesi [2]**

1 Department of Engineering, University of Ferrara, 44122 Ferrara, Italy; stefano.alvisi@unife.it (S.A.); francesco.casellato@unife.it (F.C.); marco.franchini@unife.it (M.F.); marco.govoni@unife.it (M.G.); chiara.luciani@unife.it (C.L.); giulio.riberto@unife.it (G.R.); cesare.stefanelli@unife.it (C.S.)

2 Department of Mathematics and Computer Science, University of Ferrara, 44121 Ferrara, Italy; mauro.tortonesi@unife.it

* Correspondence: filippo.poltronieri@unife.it; Tel.: +39-0532-97-4114

Received: 14 January 2019; Accepted: 16 April 2019; Published: 18 April 2019

**Abstract:** While smart metering applications have initially focused on energy and gas utility markets, water consumption has recently become the subject of increasing attention. Unfortunately, despite the large number of solutions available on the market, the lack of an open and widely accepted communication standard means that vendors typically propose proprietary data collection solutions whose adoption causes non-trivial problems to water utility companies in term of costs, vendor lock-in, and lack of control on the data collection infrastructure. There is the need for open and interoperable smart water metering solutions, capable of collecting data from the wide range of water meters on the market. This paper reports our experience in the development and field testing of a highly interoperable smart water metering solution, which we designed in collaboration with several water utility companies and which we deployed in Gorino Ferrarese, Italy, in collaboration with CADF (Consorzio Acque Delta Ferrarese), the water utility serving the city. At the core of our solution is SWaMM (Smart Water Metering Middleware), an interoperable wireless IoT middleware based on the Edge computing paradigm, which proved extremely effective in interfacing with several types of smart water meters operating with different protocols.

**Keywords:** Internet-of-Things; smart metering; water consumption

## 1. Introduction

The efficient use of natural resources is increasingly important, especially with the current trends towards a larger population mainly concentrated in urban environments (smart cities and megacities) [1,2]. Several impact analyses have proven that smart metering solutions are highly beneficial to society, as they are very effective both in engaging citizens to reduce excessive consumption and in detecting resource waste [3–5].

There is a growing interest in smart metering applications [6], that have initially focused on the energy [7] and gas [8] utility markets, which deal with the more expensive resources. Recently, a growing interest has focused on smart metering of water consumption [5]. Initial smart water metering solutions have proved very useful in order to identify and prevent water leakages, which is an increasingly important concern. In fact, while water represents a cheaper commodity than gas in many countries and its leakage does not present any hazard, water leakages in the residential and distribution network are very frequent and difficult to detect. The growing interest in sustainability in modern society has been stimulating and fostering the development of the smart water metering market to fight water leakages [9–11].

However, so far, research has dedicated a limited effort to the investigation of smart water metering solutions. This is unfortunate, because smart water metering presents interesting challenges both at the scientific and at the engineering levels. More specifically, we can identify three main problems to address.

First, lacking an open and widely accepted communication standard in the smart water metering market, most companies manufacturing smart water meters propose proprietary solutions. In addition, the smart metering market is quickly evolving from the communications perspective: while a few years ago smart meters only used the Wireless M-Bus communication protocol, more recent meter also adopt LoRa, and the first NarrowBand IoT (NB-IoT) solutions are emerging. This results in vendor and protocol lock-in issues for water utilities, which might lead to high expenditures (both CAPEX and OPEX) and ultimately hinder innovation.

Second, a significant part of the water meters installed at the end user level are of the traditional mechanical type, designed to display water consumption visually and read by a human operator. These "dumb" meters have a very large installation base and still represent a cheap and well tested solution. As a result, they are likely to play a non-negligible role in the water utility market for (at least) the near future. This means that smart water metering solutions designed for large scale deployment need to have the capability to operate with dumb meters, possibly implementing sophisticated automated reading solutions.

Finally, there is the need to address practical aspects connected to real-life deployment of large scale smart water metering solutions. More specifically, how to address radio frequency propagation issues and how to set the reading frequency (a trade off between observability and energy consumption) represent issues of major importance. Reports on practical knowledge developed on the field, which are unfortunately, very rare at the moment of this writing, would help designers to build improved smart water metering solutions.

This paper attempts to bridge this gap by introducing a comprehensive smart water metering solution, which we designed in collaboration with several Italian water utility companies to address specifically the peculiarities and requirements of the smart water utility market. This paper also reports on our extensive field testing experiences. Our solution leverages the Edge Computing approach to distribute data pre-processing functions in proximity of the water meters and takes advantage of modern Commercial Off The Shelf (COTS) IoT technologies, both at the hardware and at the (open source) software level, to reduce costs and facilitate interoperability.

At the core of our solution is SWaMM (Smart Water Metering Middleware), an interoperable wireless IoT middleware that is capable of interfacing with a wide range of smart water meters operating with different protocols. SWaMM Edge Gateway devices running the SWaMM Edge component collect water consumption data from the smart meters, with a dynamically tunable frequency that allows tradeoffs between observability and energy saving, preprocess the data, transcoding it if needed and aggregating it to consolidate transmissions, and forward it to the SWaMM Cloud platform, which further analyzes the data to return useful consumption pattern information both to the users and to the water utility.

Our solution was thoroughly tested in the context of two field experiments. First, we deployed SWaMM to perform the smart water metering of all residential houses in Gorino Ferrarese, Italy. In collaboration with CADF (Consorzio Acque Delta Ferrarese), the water utility serving the city, all the mechanical water meters in Gorino Ferrarese were replaced with smart ones. In addition, we deployed several gateways to retrieve consumption data from the smart water meters and forward it to the SWaMM Cloud platform.

In addition, we tested the capabilities of our solution to integrate with an existing water metering infrastructure by deploying SWaMM to implement the smart water metering of selected residential homes in Ferrara, Italy. More specifically, we selected several residences in Ferrara that represented frequent use cases (apartment in the historical city center, apartment and a single home in the suburbs) with different water meters.

From the initial experimentation to the deployment in the field, SWaMM has shown a remarkable interoperability and was capable of interfacing with a wide range of water meters. Not only SWaMM interfaced with smart meters adopting either the Wireless M-Bus or the LoRa communication protocols but also with traditional mechanical meters retrofitted with an Optical Reader Kit that we designed to integrate SWaMM with older water metering installations. Finally, the analysis of the data collected implemented by the SWaMM platform has proven very effective in detecting leakages.

The rest of the paper is organized as follows. Section 2 describes the smart water metering field, discussing proprietary and open solutions to address water metering data collection, also including a description of different types of water meters available on the market. Section 3 presents the design of our comprehensive and interoperable solution for smart water metering based on the SWaMM wireless IoT middleware. Section 4 discusses the architecture of SWaMM and describes (in detail) the SWaMM Edge Gateway and the SWaMM Optical Reader Kit components, that respectively deal with the data collection and elaboration at the edge and with the retrofitting of dumb water meters. Section 5 provides a thorough evaluation of our solution by reporting the results of several field experiments, including a relatively large deployment in Gorino Ferrarese, Italy, for the monitoring of water consumption in domestic residences. Section 6 discusses the advantages of open and interoperable smart water metering solutions such as SWaMM and their potential impact in a growing and competitive market. Section 7 illustrates related work in the smart metering field, highlighting the differences between those solutions and SWaMM. Finally, Section 8 concludes the paper and discusses future work.

## 2. Smart Water Metering: From Proprietary to Open Solutions

Both providers and consumers present common interests in the smart water metering market [3]. On the one hand, providers are interested in having a fully automated data collection to reduce operational costs associated to human interventions, to produce billing information based on effective/accurate data, and to have a larger control over distribution networks and delivery points. On the other hand, consumers have the opportunity to have frequent water metering data to increase consumption awareness and adapt their behaviour accordingly [4], i.e., avoiding waste of resources and being notified about possible warnings and leakages.

With respect to the energy and gas markets, smart water metering presents important criticalities. In fact, water usually tends to be much cheaper than other commodities (gas, electricity), and thus, resulting in less investments and efforts in the market. Another interesting criticality is that the water meters are usually located in underground and/or protected areas (inspection pits), which do not have an electricity coverage for obvious safety reasons. This results in systems that need to be battery powered and extremely energy efficient to provide at least an acceptable average lifetime period. In addition, these underground areas may also present radio frequency propagation issues, which significantly reduces the strength and the range of wireless transmissions. However, the increasing reduction of water availability [12] due to several factors such as climate changes and population growth [13] and significant technological advances in embedded systems fostered important innovations in smart metering solutions for the water utility market.

### 2.1. Smart Water Meters

Despite smart water metering being a relatively recent practice, there are several smart water meters available on the market. The first generation of smart water meters adopted low power short-range wireless protocols, such as Wireless M-Bus, that operates over the unlicensed spectrum (169 or 868 MHz in Europe). These meters were designed to operate in Remote Meter Reading (RMR) systems, in which data collection can be performed without a dedicated networking infrastructure: operators equipped with portable receivers collect data in the proximity of the smart meters either in walk-by or drive-by mode. RMR systems eliminate the need for physical access or visual inspection of smart meters but do not allow either real-time or automated consumption monitoring.

More recently, a second generation of smart water meters that leverage on low-power long-range wireless protocols, such as LoRa (Long Range), which is designed for a wider communication range both in urban and extra-urban environments, hit the market [14,15]. These meters were specifically designed for Automatic Meter Reading (AMR) systems, in which data collection is fully automated: smart water meters periodically transmit their consumption information to gateway devices that gather the data from the in-range smart meters and retransmit it to the utility management typically using mobile (3G/LTE/4G) communications. To guarantee user privacy, smart water meters encrypt consumption information before the transmission.

Unfortunately, despite the large number of solutions available on the market, the lack of an open and widely accepted communication standard pushed vendors to propose proprietary data collection solutions. The adoption of proprietary solutions presents water utility companies with several significant problems in term of costs, vendor lock-in, and lack of control on the data collection infrastructure.

In fact, vendors typically propose proprietary and often very expensive gateways, that need to be acquired in addition to smart meters and whose cost might account for a non negligible portion of capital expenditure for water utility companies that want to implement an AMR system. This is further exacerbated in case short-range water meters are employed. In fact, while nothing prevents short-range smart meters to be adopted within AMR systems, their limited communication radius would require the deployment of a large number of gateways. Thorough planning and field testing would be needed in order to minimize the number of gateways required to collect data from all the smart meters, and in some cases the deployment of ad hoc signal repeaters to extend the radio coverage of smart meters would be required.

In addition, due to the lack of an open standard, smart meter vendors have typically adopted proprietary protocol specifications to transmit metering data over the network. As new smart meters need to be compatible with the existing monitoring solution, this may easily result in vendor and communication protocol lock-in problems. Lock-in risks are particularly high at the moment of this writing, in which smart water metering could arguably be classified at the "early adopters" stage of the technology life cycle. In fact, solutions on the market are rapidly evolving with respect to lower energy consumption and longer communication range, and the first NarrowBand IoT (NB-IoT) proposals are emerging. As a result, choosing proprietary solutions offered by a vendor now might hinder or preclude the adoption of significantly more convenient water meters from a different vendor in a few years.

Finally, some manufacturers have developed smart meters that are able to communicate both through an open OMS (Open Meter Standard) (https://oms-group.org/en/download4all/oms-specification/) radio protocol and a proprietary protocol, which usually can provide some additional information. Furthermore, on the metering market it is common for manufacturers to provide proprietary platforms designed to receive data from different types of smart meters using proprietary radio protocol (for their own smart meters) and also to collect data using the OMS protocol, and thus, enabling data collection from smart meters that support open standard. A representative example of a similar solution is the one provided by Sensus, which makes use of the proprietary radio protocol SensusRF to collect metering data from Sensus and other types of smart meters communicating via the OMS protocol and operate on frequencies on the 868 MHz band. However, such solution, equipped only with 868 MHz antennae, cannot receive messages from smart meters that operates on different frequencies, such as the WaterTech meters that communicates on frequencies on the 169 MHz band.

The proprietary nature of most smart water metering solutions on the market also has an impact in terms of a limited amount of control for adopters. The lack of access to the software running on gateway devices in particular represents a major obstacle to the implementation of innovative features for customers. For instance, smart metering systems are typically configured to operate with a specified data collection frequency, configured either at manufacturing or at installation time. The possibility to control the behavior of gateways could, for instance, allow to dynamically tune the frequency of data acquisitions, raising it in case a leakage is suspected.

## 2.2. Traditional "Dumb" Water Meters

For many decades, water meter manufactures have used only basic physical principles of measurements in the design of water meters. For instance, the Italian legislation forced, until a few years ago, water utilities and meter manufactures to adopt and develop metering instrumentation based exclusively on mechanical/dynamic measurement principles. This regulations and the low manufacturing costs of this type of water meters reflected in water metering parks mainly composed of traditional "dumb" measuring instruments. Most of the time, this instrumentation is composed of a turbine and mechanical gears. The water passing through the meter makes move the turbine, which is connected to mechanical gears that transform the rotational speed of the turbine in a flow measurement. This measure is then expressed in cubic meters ($m^3$), or fractions of $m^3$, on the consumption wheel of the water meter's dial. Therefore, this information must be read on the water meter's dial and it refers to the total volume (cumulative consumption) of water transited since the installation of the water meter.

Specifically, there are different models of mechanical water meters that differ in terms of the mechanism the meter uses and the typology of account it was designed for (residential or industrial). For example, single-jet meters are designed for residential accounts, multiple jet meters are used both on residential and industrial accounts, and Woltmann meters are used mainly for industrial accounts.

Since the smartness of a metering instrumentation is a characteristic associated to the ability of transmitting metering data and not to the instrumentation itself, it becomes possible to enable "smartness" even on traditional mechanical meters. In fact, in recent years, water metering manufactures have introduced communication modules capable of reading data from a pulse emitter connected to the mechanical meter and transmitting over a wireless radio protocol, e.g., Wireless M-Bus. For example, Sensus produces a mechanical water meter capable of integrating a pulse emitter (https://sensus.com/products/hri-mei/), which is wire connected to a battery-powered Wireless M-Bus radio module (https://sensus.com/products/sensus-pulse-rf/).

However, not every traditional dumb water meter can be equipped with a pulse emitter. In these cases, to enable an automated data collection from the whole metering park, water utilities can adopt two different strategies. The first one consists in massive water meter replacement campaigns, but such solution requires huge investments and operational costs. Finally, the other possible strategy is to equip traditional dumb water meters with an optical reading system that can analyze an image of a water meter's dial and transform it into a computer-processable consumption value.

The two strategies present different costs. Not considering the costs of adopting a proprietary data collection infrastructure, the cost of a modern smart water meter equipped with a radio module is about EUR 100. Instead, we calculated that retrofitting a traditional mechanical dumb water meter with an optical reader kit requires approximately EUR 30. However, the adoption of one strategy instead of the other mainly depends on management and business decisions of water utilities.

## 2.3. IoT as a Foundation for Open Smart Water Metering Solutions

These considerations call for innovative and highly interoperable solutions, based on open standards and technologies instead of proprietary ones and capable of reading consumption data from a remarkably heterogeneous set of water meters, either of the smart and of the dumb type, and to enable utility companies to adopt different technologies at the same time. For instance, a similar solution would allow companies to adopt new smart meters or communication protocols in a part of their distribution network while the rest of the network would continue to operate with previous technologies.

Fortunately, the astounding progresses achieved by the Internet-of-Things (IoT) revolution have provided inexpensive and high quality tools that represent a terrific foundation upon which to build a comprehensive and interoperable smart water metering solution capable of operating with different types of meters using different open wireless protocols and data formats.

With this regards, wireless IoT communication solutions represent compelling key enabling technologies for smart metering applications. In addition to the already mentioned Wireless M-Bus and LoRa protocols, which are being increasingly adopted by smart water metering manufacturers, there are a range of wireless IoT protocols that might be relevant for smart metering applications, with different capabilities in term of allowed bitrate, range of operation (distance), number of supported devices, and power consumption [16].

Within those, IEEE 802.15.4, and its full stack proprietary extension ZigBee, is widely adopted in low power networks and provides transmissions capabilities over a limited range (10–50 m) and it supports different network topologies: star, peer-to-peer, and cluster [17]. Bluetooth LE is a relatively recent and non backwards compatible version of Bluetooth specification, featuring increased communication range (up to 100 m) and lower power consumption, albeit at the expense of a lower maximum bitrate (1 Mbps) [18]. Another interesting wireless IoT protocol is Sigfox, a proprietary communication technology developed by the eponymous Sigfox company. Unlike LoRa, which is also proprietary but infrastructureless and free to use, Sigfox adopters need to leverage on a communication infrastructure provided by Sigfox, and pay corresponding licensing fees. On the top of some of these protocols it is possible to build larger wireless network infrastructures. For instance, LoRaWAN (Low Power WAN Protocol for Internet-of-Things) provides WAN communications on top of LoRa, which end-devices can use to communicate with gateways [19].

With the only notable exception of Sigfox, all these solutions allow to perform information exchange at the edge of the network on the unlicensed spectrum, and consequently to realize wireless smart metering solutions without depending on the support of network providers/operators or paying licensing fees [20].

In addition, modern Commercial Off-The-Shelf (COTS) hardware solutions are inexpensive and battle tested, ranging from powerful Single Board Computers (SBC) based on ARM microprocessors, which can run full operating systems such as Linux, to very low power ARM microcontrollers. The former could be used as a platform for the development of interoperable gateways; the latter can be used for energy constrained applications, for instance for signal repeating or optical reading on the smart water meter side. From the communication perspective, many messaging protocols, such as Message Queue Telemetry Transport (MQTT) and Advanced Message Queuing Protocol (AMQP), have been developed as open standards and have been thoroughly designed and tested in a plethora of real life applications. They could be employed for reliable communications in smart water metering applications. Finally, a plethora of COTS open source software components, including implementations of MQTT and AMQP messaging systems, can be easily reused for IoT applications. Those components would be particularly well-suited for the realization of smart water metering systems, enabling a rapid application development approach and an impressive reduction in time to market.

## 3. Design of SWaMM

Edge Computing is a relatively recent paradigm that allows IT service developers and providers to allocate some data-processing tasks at the edge of the network on the top of different edge devices (IoT gateways, Cloudlets, Micro-Cluds, etc.), with the potential of significantly reducing services' latencies and improve the quality of those services, and thus, enabling a better utilization of both hardware resources and network bandwidth by limiting the communications with Cloud Computing platforms. We designed SWaMM in accordance with the Edge Computing paradigm to enable data processing directly at the edge of the network, thus, exploiting the available computational resources used for the collection of the metering data and also reducing the number and the size of data transmissions over the cellular network.

According to the Edge Computing paradigm, IoT applications run on devices at the edge (IoT gateways, cloudlet, micro-cloud, etc.) to exploit the proximity of users and data and provide better

QoS and QoE to the users of those applications. Exploiting the proximity of users and devices is one of the key ideas behind the realization of Edge Computing solutions. In fact, by processing partially or completely the data gathered by IoT sensors and devices at the edge is possible to create more effective and responsive services, thus, enabling a better utilization of both hardware resources and network bandwidth by limiting the communications with Cloud Computing platforms.

Figure 1 depicts the design of SWaMM and the information workflow within the middleware. At the lowest level we have the water meters (smart and dumb), which measure the consumption of the commodity (water) of the monitored accounts. As shown in Figure 1, SWaMM is capable of interfacing with different types of water meters that communicate at the edge of the network with different wireless protocols (in Figure 1 LoRa and Wireless M-Bus) and different ad hoc data formats. More specifically, Figure 1 depicts three different types of water meters: smart water meters, which were designed with radio capabilities, dumb water meters retrofitted with optical reading kits, and dumb water meters equipped with pulse emitters. Smart water meters transmit the cumulative consumption, recorded up to a certain time instant, (expressed as volume in m$^3$) using different wireless protocols on configured radio frequency channels. Instead, the dumb water meters illustrated in Figure 1 represent the case of already deployed water meters not provided with automatic reading capabilities by design, but integrated with optical reader kits or pulse emitters to provide them the smart capabilities of transmitting consumption information to the SWaMM Edge Gateways located at the edge in proximity of those metering devices.

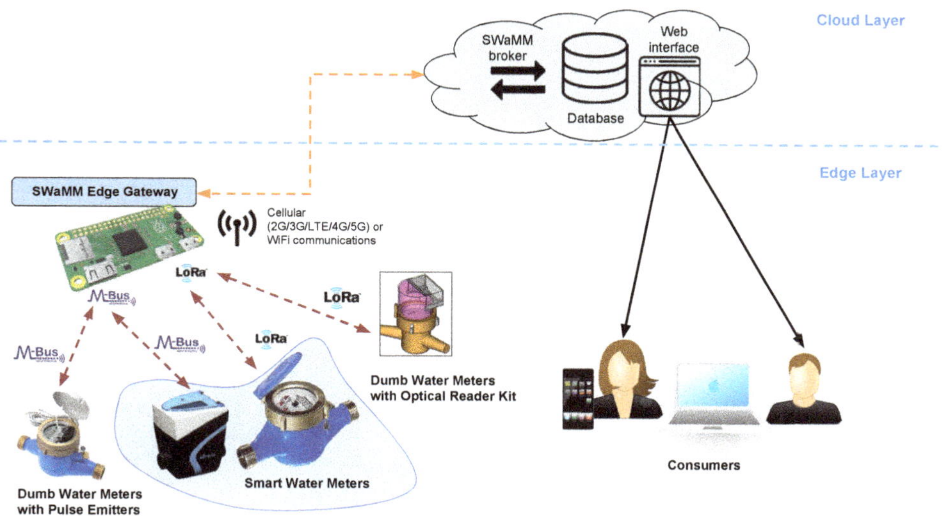


**Figure 1.** Holistic view of the SWaMM solution: from the smart and traditional water meters to the end-users (consumers and water utilities).

At the other end, we have multiple SWaMM Edge Gateways running at the edge (as shown in Figure 2) that we specifically designed to receive, concentrate, and process data from the in-range available wireless meters. With regards to the smart meters currently available on the market, we developed the SWaMM Edge Gateways to receive both LoRa and Wireless M-Bus meters by carrying one or more antennae capable of receiving meters from the two different protocols. Considering the transmission power of these smart meters, the gateways are usually located in strategic points in order to maximize the number of covered smart meters and to minimize the number of required gateways.

However, we designed the SWaMM Edge Gateway without coupling its components to a specific wireless protocol, thus, enabling the easily integration of its sensing capabilities to different type of water meters and different IoT wireless protocols that would be available in the future, and thus, allowing to avoid vendor and protocol lock-in. For instance, to allow the OCR (Optical character recognition) processing at the edge, we used relatively powerful SBCs (Raspberry Pi 3) as a hardware platform for the realization of the gateways. In fact, SWaMM Edge Gateways can also analyze data from dumb water meters using OCR techniques directly at the edge to enable data collection from old meters, which do not have automated reading capabilities, thus, avoiding to transmit pictures over the cellular network to a Cloud Computing platform for elaboration.

The gateways store and pre-process the metering data at the edge before uploading via WiFi or the cellular network (3G/4G/LTE) the aggregated data to an application running on a Cloud Computing platform, which makes further elaborations of the data by means of algorithms for consumer profiling, leakages and fault detection, and so on. Data about consumption, alarms, and profiling are then made available to the end-users of the monitoring system (utility management and consumers) via a personalized web application.

With this design, we tried to achieve a comprehensive solution, which enables an automated metering data collection from different water meters available on the market and enable interoperability from a water utilty perspective that need to collect and elaborate data from its entire metering park. SWaMM offers capabilities that result in a detailed monitoring of the water distribution system capable of detecting district and/or accounts leakages, and thus, lets water utilities to optimize the water distribution system and schedule maintenance interventions.

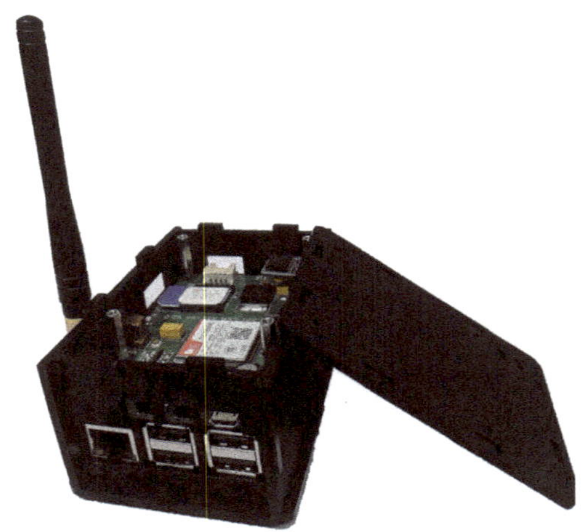

**Figure 2.** SWaMM smart metering device: SWaMM Edge Gateway, leveraging a Raspberry Pi 3 board with LTE Module.

## 4. Middleware Architecture of SWaMM

We designed SWaMM using open source technologies with the aim of creating an open solution to tackle the entire flow of a smart metering system from data collection to data elaboration. Furthermore, we exploited edge computing solutions by developing ad hoc gateways for the processing and aggregation of the metering data at the edge. In addition, having intelligent devices at the edge also enables the possibility of a granular management of the smart meters, thus, allowing to easily control all the components of the metering system.

### 4.1. SWaMM Edge Gateway Architecture

The SWaMM Edge Gateway is a data collection and elaboration kit capable of easily integrating with a wide spectrum of different types of smart meters and wireless protocols located at the edge of the network. The kit allows to collect and aggregate metering data from the nearby smart meters, to run OCR algorithms on pictures collected from old water meters, and to publish the data processed at the edge to the SWaMM elaboration platform running on the Cloud.

Figure 3 depicts the several tiers composing SWaMM Edge, the software architecture running on a SWaMM Edge Gateway. SWaMM Edge has been developed on several tiers to facilitate the development and the debugging of the system and at the same time to simplify future updates and/or extensions to the smart metering middleware. We chose to implement all the components using the Python programming language version 3.5.4 to exploit the wide range of available and open source libraries and the useful help from the international community.

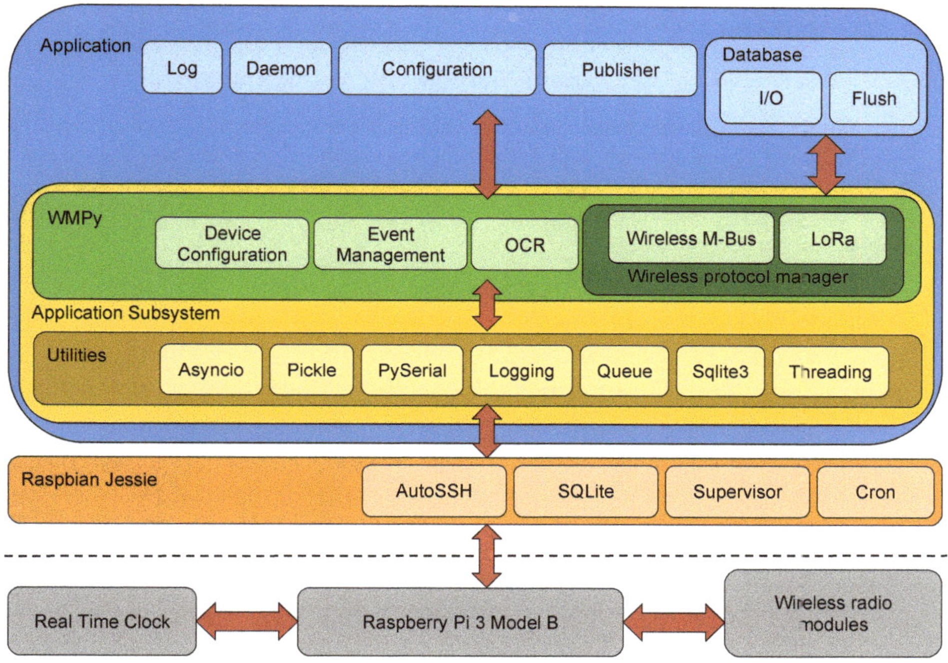


**Figure 3.** SWaMM Edge software architecture of SWaMM.

Figure 3 provides an holistic view of SWaMM Edge's architecture, including the blocks to deal with wireless modules, serial ports, and data-aggregation and elaboration. This platform runs on the top of Raspbian Jessie, the operating system we chose for the Raspberry Pi Model 3. Delving into the software architecture, the application subsystem is responsible to deal with the low level functionalities provided by the OS and the additional hardware, and at the other end to interface with the high-level functionalities (configuration, storage, etc.). In addition, the application subsystem provides several utilities used in the development of the smart water metering middleware. Specifically, the most important modules composing the application subsystem are:

- Logging for application monitoring and collection of the data;
- Threading and Asyncio, for the correct execution of the multithreaded application;
- Pyserial for interfacing with serial ports;

- Sqlite3 for interfacing with the SQLite relational database;
- Pickle for serializing and deserializing Python objects.

At the core of the application subsystem is the Water Management Python (WMPy) module, a library specifically designed to interface the SWaMM gateway with the wireless communication kits: Wireless M-Bus and LoRa. This library provides functionalities to deal with event management, configuration, devices monitoring, and communications management (sending and receiving messages over Wireless M-Bus and/or LoRa). The support for other meter types requires the development of a new plugin component for the Wireless protocol manager which provides the support for the specific communication protocol and data format used by that meter. In addition, another important piece composing the WMPy library is the OCR module. This module is capable of running OCR algorithms to detect the cumulative consumption of a water meter by processing the picture of its consumption wheel. To this end, we leveraged on the Google Tesseract OCR (https://opensource.google.com/projects/tesseract) software capabilities of analyzing and retrieving a text from an image, in this case the cumulative consumption value [21].

Then, on the top of the Application Subsystem we have several other software modules to implement data collection and aggregation. In detail, the more important modules are:

- Configuration management: this module keeps track of a list of authorized devices that determines whether or not a gateway has to analyze data sent by a specific smart meter. This module is also responsible to tune the configurations of the associated gateways;
- Daemons: processes running continuously and controlled by Supervisor, developed to listen for Wireless M-Bus and LoRa messages;
- Log: a logging module to keep track of the outgoing system's events to enable history tracking and to identify bug and system faults;
- Database management: two processes for managing the database, one for saving the received data and the other one to flush the database;
- Publisher: for sending the pre-processed data (over the cellular or WiFi network) to the Cloud Computing platform.

Finally, processing the metering data is not computationally expensive since it requires only to parse and elaborate the received message. Furthermore, messages are received and buffered/stored on an in-memory buffer waiting to be processed, and thus, avoiding to lose metering data. Finally, possible limitations regarding the number of supported smart water meters mainly depend on protocol specifications, collision and interference avoidance mechanisms, and radio frequency propagation issues, e.g., the presence of obstacles that can block the propagation of the wireless transmissions [22,23].

### 4.2. SWaMM Optical Reader Kit

To enable data collection from old water meters we specifically designed the SWaMM Optical Reader Kit (ORK). The SWaMM ORK is a sort of smart cap that can be assembled on different type of dumb water meters. The SWaMM ORK provides "smartness" to old dumb meters, which do not support other smart reader mechanisms e.g., a pulse emitter, by means of a radio module that transmits the picture of the consumption wheel visible on the water meter. In detail, Figure 4 illustrates the three dimensional model of a mechanical "dumb" water meter equipped SWaMM ORK.

To realize SWaMM ORK we developed the illustrated prototype by analyzing the factor form of the most common used water meters to make this cap suitable for collecting data on them. As illustrated in Figure 4, this cap contains a battery powered micro-controller equipped with a camera and a LoRa radio module, for wireless transmissions. More in detail, Figure 5 depicts the three dimensional model of the cap by highlighting the camera and its factor form.

SWaMM ORK can be configured to take a snapshot of the meter's consumption wheel with a desired frequency, e.g., each day, and to transmit the resulting JPEG image to the nearby SWaMM

Edge Gateway through the LoRa radio module for elaboration. Before transmitting the picture, the SWaMM ORK converts the JPEG image in gray scale in order to reduce the size of the data to be transmitted over LoRa.

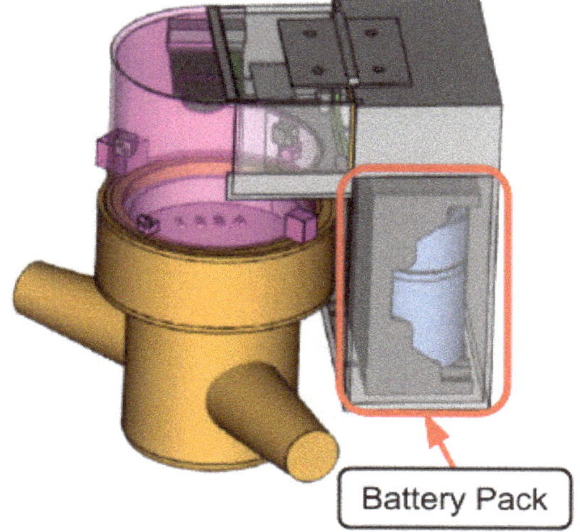


**Figure 4.** Three dimensional model of a "Dumb" water meter equipped with the SWaMM CRK module.

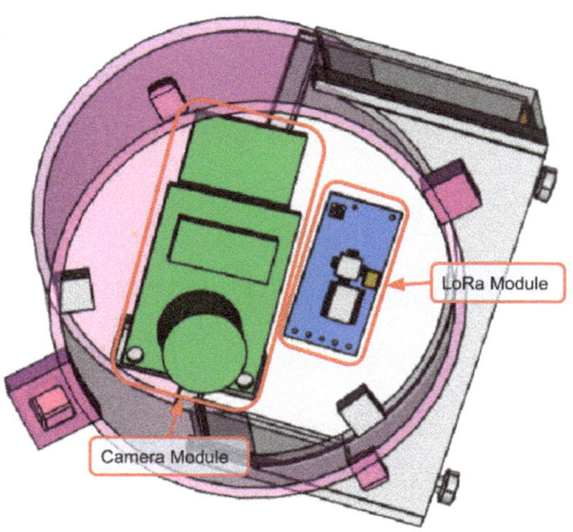


**Figure 5.** Three dimensional model of the SWaMM ORK module.

### 4.3. *SWaMM Platform*

Figure 6 illustrates the architecture of the SWaMM Platform, a comprehensive application for metering analysis and elaboration composed of seven software modules that, following the classic three-tier paradigm, uses the functions offered by a DBMS (Database Management System) for data management, and interfaces with the outside through a Web Server. The large amount of data to be stored, and subsequently analyzed, requires the adoption of tools that allow intelligent and efficient archiving, and which allow the analysis tools to quickly find the requested data. For this reason, we provided the SWaMM platform with a DBMS solution for the efficient management of data.

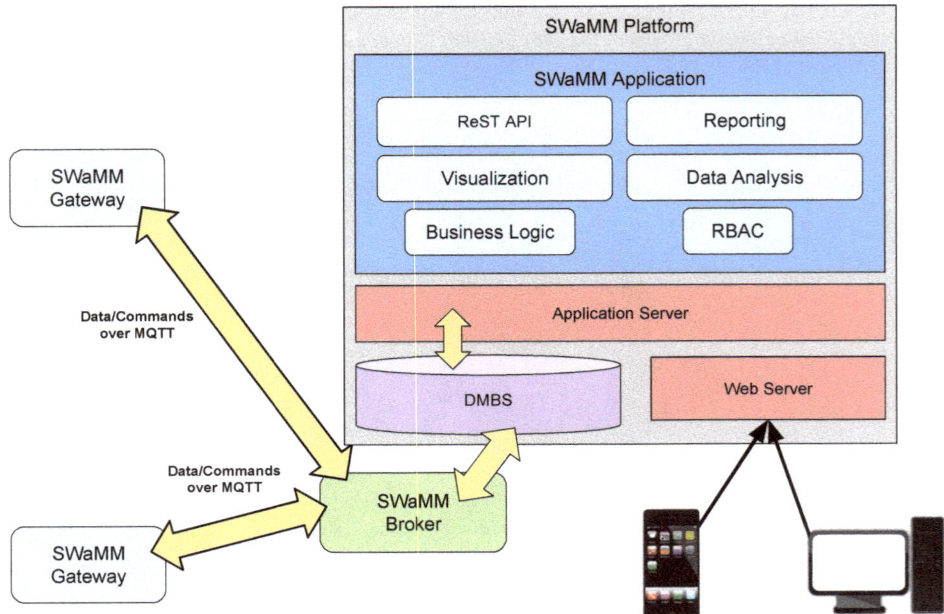


**Figure 6.** SWaMM Platform architecture.

The Business Logic, RBAC (Role Based Access Control) and CRM (Customer Relationship Management)/ERP (Enterprise Resource Planning) integration components are the business logic layer of the SWaMM platform. The Business Logic module implements the business logic of the platform and provides the management functions of gateways, including (re)configuration and firmware upgrades.

The RBAC module implements the authorization mechanisms to access the various functions offered by the SWaMM middleware, using an approach that allows the SWaMM platform to realize different access levels, e.g., SWaMM system operators, providing a specific subset of functionality depending on the user's role. This module is quite important considering the sensibility of water consumption information. In order to preserve the customers' privacy only authorized subjects can have access to sensible data.

The ReST API (Representational State Transfer Application Programming Interface) component provides an access interface to the SWaMM platform functions specially designed to allow integration with other platforms and software systems. More precisely, the functions of the SWaMM platform will be accessible through the Web using an API based on the ReST architectural paradigm. To this regard, we report some example on how to interface with the SWaMM ReST API in Table 1. Using this API is possible to retrieve the information of all accounts or of a specific account, which is identified using an identification number (id). In addition, water utilities can retrieve information about water consumption, detected leakages and push software update notifications to the SWaMM Edge gateways.

This will not only automate the export of metering data collected in an appropriate format based on XML (eXtensible Markup Language), JSON (JavaScript Object Notation) or CSV (Comma-separated values), thus, facilitating integration with sophisticated and powerful Big Data analysis tools, but also accessing "programmatically" all the functions of the platform, including the management of the SWaMM Edge Gateways. The SWaMM platform also provides advanced reporting functions, able to create, both on demand and in an automated way at scheduled deadlines, detailed technical documents that effectively highlight the state of the monitored accounts.

Finally, the Data Analysis module is the module responsible for data elaboration. In particular, this module analyzes the collected and aggregated data using fault and leakage detection algorithms that we specifically designed for the management of water distribution networks. Then, users can access the elaborated results via the Web Application, which exposes a user-friendly interface to inform about consumption and the presence of leakages on the monitored accounts.

**Table 1.** SWaMM ReST API.

| URI | HTTP Action | Behaviour |
|---|---|---|
| accounts/ | GET | retrieve the account collection |
| accounts/{data} | POST | insert a new account |
| accounts/ | PUT | replace (or create) the account collection |
| accounts/ | DELETE | delete the account collection |
| accounts/id | GET | retrieve the account with the id |
| account/id | PUT | replace the account with id |
| accounts/id | DELETE | delete the account with id |
| consumption/accounts/id | GET | retrieve the account's consumption information |
| leakages/accounts | GET | retrieve information about all detected leakages |
| leakage/accounts/id | GET | retrieve information about leakages on account with id |
| management/gateways/status | GET | retrieve gateways status information |
| management/gateways/status/id{message} | GET | retrieve status information of gateway with id |
| management/gateways/update{message} | POST | push an update notification to gateways |

SWaMM Broker

The SWaMM Broker is the component of the wireless metering middleware that deals with all aspects related to the management of interactions between the gateways and the SWaMM elaboration platform. The SWaMM Broker allows a spatial and temporal decoupling of the communications between SWaMM Edge Gateways at the edge and the SWaMM platform running on the Cloud, which makes the platform significantly more robust, since the various components can continue to work even in case of temporary component failures.

In addition, we believe that the adoption of a broker-based solution has many advantages, including the ability to transmit metering data and commands over different communication protocols, such as AMQP (Advanced Message Queuing Protocol) or MQTT (Message Queuing Telemetry Transport), and the ability to enable an efficient communication between software components with remarkably different performance. Furthermore, the SWaMM Broker also takes care of saving and storing the data by interacting with the DataBase Management System. We chose to use MQTT because of its simplicity and its small overhead. In fact, a single server can support thousands of remote clients. Furthermore, MQTT allows high scalability from the point of view of the development and management of applications. We chose RabbitMQ server side because we found it reliable and fully configurable, and thus, well suited to be installed on a Cloud Computing platform. At the same time we chose Mosquitto at the client side because is open source, well documented and supported, and it is the state of the art for implementing MQTT client applications.

In addition, the SWaMM platform makes use of the broker to interact with the SWaMM Edge Gateways by sending management commands for tuning the behaviour of the smart meters associated to the gateways. The broker exchanges both data and command messages via MQTT in order to guarantee the reliability of the platform, thus, limiting errors due to components and network failures.

Finally, MQTT topics follow the style illustrated in Table 2. According to the described architecture, a SWaMM Edge Gateway both publishes and subscribes to the SWaMM broker. First, a gateway publishes the processed data regarding the cumulative consumption value of each associated account on the consumption topic according to the format described in Table 3. This message contains all the information regarding a single meter reading needed for further elaboration. We chose to use a different topic for each account since this choice simplifies data management and elaboration (data is already organized and do not need to be divided). Second, SWaMM Edge Gateways subscribe to the management topics for receiving management commands from the SWaMM platform. Here commands can be of different types such as configuration and software update. For example, if a new functionality is required, the SWaMM platform publishes a software update message to the *gateway/management/update* topic. Then, at the reception of this message, gateways will download and install the update locally.

**Table 2.** Topic structure for MQTT messages.

| Information | MQTT Topic |
|---|---|
| cumulative consumption value related to account number | consumption/account_number |
| update commands | gateway/management/update |
| configuration commands | gateway/management/configuration |

**Table 3.** Wireless M-Bus and LoRa data delivery results.

| Location | Wireless M-Bus Delivered Packets (%) | LoRa Delivered Packets (%) |
|---|---|---|
| Location 1 | 53 packets—36.80 % | 133 packets—92.36 % |
| Location 2 | 58 packets—40.28 % | 137 packets—95.14 % |
| Location 3 | 94 packets—65.28 % | 141 packets—97.92 % |
| Location 4 | 0 packets—0 % | 121 packets—84.02 % |
| Location 5 | 76 packets—52.78 % | 135 packets—93.75 % |

## 5. Experimental Evaluation

Within the GST4Water project (Green Smart Technology for Water), we thoroughly evaluated our solution and proved the validity of SWaMM in the city of Gorino Ferrarese (Ferrara, Italy) [24,25]. As part of the GST4Water project, we installed the SWaMM solution over the residential accounts of the water district of Gorino Ferrarese in order to monitor water consumptions and identify possible leakages. In addition, we evaluated the performance of the OCR module implemented within the SWaMM Edge Gateway to recognize correctly the cumulative consumption values reported on old dumb water meters.

### 5.1. Field Testing with W-Mbus and LoRa Smart Meters

Part of these tests focused on the evaluation of Wireless M-Bus smart water meters to identify protocol performances, the number of required gateways, and other configurations. In detail, we used the Sensus iPerl smart water meter (https://sensus.com/products/iperl-international/), which has a built-in Wireless M-Bus module that we configured to transmit on the 868 MHz frequency a meter every 15 min. Instead, with regards to the SWaMM Edge Gateways, we configured the Raspberry Pi 3 with the IMST (iM871A-USB) USB Wireless M-Bus dongle to receive the Wireless M-Bus messages from the smart meters. In particular, we expected to experience radio frequency propagation issues due to the locations of these smart meters. In fact, as usually in Italy, these smart water meters were located in underground inspection pits as the one shown in Figure 7. At the other end, we installed (for testing purposes) the gateway inside the consumers' houses.

**Figure 7.** Example of an inspection pit in Gorino Ferrarese, Italy.

The following tests report the data collected from five different locations; each one characterized with distinct peculiarities:

- Location 1: the smart meter was located inside a cast iron covered inspection pit, which was 22 m away from the gateway installed at the third floor of the house;
- Location 2: the smart meter was located inside a cast iron covered inspection pit, which was 15 m away from the gateway installed at the third floor of the house;
- Location 3: the smart meter was located inside a concrete inspection pit, which was 10 m away from the gateway installed at the ground floor of the house;
- Location 4: the smart meter was located inside a cast iron covered inspection pit, which was 20 m away from the gateway installed at the first floor of a two-family house. In particular, this location presents another issue related to presence of an house between the smart meter and the gateway;
- Location 5: the smart meter was located inside a cast iron covered inspection pit, which was 22 m away from the gateway installed at the third floor of the house;

The results reported in Table 3 regard two 36 h periods (one for Wireless M-Bus and the other one for LoRa, which correspond to 144 Wireless M-Bus packets.

As depicted in Table 3, the Wireless M-Bus protocol (configured on the 868 MHz) did not guarantee a reliable transmission of data in the above described configurations. In detail, we achieve a 65.28% delivery ratio in location 3, when the smart meter was located inside a concrete inspection pit relatively close to the gateway (10 m). On the other hand, the characteristics of location 4 resulted in a 0% delivery ratio. The other results indicate that only part of the sensed data was delivered to the SWaMM Edge Gateways (from 36.80% to 52.78%). Possible solutions to overcome these results are to increase the data collection frequency, to guarantee the collection of a sufficient amount of data, or to tune Wireless M-Bus protocol specific configurations, e.g., increase the transmission power. However, both solutions present drawbacks related to an increased battery consumption, which results in a reduced smart meters lifetime.

To solve these data delivery issues and increase the reliability of the data collection system, we used an ad-hoc designed bridge to transmit over LoRa the Wireless M-Bus messages. During these tests, we noticed a clear delivery ratio improvement, as depicted in Table 3.

According to these results, we believe that the LoRa protocol outperforms the Wireless M-Bus protocol in all those situations where the outdoor smart meters are installed in underground inspection pits or they are particularly distant from the gateways where the presence of natural shielding conditions considerably reduce the range of those smart meters.

### 5.2. Water Leakage Detection

We collected and analyzed the consumers' consumption data to demonstrate the validity of SWaMM as smart water metering solution. Using the collected data, the SWaMM elaboration platform running on the Cloud can identify leaks of different sizes and behavior at household level: from large leakages due to broken pipes, to small leakages due to the faulty operation of sanitary appliances. Furthermore, the collected data can be used to characterize the leak level in the distribution network. Indeed, leaks, both at pipe network and household level, nowadays represent an important problem and huge effort is required to characterize and reduce these leakages through development and application of new approaches [26–29]. In particular, in the case study considered at household level a total of about 200 leakages were identified, of which the 25% corresponding to large water leakages of more than 10 L/h and the 75% corresponding to small water leakages in the range from 1 to 10 L/h.

The SWaMM elaboration platform detects water leakages through an empirical algorithm based on the analysis of the hourly flow rates time series. This criterion has been defined taking into consideration the typical water consumption pattern of residential accounts. With this regard, a residential account is characterized by periods of higher consumption (peaks), which mainly occurs during the day, alternating with periods when water consumption decreases to zero (mostly during the night). Instead, if an account is characterized by a leak, the smart water meter will detect a continuous consumption, which will be marked as leakage by the monitoring system (incremental consumption that never stops during the whole day). Thus, the algorithm is set up to assess the presence/absence of leakages inside residential houses by looking for non-consumption in certain periods of the day. Examples of detected leakages are the ones depicted in Figures 8 and 9, which show a large water leakage due to a damage of the hydraulic system and a small water leakage due to a failure of a toilet box, respectively. In both Figures the water leakage is identified by the incremental consumption that never goes to zero.

On the other hand, water utility managers have the opportunity to analyze the collected data to optimize the management of the water supply network and to reduce real water loss levels, by applying the water balance method using the high resolution data. In fact, managers can estimate the leak level of the distribution network by monitoring the incoming and outgoing discharges from well-defined portions of the distribution network (using Supervisory Control And Data Acquisition (SCADA) systems) and by combining this information with the consumption trend of all users belonging to the same area. For example, Figure 10 shows the water balance with reference to the water distribution network of Gorino Ferrarese (FE). In detail, the blue line indicates the consumption of all users belonging to the monitored area, while the red line (dashed line) indicates the result of the combination of the incoming and outgoing discharges from the same area, measured through the proprietary SCADA system. If the distribution network does not present leakages, the two lines (illustrated in red and blue) should overlap. However, as shown in Figure 10, the distribution network is affected by water leak, since the combination of incoming and outgoing discharges at the district is always greater than the monitored consumption of all users during the illustrated period, and thus, identifying a water leakage in the water distribution network.

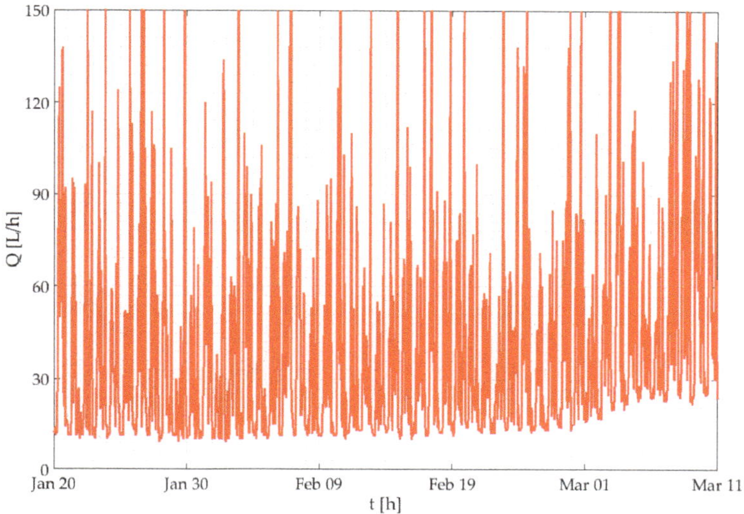


**Figure 8.** Hourly incremental consumption trends of a user affected by a large leak.

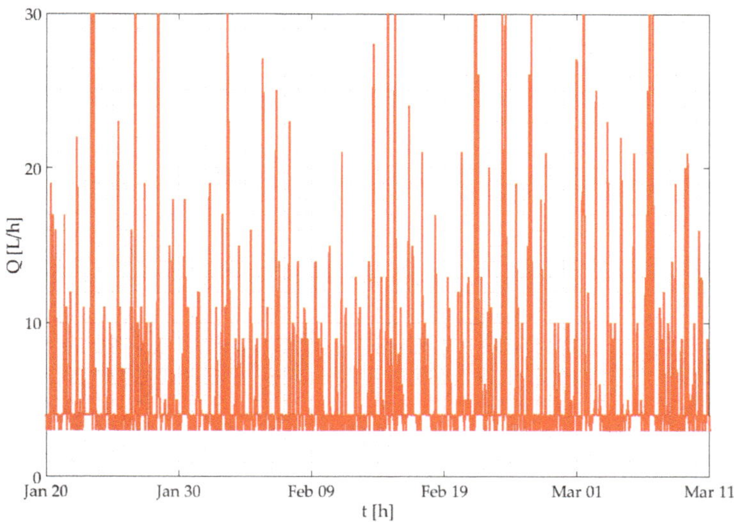


**Figure 9.** Hourly incremental consumption trends of a user affected by a small leaks.

In addition, the water utility has the opportunity to identify faulty smart meter devices immediately, reducing the apparent water losses due to measurement errors. For example, Figure 11a illustrates the total consumption trend of a residential account starting to decrease, and thus, indicating faulty data (the total consumption cannot decrease over time). This result is also confirmed in

Figure 11b, where the incremental consumption values are negative, which is impossible since the discharge cannot assume negative values.

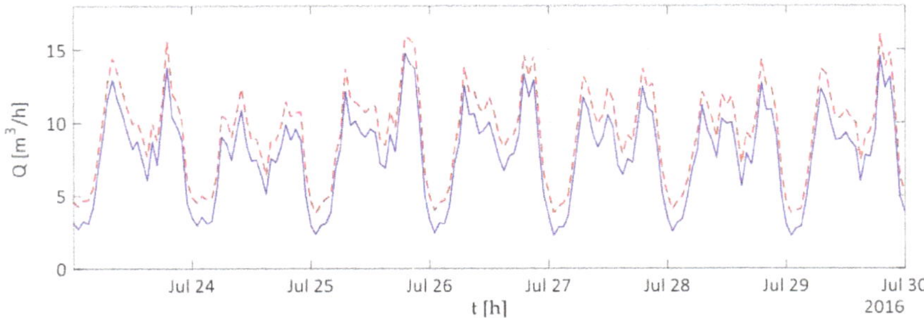


**Figure 10.** Water balance with reference to the district metered area of Gorino Ferrarese (FE) in July 2016.

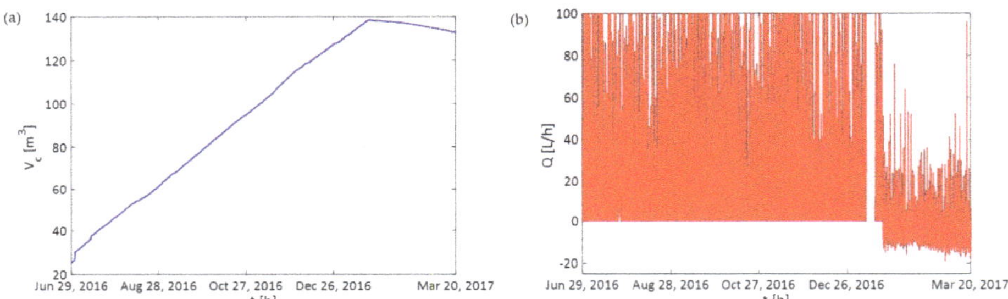


**Figure 11.** Totalized consumption trend measured by a faulty smart meter (**a**) and incremental consumption trend reported by the Cloud platform (**b**).

### 5.3. *Interfacing with Dumb Water Meters*

We conducted several tests to evaluate the performance and the quality of the OCR module in detecting the cumulative consumption on pictures taken from different types of water meters. Considering the wide variety of water meters deployed on the field, we consider for the scope of these tests only three types of water meters, which are the most common used in the city of Ferrara, Italy. Figure 12 depicts three different types of mechanical water meters and how each water meter is characterized with a particular form-factor and appearance, and thus, tending to influence the text recognition results of the OCR processing.

During these tests, we evaluated SWaMM ORK and the efficiency of the OCR module on dumb water meters put into operation, to verify its functionality during the time. To this end, we realized a simplified hydraulic network, as the one illustrated in Figure 13. Into this hydraulic network, an hydraulic pump puts the water meters at work, meanwhile the 'reader' components works continuously taking snapshots and transmitting pictures to the SWaMM Edge Gateway, located nearby the hydraulic network. Figure 14 depicts an example of images collected using the SWaMM ORK on the gateway side. As illustrated in Figure 14, SWaMM ORK converts light compressed JPEG into grayscale pictures before the transmission, in order to reduce the size of the data to be transmitted over LoRa and transmission time consequently.

On the SWaMM Edge Gateway side, the OCR module is initially instructed on the position of the digits to process each one individually. Initially, we used the standard Tesseract engine of Google

Tesseract 4.0. Unfortunately, applying the standard Tesseract engine on the meter's digits did not produce extremely brilliant results. A first cause was due mainly to the fact that Tesseract's neural network is trained to identify all the possible characters commonly used for writing (alphanumeric) for a multitude of fonts. In our case, instead, the interest have to be focused only for the specific fonts used on the water meters' consumption wheels. Another fundamental motivation is depicted in Figure 14b. The consumption wheel does not always show "entire" digits, but sometimes it could show a "digit transaction" state, which is a digit between two consecutive values. For example, in the least significant digit of Figure 14b, we find a transaction between value 2 and 3, meanwhile the remainder digits are all "entire" numbers.

**Figure 12.** Three types of mechanical dumb water meters.

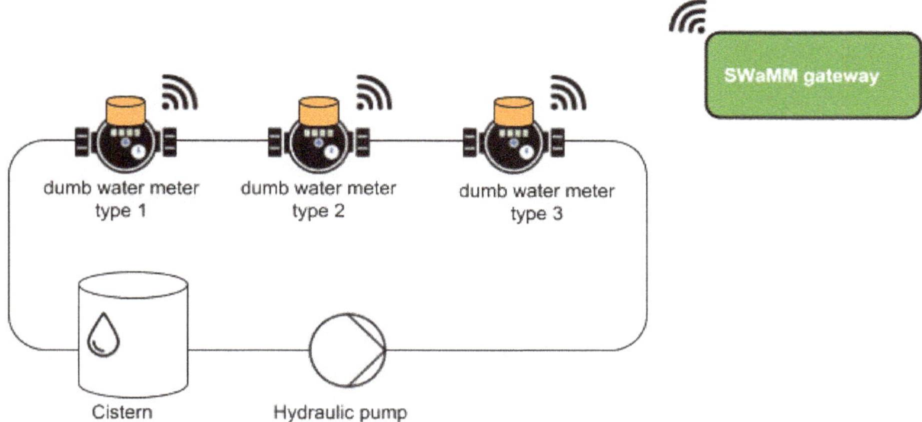


**Figure 13.** The hydraulic network setup.

For this reason, we decided to re-train the Tesseract LSTM neural network, using the pictures collected from the dumb water meters as a training dataset. After two days of system operation, we collected about 1400 images for each dumb meter, which included both entire digits and "digit transactions". Using the re-trained neural network, we achieved an accuracy of the results between 80 and 90 percent.

However, we tried to further improve the text recognition quality results. To this end, we designed a specific software component that works following the Tesseract processing, implementing an algorithm we called "validation process", which tries to improve the precision of the results obtained using Tesseract. We developed this validation algorithm according to the suggestions of the water utilities we collaborated with during the development and field testing of SWaMM. This algorithm tries to improve the text recognition accuracy by limiting the identification of result not corresponding to the reality, using the know-how of the typical average water consumption of residential accounts. In this way, the algorithm can validate or possibly correct the Tesseract results. For instance, in the event that a proposed result is not compatible with domestic consumption patterns, the validation algorithm runs a series of checks to identify what caused the possible recognition error. Errors can be differentiated in single digit recognition error (marginal) or multi-digit recognition error (substantial). In the first case, the algorithm tries to correct the possible error by leveraging on the Tesseract capabilities, while in case of substantial errors, the algorithm adopts a more conservative approach and keeps the value of the previous correct reading since it is not possible to interpret the result of Tesseract processing.

**Figure 14.** Different radials example. Dumb Meter 1 (**a**) Dumb Meter 2 (**b**) Dumb Meter 3 (**c**).

To demonstrate the functionality and utility of the validation process algorithm, we conducted a series of tests for the dumb meter of type 1, each time selecting 40 different images that depicted the consumption trends of this meter. In Table 4, we present the obtained results. The first case reports the recognition results obtained using only the re-trained Tesseract network. Instead, the second case shows the results applying our validation algorithm on the re-trained Tesseract outputs. Finally, the last two columns represent indicators that we calculated to estimate the size of the algorithms errors. In detail, deltaT represents the sum of the differences, expressed in $m^3$, between the Tesseract results and the correct ones. At the same way, deltaP represents the sum of the differences, expressed in $m^3$, between the validation process results and the correct ones.

As shown in Table 4, the validation process has a final 88.63% medium percentage of correct results of all the tests performed. Although it slightly improves the percentages of the medium re-trained Tesseract precision (87.66%), we find a good improvement on the errors size, from an average error of 2392 $m^3$ using only the re-trained Tesseract, to only 5 $m^3$ of average error for the validation process. These results are acceptable for water bill calculation. In fact, considering that billing procedures are usually performed monthly, or sometimes even bimonthly, the amount of readings pertaining to the billing time window tend to even out the error in a single cumulative water consumption reading. Therefore, an average error of 5 $m^3$ for a single cumulative water consumption reading is relatively small and well within the acceptance window.

Finally, we calculated using the *process_time* function (process_time function, https://docs.python.org/3/library/time.html#time.process_time) of the Python time module that the entire processing of the consumption wheel's picture (image elaboration and OCR processing) takes about 1261 ms (milliseconds) on the Raspberry Pi 3. This means that a SWaMM Edge Gateway can process around 7100 readings from Optical Reading Kits per day. Considering one or two readings per meter per day (sensible reading frequencies, which represent a good tradeoff between energy consumption at the Optical Reading Kit level and "observability" of the water consumption process) this means that a gateway could respectively support 7100 and 3550 water meters. However, radio frequency propagation issues and bandwidth limitations impact on the amount of data that can be transmitted and consequently reduce the number of water meters that can be served by a single gateway. These results seem to confirm the validity of the retrofitting solution in those situations in which replacing a traditional water meter with a new one is difficult. Therefore, the OCR processing does not present a limiting factor on the SWaMM Edge Gateway.

**Table 4.** Comparision between re-trained Tesseract network and validation process for dumb meter of type 1.

| Re-Trained Tesseract | | Validation Process | | | |
|---|---|---|---|---|---|
| **Errors** | **Right Results** | **Errors** | **Right Results** | **DeltaT ($m^3$)** | **DeltaP ($m^3$)** |
| 15.79% | 84.21% | 10.53% | 89.47% | 469 | 4 |
| 6.9% | 93.1% | 3.45% | 96.55% | 69 | 1 |
| 17.65% | 82.35% | 8.82% | 91.18% | 10,022 | 5 |
| 6.06% | 93.94% | 6.06% | 93.94% | 600 | 2 |
| 10.00% | 90.00% | 12.5% | 87.5% | 605 | 6 |
| 18.75% | 81.25% | 15.63% | 84.38% | 622 | 6 |
| 8.57% | 91.43% | 5.71% | 94.29% | 904 | 3 |
| 12.50% | 87.50% | 18.75% | 81.25% | 605 | 8 |
| 15.79% | 84.21% | 23.68% | 76.32% | 17 | 12 |
| 11.43% | 88.57% | 8.57% | 91.43% | 10,006 | 4 |
| **12.34%** | **87.66%** | **11.37%** | **88.63%** | **2392** | **5** |

## 6. Discussion

In the near future, the growing need for environmental sustainability is expected to push the need for smart metering solutions that avoid the waste of natural resources and increase the population awareness. To this end, smart water metering solutions provide many advantages for both consumers and water utilities. On the one hand, consumers can look over the time series of totalized and incremental consumption and the volume of water required during the last 24 h, thus, allowing real-time monitoring of their consumption and consequently to adopt more environmental friendly (virtuous) behaviors, thus, avoiding bad habits and waste of water. In addition, consumers can benefit of a more transparent water bill and avoid possible inconveniences caused by refund procedures, since the consumption is always determined on the basis of real data instead of a general forecast. Furthermore, this solution offers consumers the opportunity to receive an alarm message when the smart water metering platform detects leaks within the consumer's property, and thus, reducing waste of water even more.

Another advantage of using smart water metering platform is that the collected data can be used to implement water demand forecasting models for the entire district monitored areas. As a result, the general utility would be able to optimise the distribution network management by controlling the pressures of the distribution network and ensuring a more efficient management of the entire water district (control valves and pumping systems). In addition, a smart monitoring system also allows to reduce energy consumption and water losses, which are usually connected to an excessive amount of pressure in the distribution network. Finally, water utilities can make use of the collected data to plan marketing strategies, provide feedbacks on their water consumption or recommendations on personalized water-saving practices.

Our extensive field testing experience with SWaMM showed that the design of an efficient and automated smart water metering solution requires an intelligent analysis of the performance of wireless protocols, the optimal placement of gateways, and a thorough testing and evaluation phase. As we reported in the previous sections, some wireless protocols may perform significantly better than others when physical limitations can interfere with data collection procedures. On the other hand, protocols that enable longer ranges usually require a bigger power consumption, and thus, limiting the average lifetime of smart meters.

In the smart water metering market, interoperable wireless middleware solutions such as SWaMM, based on open technologies, both at the software and the hardware levels, and capable of dealing with different communication protocols and data formats, represent a key enabling technology. In fact, unlike stovepiped closed proprietary systems, SWaMM does not force to use a specific type of water meter, neither at the radio communication protocol nor at the meter type (smart or dumb) level, and allows water utilities to choose among different water metering technologies and to switch to different vendors. We believe that this capability will gain importance over the next years when new manufacturers will enter into the smart metering market, thus, influencing the heterogeneity in metering parks composition.

Finally, the use of SWaMM might open the market to 3rd party companies that operate between consumer and water utilities. These companies can operate as intermediary and offer metering solutions for collecting and elaborating metering consumption data, and thus, opening scenarios where the metering infrastructure is sold as a service.

## 7. Related Work

Smart metering is an active topic in research literature. Many efforts have been done to propose open and interoperable solutions for the smart metering of water and other commodities. This work [30] analyzes the applicability of the LoRa protocol for data collection purposes in a smart electricity metering system.

With focus on wireless protocols for smart metering solutions, [31] provides an accurate study of LP-WAN protocols for IoT, by describing performances of (LoRaWAN, NB-IoT, etc.) in different conditions such as small and large cities, rural areas, and so on.

In [32], Simmhan et al describe a service-oriented and data-driven architecture for Smart Utilities. In particular, this work describes, proposes, and evaluates protocols and solutions for IoT infrastructures in Smart Cities, also presenting real applications developed within a campus in Bangalore, India. Another interesting work is the one described in [33], in which the authors propose a smart metering infrastructure for water, electricity, and gas, by discussing possible wireless protocols for data collection, clustering techniques and prediction models for data elaboration and forecasting.

In [34], the author presents an interesting survey on smart grid and smart grid communication. The survey discusses smart grid and related technologies also by giving an exhaustive example of a smart energy infrastructure. Furthermore, the author analyzes wireline and wireless smart metering communication technologies needed to implement a measurement system, and the possible security threats and issues related to a smart grid infrastructure.

A smart water metering architecture is discussed in [35]. In this work, the authors describe an IoT architecture for smart water management by proposing a solution called the MEGA model. This work focuses mainly on the architecture and the interactions within the subsystem, but it also presents an implementation scenario in which the proposed solution is evaluated. [36] presents instead an IoT smart water monitoring system based on COTS hardware (Arduino and Raspberry Pi) to monitor the water level of tanks located across a campus.

SWaMM differs from the other solutions by proposing a comprehensive wireless middleware for smart water metering, which tackles the entire information flow of the metering data, from the sensing to the elaboration. Furthermore, we designed SWaMM to be open and interoperable with a wide range of wireless protocols, thus, enabling the integration of more reliable and innovative solutions.

## 8. Conclusions

We presented a comprehensive smart water metering solution that aims at tackling the heterogeneity of protocols and standards in a competitive and proprietary market and to avoid vendor and protocol lock-in. The core of this solution is SWaMM, a wireless IoT middleware for the collection and elaboration of metering data from a wide range of smart water meters. In addition, purposely developed Optical Reading Kits allow to integrate SWaMM with dumb water meters. SWaMM was thorougly tests on the field, also in the context of a full scale deployment in Gorino Ferrarese, Italy as part of the GST4Water project. The results obtained demonstrate the effectiveness of our middleware in reducing the waste of commodities over the water distribution network, by means of an accurate tracking and monitoring of consumptions.

Finally, let us note that, while SWaMM was specifically designed for smart water metering applications, its flexible design allows the middleware to be easily adopted for the smart metering of other commodities. In fact, SWaMM Edge Gateways could be used to collect data from other smart meters such as electricity and gas, as far as these smart meters will use supported wireless protocols. We believe that other IoT low power wireless protocol such as NB-IoT should be investigated and proved on the field. With this regard, we specifically designed SWaMM to be open and interoperable to future protocol extensions.

In addition to the experimentation with new water meter models and more sophisticated water leakage detection algorithms, future works will include the development of a dedicated OCR solution for dumb water meters that could provide better accuracy with respect to Tesseract, and the realization of a significantly improved version of the Optical Reader Kit, based on a low-power computing platform that allows it to operate for around 2 years on a couple of AA batteries.

**Author Contributions:** Conceptualization: all the authors; methodology: all the authors, formal analysis: C.L. and S.A.; software and validation: F.C., M.G., and G.R.; writing—original draft preparation, F.P. and M.T.; writing—review and editing: all the authors; visualization: F.C., M.G., C.L., F.P., and G.R.; supervision: S.A., C.S., and M.T.; project administration and funding acquisition: M.F.

**Funding:** This research was partially funded by Regione Emilia-Romagna within the context of the POR FESR 2014-2020 project "Green Smart Technology for water (GST4Water)".

**Conflicts of Interest:** The authors declare no conflict of interest.

## Abbreviations

The following abbreviations are used in this manuscript:

| | |
|---|---|
| IoT | Internet-of-Things |
| CADF | Consorzio Acque Delta Ferrarese |
| LoRa | Long Range |
| OCR | Optical character recognition |
| SWaMM | Smart Water Metering Middleware |
| DBMS | database management system |
| RBAC | Role Based Access Control |
| MQTT | Message Queuing Telemetry Transport |
| LoraWAN | Low Power WAN Protocol for Internet of Things |
| AMQP | Advanced Message Queuing Protocol |
| AMR | Automatic Meter Reading |
| RMR | Remote Meter Reading |
| CAPEX | Capital expenditure |
| OPEX | Operating expense |
| NB-IoT | Narrowband IoT |
| COTS | Commercial Off The Shelf |
| SBC | Single-board Computer |
| WMPy | Water Management Python |
| ReST | Representational State Transfer |
| API | Application Programming Interface |
| SCADA | Supervisory Control And Data Acquisition |

## References


1. Santos, J.; Vanhove, T.; Sebrechts, M.; Dupont, T.; Kerckhove, W.; Braem, B.; Van Seghbroeck, G.; Wauters, T.; Leroux, P.; Latre, S.; et al. City of Things: Enabling Resource Provisioning in Smart Cities. *IEEE Commun. Mag.* **2018**, *56*, 177–183. [CrossRef]
2. Willis, R.M.; Stewart, R.A.; Panuwatwanich, K.; Williams, P.R.; Hollingsworth, A.L. Quantifying the influence of environmental and water conservation attitudes on household end use water consumption. *J. Environ. Manag.* **2011**, *92*, 1996–2009. [CrossRef] [PubMed]
3. Romer, B.; Reichhart, P.; Kranz, J.; Picot, A. The role of smart metering and decentralized electricity storage for smart grids: The importance of positive externalities. *Energy Policy* **2012**, *50*, 486–495. [CrossRef]
4. Kaufmann, S.; Künzel, K.; Loock, M. Customer value of smart metering: Explorative evidence from a choice-based conjoint study in Switzerland. *Energy Policy* **2013**, *53*, 229–239. [CrossRef]
5. Boyle, T.; Giurco, D.; Mukheibir, P.; Liu, A.; Moy, C.; White, S.; Stewart, R. Intelligent Metering for Urban Water: A Review. *Water* **2013**, *5*, 1052–1081. [CrossRef]
6. Sharma, K.; Saini, L.M. Performance analysis of smart metering for smart grid: An overview. *Renew. Sustain. Energy Rev.* **2015**, *49*, 720–735. [CrossRef]
7. de Araujo, P.; Filho, R.; Rodrigues, J.; Oliveira, J.; Braga, S. Infrastructure for Integration of Legacy Electrical Equipment into a Smart-Grid Using Wireless Sensor Networks. *Sensors* **2018**, *18*, 1312. [CrossRef]
8. di Castelnuovo, M.; Fumagalli, E. An assessment of the Italian smart gas metering program. *Energy Policy* **2013**, *60*, 714–721. [CrossRef]
9. Britton, T.C.; Stewart, R.A.; O'Halloran, K.R. Smart metering: Enabler for rapid and effective post meter leakage identification and water loss management. *J. Clean. Prod.* **2013**, *54*, 166–176. [CrossRef]
10. Loureiro, D.; Vieira, P.; Makropoulos, C.; Kossieris, P.; Ribeiro, R.; Barateiro, J.; Katsiri, E. Smart metering use cases to increase water and energy efficiency in water supply systems. *Water Sci. Technol.* **2014**, *14*, 898–908. [CrossRef]
11. Giurco, D.; White, S.; Stewart, R. Smart Metering and Water End-Use Data: Conservation Benefits and Privacy Risks. *Water* **2010**, *2*, 461–467. [CrossRef]
12. Cosgrove, C.E.; Cosgrove, W.J. The Dynamics of Global Water Futures: Driving Forces 2011–2050. In *Volume 2 of The United Nations World Water Development Report*; UNESCO: Paris, France, 2012.
13. McDonald, R.; Douglas, I.; Grimm, N.; Hale, R.; Revenga, C.; Gronwall, J.; Fekete, B. Implications of fast urban growth for freshwater provision. *Ambio* **2011**, *40*, 437. [CrossRef] [PubMed]
14. Sanchez-Iborra, R.; Sanchez-Gomez, J.; Ballesta-Viñas, J.; Cano, M.-D.; Skarmeta, A.F. Performance Evaluation of LoRa Considering Scenario Conditions. *Sensors* **2018**, *18*, 772. [CrossRef] [PubMed]
15. Augustin, A.; Yi, J.; Clausen, T.; Townsley, W.M. A Study of LoRa: Long Range & Low Power Networks for the Internet of Things. *Sensors* **2016**, *16*, 1466.
16. García-García, L.; Jimenez, J.M.; Abdullah, M.T.A.; Lloret, J. Wireless Technologies for IoT in Smart Cities. *Netw. Protoc. Algorithms* **2018**, *10*, 23–64. [CrossRef]
17. Zigbee Specfication, September 2012. Available online: http://www.zigbee.org/wp-content/uploads/2014/11/docs-05-3474-20-0csg-zigbee-specification.pdf (accessed on 14 January 2019).
18. Gomez, C.; Oller, J.; Paradells, J. Overview and Evaluation of Bluetooth Low Energy: An Emerging Low-Power Wireless Technology. *Sensors* **2012**, *12*, 11734–11753. [CrossRef]
19. de Carvalho Silva, J.; Rodrigues, J.J.P.C.; Alberti, A.M.; Solic, P.; Aquino, A.L.L. LoRaWAN—A low power WAN protocol for Internet of Things: A review and opportunities. In Proceedings of the 2017 2nd International Multidisciplinary Conference on Computer and Energy Science (SpliTech), Split, Croatia, 12–14 July 2017; pp. 1–6.
20. Oliveira, L.; Rodrigues, J.J.P.C.; Kozlov, S.A.; Rabêlo, R.A.L.; de Albuquerque, V.H.C.D. MAC layer protocols for internet of things: A survey. *Future Internet* **2019**, *11*, 16. [CrossRef]
21. Smith, R. An Overview of the Tesseract OCR Engine. In Proceedings of the Ninth International Conference on Document Analysis and Recognition, Parana, Brazil, 23–26 September 2007; pp. 629–633.
22. Masek, P.; Zeman, K.; Kuder, Z.; Hosek, J.; Andreev, S.; Fujdiak, R.; Kropfl, F. Wireless M-BUS: An Attractive M2M Technology for 5G-Grade Home Automation. In *Internet of Things. IoT Infrastructures*; IoT360 2015; Lecture Notes of the Institute for Computer Sciences, Social Informatics and Telecommunications Engineering; Springer: Cham, Switzerland, 2016; Volume 169.

23. Mahmood, A.; Sisinni, E.; Guntupalli, L.; Rondón, R.; Hassan, S.A.; Gidlund, M. Scalability Analysis of a LoRa Network Under Imperfect Orthogonality. *IEEE Trans. Ind. Inform.* **2018**, *15*, 1425–1436. [CrossRef]
24. GST4Water, Project Website. Available online: https://www.gst4water.it/ (accessed on 14 January 2019).
25. Luciani, C.; Casellato, F.; Alvisi, S.; Franchini, M. From Water Consumption Smart Metering to Leakage Characterization at District and User Level: The GST4Water Project. *Proceedings* **2018**, *2*, 675. [CrossRef]
26. Puust, R.; Kapelan, Z.; Savic, D.A.; Koppel, T. A review of methods for leakage management in pipe networks. *Urban Water J.* **2010**, *7*, 25–45. [CrossRef]
27. Farley, M.; Trow, S. *Losses in Water Distribution Networks—A Practitioner's Guide to Assessment, Monitoring and Control*; IWA Publishing: London UK, 2003; pp. 146–149, ISBN 13 9781900222112.
28. IWA, Water Loss Task Force. *Best Practice Performance Indicators for Non-Revenue Water and Water Loss Components: A Practical Approach*; International Water Association: London, UK, 2005.
29. Britton, T.; Cole, G.; Stewart, R.; Wiskar, D. Remote diagnosis of leakage in residential households. *Water* **2008**, *35*, 56–60.
30. de Castro Tomé, M.; Nardelli, P.H.J.; Alves, H. Long-Range Low-Power Wireless Networks and Sampling Strategies in Electricity Metering. *IEEE Trans. Ind. Electron.* **2019**, *66*, 1629–1637. [CrossRef]
31. Ikpehai, A.; Adebisi, B.; Rabie, K.M.; Anoh, K.; Ande, R.E.; Hammoudeh, M.; Gacanin, H.; Mbanaso, U.M. Low-Power Wide Area Network Technologies for Internet-of-Things: A Comparative Review. *IEEE Internet Things J.* **2018**. [CrossRef]
32. Simmhan, Y.; Ravindra, P.; Chaturvedi, S.; Hegde, M.; Ballamajalu, R. Towards a data-driven IoT software architecture for smart city utilities. *Softw. Pract. Exper.* **2018**, *48*, 1390–1416. [CrossRef]
33. Lloret, J.; Tomas, J.; Canovas, A.; Parra, L. An Integrated IoT Architecture for Smart Metering. *IEEE Commun. Mag.* **2016**, *54*, 50–57. [CrossRef]
34. Kabalci, Y. A survey on smart metering and smart grid communication. *Renew. Sustain. Energy Rev.* **2016**, *57*, 302–318. [CrossRef]
35. Robles, T.; Alcarria, R.; Andrés, D.M.; Cruz, M.N.; Calero, R.; Iglesias, S.; López, M. An IoT based reference architecture for smart water management processes. *JoWUA* **2015**, *6*, 4–23.
36. Towards a data-driven IoT software architecture for smart city utilities Framework for a Smart Water Management System in the Context of Smart City Initiatives in India. *Procedia Comput. Sci.* **2016**, *92*, 142–147. [CrossRef]

*Article*

# Container Migration in the Fog: A Performance Evaluation †

**Carlo Puliafito [1,2,*], Carlo Vallati [2], Enzo Mingozzi [2], Giovanni Merlino [3], Francesco Longo [3] and Antonio Puliafito [3]**

1 DINFO, University of Florence, Via di S. Marta 3, 50139 Florence, Italy

2 Department of Information Engineering, University of Pisa, Largo Lucio Lazzarino 1, 56122 Pisa, Italy; carlo.vallati@iet.unipi.it (C.V.); enzo.mingozzi@unipi.it (E.M.)

3 Department of Engineering, University of Messina, Contrada di Dio, Sant'Agata, 98166 Messina, Italy; gmerlino@unime.it (G.M.); flongo@unime.it (F.L.); apuliafito@unime.it (A.P.)

* Correspondence: carlo.puliafito@ing.unipi.it

† This paper is an extended version of our paper published in the IEEE 4th International Workshop on Sensors and Smart Cities (SSC), which was co-located with the IEEE 4th International Conference on Smart Computing (SMARTCOMP).

Received: 15 January 2019; Accepted: 22 March 2019; Published: 27 March 2019

**Abstract:** The internet of things (IoT) is essential for the implementation of applications and services that require the ability to sense the surrounding environment through sensors and modify it through actuators. However, IoT devices usually have limited computing capabilities and hence are not always sufficient to directly host resource-intensive services. Fog computing, which extends and complements the cloud, can support the IoT with computing resources and services that are deployed close to where data are sensed and actions need to be performed. Virtualisation is an essential feature in the cloud as in the fog, and containers have been recently getting much popularity to encapsulate fog services. Besides, container migration among fog nodes may enable several emerging use cases in different IoT domains (e.g., smart transportation, smart industry). In this paper, we first report container migration use cases in the fog and discuss containerisation. We then provide a comprehensive overview of the state-of-the-art migration techniques for containers, i.e., cold, pre-copy, post-copy, and hybrid migrations. The main contribution of this work is the extensive performance evaluation of these techniques that we conducted over a real fog computing testbed. The obtained results shed light on container migration within fog computing environments by clarifying, in general, which migration technique might be the most appropriate under certain network and service conditions.

**Keywords:** fog computing; internet of things; mobility; container; migration; CRIU; pre-copy; post-copy

---

## 1. Introduction

The internet of things (IoT) [1] is a paradigm where each object, from a "smart" one (e.g., a smartphone, a wearable device) to a "dumb" one (e.g., a lamp post, a dumpster), can exchange data over the internet. Such objects may also store and process data, use sensors to perceive the surrounding environment, and modify the latter through actuators. In this scenario, people are active players in this ecosystem, as they consume and produce data through their smartphones and wearable devices. The number of smart objects connected to the internet exceeded the world human population in 2010 [2], and the McKinsey Global Institute estimates a potential economic impact for IoT applications of as much as $11.1 trillion per year in 2025 [3]. The great popularity that the IoT has been gaining is due to the fact that this paradigm is necessary for the implementation of many information and communications technology (ICT) services. However, IoT devices, and especially the wireless ones, typically have limited compute, storage, and networking resources and can be battery-powered [4]. As such, they are not sufficient in most cases to directly host resource-intensive services that perform complex computations over the big data [5] collected by sensors.

IoT devices often need to be integrated with more powerful resources, and fog computing [6] is emerging as a paradigm to provide them. Fog computing extends the cloud towards the network edge, distributing resources and services of computing, storage, and networking anywhere along the cloud-to-things continuum, close to where data are sensed and actions need to be performed. Depending on its requirements, a fog service may be deployed on: (i) resource-rich end devices, such as video surveillance cameras; (ii) advanced nodes at the network edge (e.g., gateways, cellular base stations); and (iii) specialised routers at the core network. Besides, fog services may be either stand-alone or can interact with one another and with cloud services to fulfil more complex use cases. As a result, IoT devices may perform only the lightest tasks and count on the fog for the most intensive ones [7]. In addition, the closer proximity of fog services to the end devices permits a set of advantages with respect to the exclusive dependence on the distant cloud, such as [8–11]: (i) low and predictable latencies; (ii) reduction in bandwidth consumption; (iii) better privacy and security; (iv) improved context awareness; and (v) uninterrupted services in the presence of intermittent or no network connectivity to the cloud. These benefits are indispensable to many emerging applications in several domains (e.g., smart healthcare, smart surveillance, smart transportation).

Virtualisation is an essential aspect of cloud as well as fog computing. Providing a cloud or fog service as a virtual environment (e.g., a virtual machine, a container) grants great elasticity and isolation, which in turn allows multi-tenancy and thus resource efficiency. Furthermore, migrating a service among nodes is another major feature that brings high flexibility and adaptability, as it is described in [12,13], where the authors mainly focus on virtual machine migration within cloud data centres. However, migration in the fog is influenced by some aspects that are not present in cloud-only environments. These aspects, which are as follows, occur when extending the cloud towards the network edge, indeed:

- Fog environments are characterised by a high **heterogeneity** of nodes in terms of hardware capabilities, architectures, and operating systems. Hence, there is the need for a virtualisation technology that is generic and lightweight enough to run on as many different types of fog nodes as possible;
- Fog nodes are interconnected through a wide area network (**WAN**) and therefore experience higher latencies and lower throughputs than those present within a cloud data centre. Based on this, it is beneficial during migration to transmit the lowest possible amount of data;
- In the cloud, the **total migration time** is only a secondary consideration. In the fog, instead, limiting it may be of paramount importance as there are situations in which protracted total migration times may lead to overall degraded performances [14];
- Most fog services, especially those deployed and running at the network edge, typically perform transient data analysis and time-critical control and are thus not supposed to write to any **persistent memory** (e.g., the disk), unlike cloud services. As a consequence, it is in general not necessary to transfer any persistent data during migration; what typically happens is that only the runtime (i.e., volatile) state is migrated and applied at destination to an available base service image representing the default disk state.

The focus of this paper is on **container migration in the fog**. This is a topical issue that promises to attract great attention in the near future for several reasons. Firstly, fog computing has been gaining much popularity as an enabler of emerging applications and services, especially in the IoT context. Secondly, containers are becoming a common virtualisation approach, within fog computing environments in particular, for their lightweight nature and overall performances [15–17]. Thirdly, container migration in the fog enables various use cases, both traditional ones (i.e., already present in cloud contexts) and emerging ones that are instead specific to fog computing, as we will further discuss. Nonetheless, container migration is still in its infancy, and hence only few works in literature provide knowledge in the field [17–20]. Besides, none of them evaluates and compares the existing container migration techniques. In our previous conference paper [21], we surveyed these techniques and attempted to point out their advantages and disadvantages, with specific attention to fog computing contexts. However, our claims were not supported by any experimentation. This article, instead, reports the extensive performance evaluation of the state-of-the-art container migration techniques that we carried out over a real fog computing testbed. The obtained results confirm most of the statements in [21], although we found some unexpected results and were able to analyse each aspect of container migration in depth. To the best of our knowledge, there is only one work in literature that makes a similar contribution [22]. Nonetheless, it only assesses a subset (i.e., cold and pre-copy migration) of the available techniques and concludes that container migration results in unexpected errors.

The rest of the paper is organised as follows. Section 2 describes the fog computing use cases where container migration may play an important role. Then, Section 3 points out the main virtualisation techniques, with specific attention to containerisation. In Section 4, we detail the characteristics of the existing container migration techniques. Next, Section 5 reports the experiments that we carried out along with the analysis of the obtained results. Finally, in Section 6, we draw the conclusions and outline the future work.

## 2. Motivation and Use Cases

Sensors and actuators can be connected to the internet either directly or through a sensor network that creates the necessary path. In any case, physical sensors and actuators exploit the concept of gateway, i.e., a device that on one side communicates with them and on the other side with the internet and the available services. Such a gateway is usually provided with computing and storage capabilities that allow code execution. Containers represent a valid solution for code execution (see Section 3), and their controlled migration is expected to play an important role in future fog-IoT deployments. Container migration will be mandatory in two different use cases: (i) mobility use case, i.e. to support use cases that natively require the service to migrate from one fog node to another for instance to support mobility of users and IoT devices [23]; (ii) management/orchestration use case, i.e. to enable dynamic management of resources and orchestration of services to ensure proper management of the fog computing infrastructure at large, in a similar manner cloud computing platforms operate [24].

**Mobility.** In specific use cases, migration will represent a mandatory function, required to properly support mobility of users or IoT devices [17]. In particular, migration will be crucial to handle scenarios in which sensors and actuators are natively characterised by mobility, such as when worn or brought by mobile users (e.g., wearables, smartphones) or carried by vehicles (e.g., car sensors). One scenario is the vehicular use case in which applications implementing real-time situation awareness or cooperative functionalities will require fog computing and storage capabilities to implement time-critical and data-intensive tasks [25]. For instance, let us consider an autonomous driving vehicle and a supporting service that collects context data from the environment (e.g., roadside sensors) and other vehicles to notify possible hazardous situations by applying mining techniques on the collected data. In this case, service migration is mandatory in order to keep the communication latency between the vehicle and the supporting service as low as possible and allow the vehicle to react promptly.

Another scenario pertains applications using augmented reality (AR). In an industrial environment, portable AR devices (e.g., smart glasses, tablets) could be used by operators to obtain real time information on aspects such as: products, production procedures, instructions. Real-time AR systems exploit fog computing services to analyse and augment images with bounded low latency and without the need to offload them to the cloud [26]. In this case, for instance, migration could be exploited to support AR functionalities even when mobility is involved, as showed in Figure 1a. Let us consider an operator moving through the production plant for inspection equipped with smart glasses that show status information on the various equipment. In this case, the migration of the AR service is mandatory to ensure that images are always collected and analysed in due time.

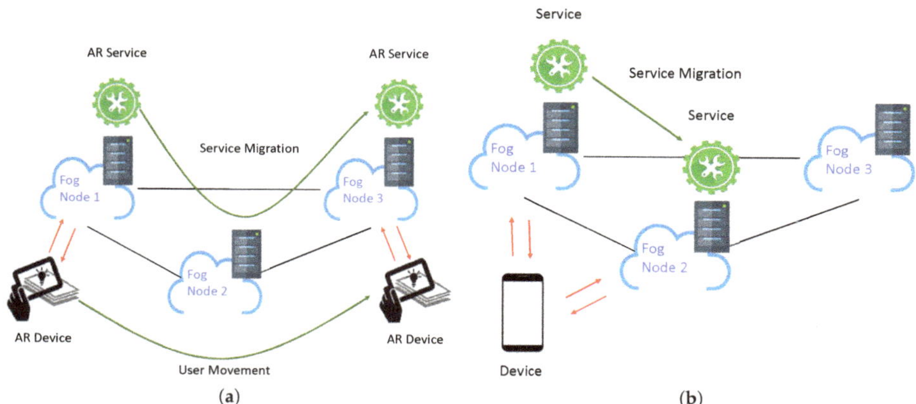


(a) (b)

**Figure 1.** Container migration use cases: (**a**) mobility support; (**b**) platform management/orchestration.

**Management/orchestration.** The ultimate fog computing vision foresees an infrastructure in which single fog nodes or edge data centres (EDC) are seamlessly integrated with the cloud infrastructure. Such multi-layer infrastructure will require support for migrating services from the cloud to the fog and vice versa with the objective to accommodate application requirements that vary over time [27]. For instance, a service running in the cloud that requires low-latency interactions with cyber-physical systems for a certain period of time could be migrated from the cloud to a fog node, while a service running on a fog node that at a certain point demands for more resources can be offloaded to the cloud.

In addition to this feature, migration of services within the fog will represent an important feature to ensure the proper functionality of the fog infrastructure (see Figure 1b). Migration between different fog nodes or EDCs could be exploited to implement load balancing policies that aim at improving fog resource utilisation and at ensuring that application requirements are constantly met [28]. A management/orchestration service that controls load distribution on a fog infrastructure, for instance, could decide to migrate a service from one overloaded fog node to another in order to avoid service degradation. Conversely, the same management/orchestration service could decide to move out all the services from one fog node to others to enable its power saving mode. Such adaptive energy-aware computation offloading [29] could be implemented to mitigate the energy consumption of fog computing systems, which are expected to represent a significant share of the overall energy consumption of fog/cloud systems [30].

## 3. Virtualisation in the Fog

In this section, we give some background knowledge about the most popular virtualisation approaches, with specific attention to containerisation.

### *3.1. Virtualisation Approaches*

Hardware virtualisation (also known as platform virtualisation) [31] is a set of techniques to support the creation of a so-called virtual machine (VM), which acts like a real computer able to run an operating system. More in general, software, when executed on VMs, does not have any access to, or visibility into, the underlying hardware resources of the host machine [32]. In platform virtualisation then, guest is another name for a VM, whereas the software (or firmware) in charge of VM instantiation on the host is called a hypervisor, or virtual machine monitor (VMM) [33]. Hardware-assisted virtualisation is a way for CPUs and other computer components to support virtualisation in hardware and thus improve virtualisation performance.

Operating system-level virtualisation [34], better known as containerisation, is a virtualisation approach enabled by a set of operating system features where the kernel itself allows the coexistence of multiple and isolated instances of user-space environments, leaving the hardware abstraction layer as well as the enforcement for process sandboxing to the shared kernel co-hosting them. Still, no limitations are in place in terms of choice for user-space, e.g., down to the C standard library if needed. Virtualisation instances of this kind, the so-called containers, still look like real computers from the point of view of programs running in them, as the latter can see select resources (file systems, networks, hardware devices and capabilities), i.e., those exposed by an ordinary operating system. However, software running inside a container can only see a subset of the resources made available by the kernel and specifically those assigned (intentionally) to the container.

The main difference between (platform) virtualisation and containerisation (see Figure 2) thus lies in the level at which abstraction and resource partitioning occurs, i.e., at the kernel level for the latter, below it for the former, and in the corresponding breadth of scope in reproducing the machine abstraction exposed to the guest environment, e.g., including emulation of low-level (firmware-based) hardware management interfaces for the former only. The requirements of virtualisation lead to higher-complexity implementations and higher storage requirements and are thus prone to substantial overhead, especially in case of unavailability of extensions for hardware-assisted virtualisation. This aspect, coupled with constraints of fog computing, where single-board computers (SBCs) or embedded systems are often involved whilst hardware-assisted virtualisation cannot always be taken for granted, makes VMs less suitable than containers for the fog computing domain in terms of overall footprint and support of workload mobility (i.e., migration) [15–17,35–37].

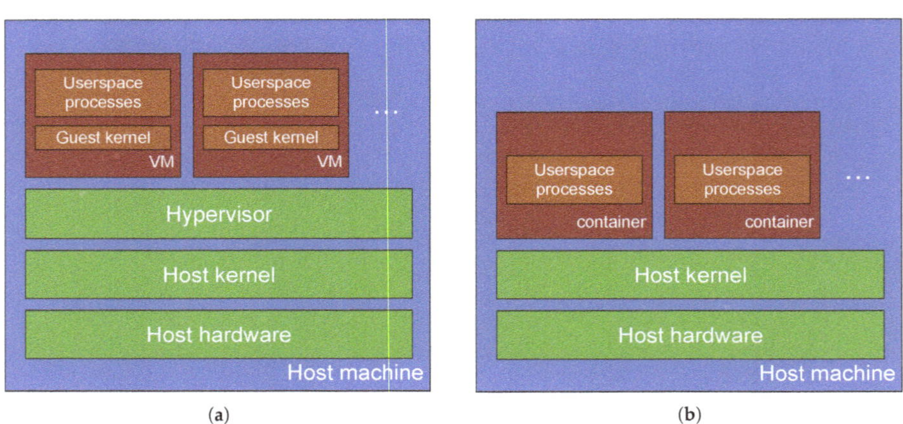


**Figure 2.** Hardware virtualisation (**a**) vs. containerisation (**b**).

### *3.2. Containerisations*

Containerisation takes many forms, according to the combination of techniques employed (and thus the level of control exerted on the guest environment) as well as the scope and usage [38]. Given that Linux is the most advanced example of an operating system featuring full support for containerisation, from now onwards the overview will refer to Linux-specific techniques, even if the same concepts have been applied elsewhere and the discussion does not otherwise lose generality. In terms of container-enabling techniques, the Linux kernel provides in particular control groups (typically abbreviated as cgroups), secure computing mode (typically abbreviated as seccomp), and namespacing. The latter implements kernel resource partitioning such that a set of processes is able to enumerate and access certain resources while another set of processes is given visibility on a different set of resources. This works by having the same namespace for the resources in the different process sets, while having those names actually point to distinct resources. Namespacing [39] is an approach first and foremost, and the Linux implementation provides interfaces to namespace a number of resources and subsystems: process/user/group IDs, filesystem resources, hostnames, routes, network interfaces, and interprocess communication descriptors, to name a few. Cgroups [40] is a kernel feature to limit, account for, and isolate usage of resources (e.g., CPU, memory, disk I/O, bandwidth) by a collection of processes. The seccomp [40] feature is a sandboxing facility first devised for grid computing users and originally intended as a technique to safely run untrusted compute-bound code. As such, it allows a process to make a one-way transition into a "secure" state from where system calls are disallowed, except for a very limited subset, restricted to (already) open file descriptors. In contrast to namespacing, seccomp does not virtualise (assigned) system resources but isolates the process from them entirely.

A middle ground between VMs and containers is represented by extremely lightweight and novel VMs that seamlessly plug into the containers ecosystem. An example of these are Kata Containers (see https://katacontainers.io/, accessed on 15 January 2019), which are managed by the OpenStack Foundation and merge technologies from Intel Clear Containers and Hyper runV. Specifically, their rationale is providing a user experience very similar to that of containers, while ensuring the portability and security advantages of VMs. This is achieved by featuring lightweight operating systems and kernels where isolation is enforced with the help of hardware-assisted (i.e., processor) extensions for hypervisors. Such extensions have been historically available on x86-architecture processors but are recently making their appearance on the ubiquitous ARM64 hardware.

As another example of performance enhancement technique, nvdimm (see https://github.com/qemu/qemu/blob/master/docs/nvdimm.txt, accessed on 15 January 2019) enables the guest to directly access (see https://git.kernel.org/pub/scm/linux/kernel/git/torvalds/linux.git/tree/Documentation/filesystems/dax.txt, accessed on 15 January 2019) the root file system from host memory pages, bypassing the guest page cache, by providing a memory-mapped persistent memory device to the VM hosting the container(s). Other so-called machine accelerators, at the moment x86-specific only, include skipping the firmware, e.g., basic input-output system (BIOS) or extensible firmware interface (EFI), in the guest when booting an ELF-format kernel (nofw) and reducing the interpretation burden for guest advanced configuration and power interface (ACPI) component (static-prt). These customisations together improve performance and boot time significantly. Finally, virtcontainers are highly optimised/accelerated VMs hosting (typically, but not solely) single-container workloads and an agent guest-side that enables interaction with the guest environment, e.g., command relaying for container life cycle management.

## 4. Container Migration Techniques

Container migrations may be either stateless or stateful. The former is so called because it causes the loss of the whole container state. Therefore, once on the destination node, the container restarts from scratch. As such, stateless migration simply consists in the following two steps: (i) the start of a new container on the destination node; and (ii) the deletion of the old container on the source node. **Stateful migration**, instead, is such that both the volatile and persistent states of the container are made available at destination once migration is completed. This paper focuses on stateful container migration. More specifically, in this section we provide an overview of the state-of-the-art techniques that may be adopted to statefully migrate containers, distinguishing between cold and live migration techniques. At the end of this section, Table 1 compares the discussed techniques in terms of what is transmitted in each migration phase.

**Table 1.** What is transferred in each phase of the migration techniques.

| Technique | Pre-Dump | Dump | Faulted Pages |
|---|---|---|---|
| Cold | | Memory pages and execution state | |
| Pre-copy | Memory pages and execution state | Dirty pages and changes in execution state | |
| Post-copy | | Execution state | Memory pages |
| Hybrid | Memory pages and execution state | Changes in execution state | Dirty pages |

### 4.1. Cold Migration

The steps for **cold migration** are depicted in Figure 3, where the source node performs the blue steps, the destination node performs the green one, and the grey step involves both nodes. This migration technique is said to be "cold" because it: (i) first freezes/stops the container to ensure that it no longer modifies the state; (ii) then dumps the whole state and transfers it while the container is stopped; and (iii) finally resumes the container at destination only when all the state is available. As such, cold migration features a very long **downtime**, namely the time interval during which the container is not up and running. As shown in Figure 3, downtime even coincides with the **total migration time**, which is the overall time required to complete the migration. This aspect of cold migration goes against one of the main objectives of service migration in both the cloud and the fog, i.e., the limitation of the downtime. However, we highlight that this technique transfers each memory page only once, and this should significantly reduce both the total migration time and the overall amount of data transferred during migration.

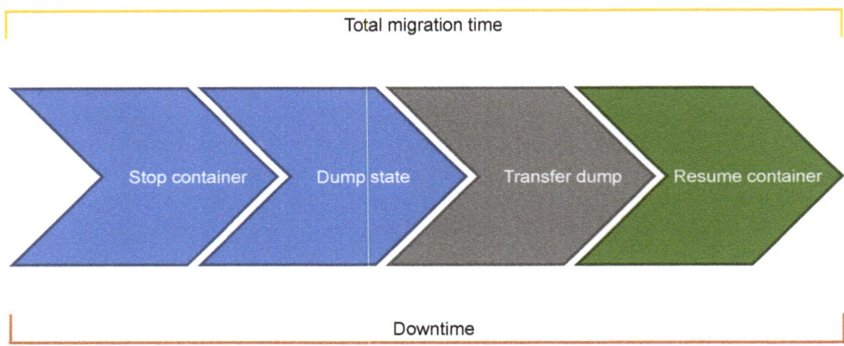


**Figure 3.** Cold migration.

Cold migration of containers, like all the other container migration techniques, is strongly based on checkpoint/restore in userspace (*CRIU*) (see https://criu.org/Main_Page, accessed on 14 January 2019). This started as a project of Virtuozzo (see https://virtuozzo.com/, accessed on 14 January 2019) for its OpenVZ containers but has been getting so much popularity over time that now it is used by all the most prominent container runtimes, such as runC. In detail, CRIU is a software tool for Linux that is written in C and mainly allows to: (i) freeze a running container; (ii) checkpoint/dump its state as a collection of files on disk; and (iii) use those files to restore the container and run it exactly as it was before being stopped. With respect to Figure 3, CRIU misses the third step, namely the state transfer to the destination node. Indeed, CRIU by itself only allows to restore a checkpointed container on the same host node; therefore, the actual state transfer has to be performed by exploiting tools such as *rsync* or secure copy protocol (*SCP*). Last but not least, we highlight that CRIU captures the runtime state of a container, namely all the memory pages and the execution state (e.g., CPU state, registers), but not the persistent one. However, this is in general not an issue within a fog computing environment, as we pointed out in the introduction of this paper, and might even represent a strength since less data are transmitted.

### 4.2. Live Migration

The main concern of live migration is the limitation of service downtime. Indeed, "live" means that the container keeps on running while most of its state is being transferred to the destination node. The container is typically suspended only for the transmission of a minimal amount of the overall state, after which the container runs at destination. When the downtime is not noticeable by the end user, live migration is said to be "seamless". Three different live migration techniques exist for containers: pre-copy, post-copy, and hybrid.

#### 4.2.1. Pre-Copy Migration

**Pre-copy migration** is so called because it transfers most of the state prior (i.e., *pre*) to freezing the container for a final dump and state transfer, after which the container runs on the destination node. It is also known as *iterative migration*, since it may perform the pre-copy phase through multiple iterations such that each iteration only dumps and retransmits those memory pages that were modified during the previous iteration (the first iteration dumps and transfers the whole container state as in cold migration). The modified memory pages are called **dirty pages**. Typically, iterations in the pre-copy phase are convergent, i.e., of shorter and shorter duration. If iterative, the pre-copy phase generally concludes when a predetermined number of iterations is reached. The container is then suspended on the source node in order to capture the last dirty pages along with the modifications in the execution state (e.g., changes in the CPU state, changes in the registers) and copy them at destination without the container modifying the state again. Finally, the container resumes on the destination node with its up-to-date state. Figure 4 shows the steps of pre-copy migration with a one-iteration pre-copy phase. The reason for this is that our current implementation of pre-copy migration, which is then evaluated in Section 5, presents only one pre-copy iteration. This implementation is based on CRIU, which provides all the basic mechanisms (e.g., the *-pre-dump* option) that are necessary to pre-dump the runtime state of a container and restore it afterwards (see https://criu.org/Iterative_migration, accessed on 15 January 2019). All the following considerations on pre-copy migration and hybrid migration (see Section 4.2.3) assume a one-iteration pre-copy phase.

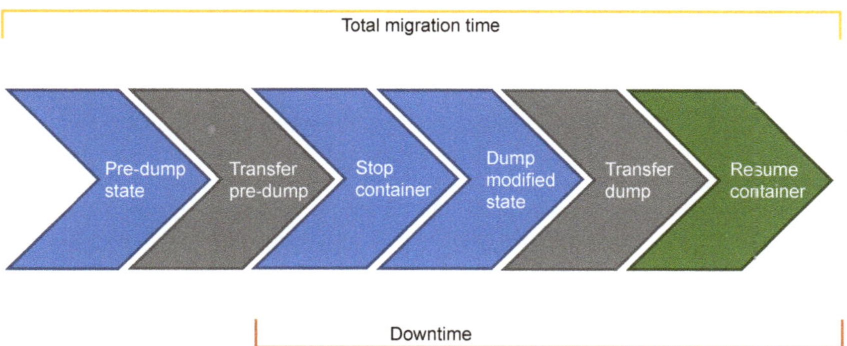


**Figure 4.** Pre-copy migration.

The main difference between cold and pre-copy migrations lies in the nature of their dumps. The dump in cold migration represents the whole container state (as the pre-dump in pre-copy migration) and thus always includes all the memory pages and the execution state. The dump in pre-copy migration, instead, only includes those memory pages that were modified during the pre-copy phase, together with the changes in the execution state. As such, downtime for pre-copy migration should be in general shorter than that for cold migration because less data are transferred while the container is stopped. However, downtime for pre-copy migration is not deterministic, as it significantly depends on the number of dirty pages. Therefore, we expect pre-copy migration to be afflicted by the two factors that may increase the number of dirty pages: (i) the **page dirtying rate** featured by the container-hosted service, namely the speed at which the service modifies memory pages; (ii) the **amount of data** that are transferred during the pre-copy phase, since the more data are transferred in that phase, the more time the service has to modify pages. It is worth noting that both these factors should be always considered against the available **throughput** between the source and the destination node. Specifically, this is more obvious for the second factor, since this needs to be considered against the available throughput in order to estimate the time that the service has to modify pages during the pre-copy phase. With regard to the first factor, we expect that pre-copy migration performances are impaired by a page dirtying rate that starts approaching the available throughput (i.e., same order of magnitude), as in that case memory pages are modified at a rate which is comparable to that of page transfer. This would result in a state dump that is in general comparable to the state pre-dump or, in other words, to a downtime which is comparable to that for cold migration. As a final remark, we highlight that, unlike cold migration, pre-copy migration might transfer each memory page several times, with possible negative consequences on the overall amount of data transferred during migration and thus on the total migration time.

#### 4.2.2. Post-Copy Migration

**Post-copy migration** is the exact opposite of pre-copy migration. Indeed, it first suspends the container on the source node and copies the execution state to the destination so that the container can resume its execution there. Only after that (i.e., *post*), it copies all the remaining state, namely all the memory pages. Actually, there exist three variants of post-copy migration, which differ from one another on how they perform this second step. In this paper, we only describe the *post-copy migration with demand paging* variant, better known as **lazy migration** (see Figure 5), which is the only one that may be currently implemented using the functionalities provided by CRIU (see https://criu.org/Lazy_migration, accessed on 15 January 2019), e.g., the *-lazy-pages* and *-page-server* options. With lazy migration, the resumed container tries to access memory pages at destination, but, since it does not find them, it generates **page faults**. The outcome is that the *lazy pages daemon* at destination contacts the *page server* on the source node. This server then "lazily" (i.e., only upon request) forwards the faulted pages to the destination.

Post-copy migration copies each memory page only once. Therefore, it should transfer a data volume that is comparable with that of cold migration and with that of the pre-copy phase of pre-copy migration. Besides, similarly to cold migration, downtime for post-copy migration is irrespective of the page dirtying rate featured by the container-hosted service and of the overall amount of data that need to be transferred. This is due to the fact that the dump in post-copy migration is simply the execution state and does not contain any dirty memory pages. However, post-copy migration is afflicted by two drawbacks that are worthy of consideration. Firstly, **page faults degrade service performances**, as memory pages are not immediately available at destination once the container resumes. This could be unacceptable to the many latency-sensitive services present in fog computing environments. Secondly, during migration, this technique **distributes the overall up-to-date state** of the container between both the source and the destination node (before completion of post-copy migration, the source node retains all the memory pages, but some of them may be out-of-date because they have already been copied at destination and modified by the resumed container), whereas approaches like cold or pre-copy migrations retain the whole up-to-date state on the source node until the termination of the migration process. Therefore, if the destination node fails during migration, it may be no more possible to recover the up-to-date state of a post-copied container.

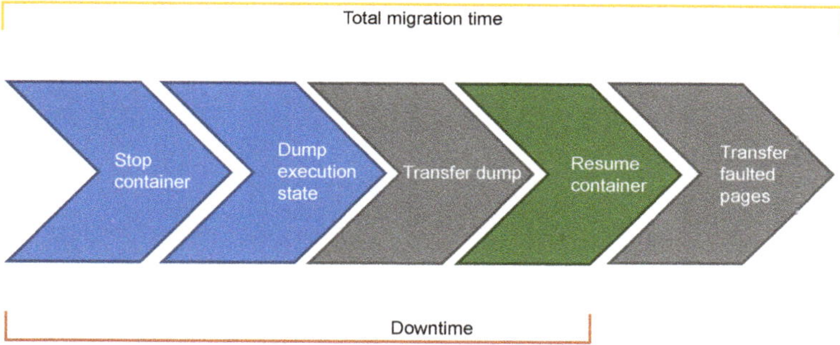


**Figure 5.** Post-copy migration.

#### 4.2.3. Hybrid Migration

As discussed in the previous sections, both pre-copy and post-copy migrations present some shortcomings: (i) pre-copy migration has a non-deterministic downtime; (ii) faulted pages in post-copy migration degrade service performances. **Hybrid migration**, which is illustrated in Figure 6, combines pre-copy and post-copy with the objective to subdue their limitations and sharpen their strengths. Going into detail, the first two steps of hybrid migration coincide with those of pre-copy migration, namely a pre-dump of the whole state and its transmission at destination while the container is still running on the source node. Then, the container is stopped, and its state is dumped in a way that combines the dumps of pre-copy and post-copy migrations. Indeed, the dump in hybrid migration is represented by the modifications in the execution state that occurred during the pre-copy phase. Once the dump is transferred to the destination node, the container can be restored. At this step, the destination node has the up-to-date container execution state along with all the memory pages. Nonetheless, some of them were dirtied during the pre-copy phase. As a result, the last step in hybrid migration consists in the page server at source lazily transmitting the dirty pages to the lazy pages daemon on the destination node. It is worth noting that the number of memory pages that are lazily transmitted in hybrid migration is generally less than that of post-copy migration since only the dirty pages are transferred. From now on, we will refer to these dirty pages as faulted pages, in line with the name used in post-copy migration to indicate the data transferred in this last phase. As pre-copy migration, we expect also hybrid migration to be affected by the page dirtying rate and the amount of data that are transmitted during the pre-copy phase. However, these factors should influence the total migration time for hybrid migration but not the downtime, as they alter the number of faulted pages. Moreover, hybrid migration is affected by the same two drawbacks of post-copy migration (see Section 4.2.2). To conclude, hybrid migration can be implemented by merging the CRIU options for pre-copy and post-copy migrations.

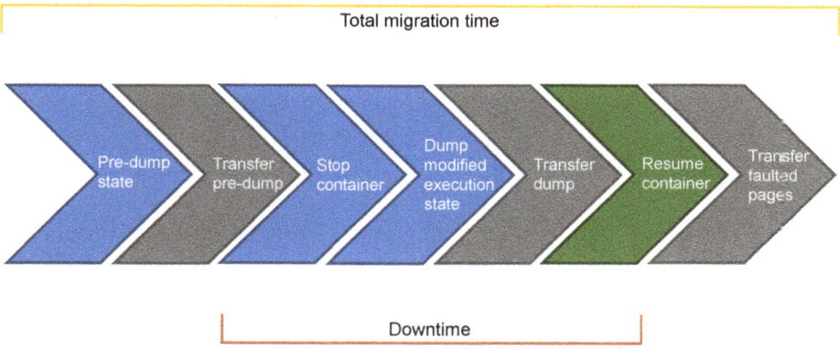


**Figure 6.** Hybrid migration.

## 5. Performance Evaluation

In this section, we evaluate and compare the performances of the container migration techniques described in Section 4. The main objective is to determine whether there exists a technique that always performs the best or, otherwise, to delineate which technique might be the most suitable under certain network and service conditions. Section 5.1 illustrates the experiment setup, which consists in a real fog computing testbed. Next, Section 5.2 analyses the obtained results.

### 5.1. Experiment Setup

The overall testbed comprises two fog nodes and one end device. The former are two *Raspberries Pi 3 Model B* (i.e., ARMv8-A architecture) with Debian 9.5 and Linux kernel 4.14.73-v8+. They both run: (i) **CRIU 3.10** for the checkpointing and restore functionalities; (ii) **rsync 3.1.2** as file transfer mechanism; and (iii) **runC 1.0.1** as container runtime. Besides, they are both deployed within an office in the Department of Information Engineering of the University of Pisa. We chose Raspberries Pi as fog nodes because, even though they are able to host and run containers [36], they are rather limited in terms of hardware capabilities and thus represent the worst-case scenario in a fog environment. On the other hand, an *Asus ZenBook* notebook with Windows 10 emulates an IoT device that forwards data to a container-hosted fog service and receives actuation commands from it.

The core part of the experiment setup consisted in the appropriate tuning of the **throughput** between the fog nodes as well as in: (i) the choice of the **runtime state size** of the container (we highlight that, given the rsync settings described at the end of this section, the runtime state size represents the actual amount of data transferred during the pre-copy phase. Therefore, throughout Section 5, we use these two terms interchangeably); (ii) the selection of the values for the **page dirtying rate** of the container-hosted service. Indeed, as explained in Sections 4.2.1 and 4.2.3, we expect pre-copy and hybrid migrations to be afflicted by these two factors, both to be considered against the available throughput. Instead, we do not expect these or any other specific factor to affect cold or post-copy migrations. By calibrating and combining the aforementioned factors, we implemented the following **four configurations**, which allow to carry out a comprehensive performance evaluation:

- **A**—this configuration presents a page dirtying rate and a throughput of different orders of magnitude, with the throughput higher than the page dirtying rate. However, given the throughput, the size of the runtime state leads to a prolonged pre-copy phase and thus gives the service plenty of time to modify memory pages. Therefore, this configuration is mainly aimed at evaluating the effects of a considerable runtime state size on the migration techniques;
- **B**—this configuration resembles configuration A in terms of runtime state size but features a page dirtying rate of the same order of magnitude of the available throughput. As such, this configuration is mainly aimed at investigating the effects of a high page dirtying rate on the migration techniques;
- **C**—this configuration shows a runtime state size that, considering the throughput, causes a shortened pre-copy phase and hence gives the service little time to modify memory pages. Besides, the page dirtying rate and the throughput are of different orders of magnitude, with the throughput higher than the page dirtying rate. Thus, the main objective of this configuration is to assess the migration techniques when both the factors are low, given the available throughput;
- **D**—this configuration resembles configuration C. The only difference is that the page dirtying rate in C and in D are of different orders of magnitude, with the former lower than the latter. However, the page dirtying rate in D is still lower than the throughput and of a different order of magnitude. The main purpose of this configuration is to estimate whether there are evident effects on the migration techniques when the page dirtying rate increases, though still being considerably lower than the throughput.

Going into detail, we achieved the previous four configurations by choosing and combining values as follows. Firstly, we selected two different throughput values that may both occur in real-life use cases within a fog environment (see Section 2). One throughput value may be present within a **mobility use case** where the destination fog node connects to the internet through 4G/LTE (e.g., a fog node within a public bus). The other throughput value, instead, may exist in a **management/orchestration use case** within the fog. Table 2 reports these values, together with the round trip time (RTT) values associated to them in the aforementioned use cases. Indeed, we also considered RTT values among the fog nodes in order to more accurately replicate the use cases network conditions in our testbed. To obtain the values presented in Table 2, we proceeded as follows. For the mobility use case, we considered a computer connected through Ethernet to the University of Pisa network as source fog node and a smartphone connected to the internet through 4G/LTE as destination. For the management/orchestration use case, instead, we employed two fixed computers belonging to a bridged LAN of the University of Pisa and installed in two different buildings placed about 1 km far apart. Throughput values were then obtained by performing 10 runs with the *iperf3* (See https://iperf.fr/, accessed on 30 December 2018.) tool, sending 50 MB each time. Similarly, RTT values were calculated over 10 runs, with 20 measurements per run. Since, in our testbed, the two Raspberries behaving as fog nodes are located in the same office, we had to emulate the values described in Table 2. Therefore, we exploited Linux Traffic Control Hierarchy Token Bucket (*tc-htb*) (see https://linux.die.net/man/8/tc-htb, accessed on 30 December 2018) to limit the throughput and Linux Traffic Control Network Emulator (*tc-netem*) (see https://www.systutorials.com/docs/linux/man/8-tc-netem/, accessed on 30 December 2018) to artificially set RTT values between the Raspberries.

**Table 2.** Considered throughput and round trip time (RTT) values (95% confidence intervals).

| Use Case | Throughput (Mbps) | RTT (ms) |
|---|---|---|
| Mobility | $11.34 \pm 2.31$ | $122.95 \pm 5.57$ |
| Management/Orchestration | $72.41 \pm 3.87$ | $6.94 \pm 0.61$ |

We then chose the other values based on the identified throughput values. Specifically, we implemented a distributed application where both the client and the server are written in Java using the *Californium* (see https://www.eclipse.org/californium/, accessed on 15 January 2019) *CoAP* framework and thus required *openjdk8* to be installed on all the devices of the testbed. The server runs within a runC container in the fog and, once started, allocates 75 MB of RAM for random data. For our purpose, 75 MB is a suitable runtime state size; indeed, it determines a pre-copy phase that lasts tens of seconds with the lowest throughput and only few seconds with the highest one. The client, instead, runs on the Asus notebook and sends a POST request to the server every second to represent sensor data. Every time the server receives a request from the client, it modifies some of the memory pages with new random values. The server may perform this task with the following two page dirtying rates, which were identified by taking the throughputs into consideration. The lowest page dirtying rate is 10 KBps, which is about two orders of magnitude lower than the mobility use case throughput and about three orders of magnitude lower than the management/orchestration use case throughput. The highest page dirtying rate, instead, is 500 KBps, which is of the same order of magnitude of the throughput in the mobility use case and about one order of magnitude lower than the throughput in the management/orchestration use case. By combining the aforementioned values as reported in Table 3, we obtained the four configurations that were previously described. We highlight that all the implemented configurations share the same value of the runtime state size.

During the experiments, we evaluated all the migration techniques that are discussed in Section 4. In particular, we tested each technique five times for each of the four configurations. For each migration technique, we observed all the metrics that characterise it. As a result, the metrics that were overall observed are the following:

- **Pre-dump time**—taken in the pre-copy phase to dump the whole state on the source node while the service is still running;
- **Pre-dump transfer time**—needed in the pre-copy phase to transfer the generated pre-dump from the source to the destination node. It is not to be confused with the pre-dump time;
- **Dump time**—necessary in the dump phase to stop the container and dump its (modified) state on the source node. As described in Table 1, each migration technique presents a different concept of state dump;
- **Dump transfer time**—needed in the dump phase to transfer the generated dump from the source to the destination node. It is not to be confused with the dump time;
- **Resume time**—taken to restore the container at destination based on the state that was transferred up to that moment;
- **Faulted pages transfer time**—required in the last phase to transfer the faulted pages from the source to the destination node. Table 1 presents the different meanings that the term "faulted pages" assumes for post-copy and hybrid migrations;
- **Pre-dump size**—transferred during the pre-dump transfer time;
- **Dump size**—sent from the source to the destination node during the dump transfer time;
- **Faulted pages size**—transferred during the faulted pages transfer time.

We exploited the Linux *time* command to measure the times (i.e., the first six metrics) and the rsync *-stats* option to collect statistics regarding the amount of data transferred through rsync (i.e., the last three metrics). It is worth noting that we disabled the rsync data compression functionality during all the experiments. This was done because compressibility depends on data, and we did not want the experiment results to be influenced by this aspect. Similarly, we deleted the dump and the eventual pre-dump from the destination fog node after every experiment run. This was done to avoid that, by finding any of them at destination the next time, the incremental data transfer performed by rsync could transmit less data, thus influencing the experiment results. To conclude, we stored raw data in .csv files for the next phase of results analysis and plotting, which was performed in Python through *Plotly*, an open-source graphing library for Python.

**Table 3.** How values were combined to obtain the four configurations.

| Throughput and RTT | Page Dirtying Rate | Configuration |
|---|---|---|
| Mobility | Low | A |
| Mobility | High | B |
| Management/Orchestration | Low | C |
| Management/Orchestration | High | D |

### 5.2. Results

We now analyse and discuss the results obtained from the experiments. More specifically, Section 5.2.1 compares the migration techniques in terms of the total migration times; Section 5.2.2 evaluates the techniques with respect to the downtimes, while Section 5.2.3 does it with regard to the amounts of transferred data. As we will clarify in the following sections, each of these three metrics is the result of adding up some of the metrics from Section 5.1. To conclude, Section 5.2.4 summarises the main lessons learnt. All the following results are presented with a 95% confidence level.

#### 5.2.1. Total Migration Times

Figure 7 depicts the total migration times, highlighting their components for each migration technique. It is evident how the times to perform local computations (i.e., pre-dump and dump the state, resume the container) are negligible with respect to those needed to transfer the state at destination (i.e., pre-dump transfer, dump transfer, and faulted pages transfer times). This is clearly more visible under configurations A and B, where the available throughput is significantly lower than that of configurations C and D.

Let us now compare the migration techniques. Cold migration presents the lowest total migration times, irrespective of the specific configuration. We were expecting this result, as cold migration transmits each memory page only once unlike pre-copy and hybrid techniques, which instead may transmit a memory page more than once (i.e., if it is dirtied). Less pages transferred result in a shorter total migration time, indeed. However, this is not always true. In Figure 7a, post-copy migration presents a longer total migration time than pre-copy, even though it transmits less data (see Section 5.2.3). Similarly, total migration times for post-copy migration are always longer than those for cold migration, even though these two techniques overall transmit similar amounts of data, as confirmed in Section 5.2.3. This unexpected result can be explained as follows. Post-copy migration is currently implemented according to the "lazy migration" variant, and hence faulted pages are transferred by the page server on the source node only upon request of the lazy pages daemon running at destination (see Section 4.2.2). Thus, **the time to perform such requests, which are not present in cold and pre-copy migrations, increases the overall total migration time**. This is particularly noticeable under configurations A and B, where RTT between the fog nodes is considerably higher than that of C and D.

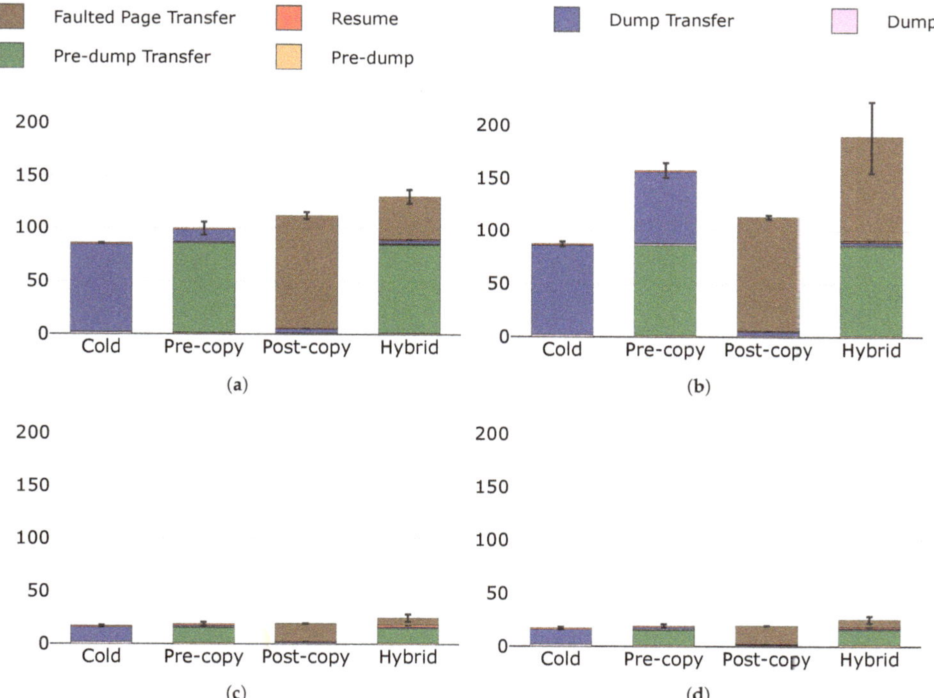


**Figure 7.** Total migration times (s) with their components in evidence under: (**a**) configuration A; (**b**) configuration B; (**c**) configuration C; and (**d**) configuration D.

Total migration times for cold and post-copy migrations are never influenced by an increase in the page dirtying rate. We expected this result as neither of these two techniques transfers dirty pages in any of its phases (see Table 1). Also pre-copy and hybrid migrations are not affected in terms of total migration time when page dirtying rate raises from configuration C to D. This important result shows how **there are no evident effects on the total migration times for pre-copy and hybrid migrations when the page dirtying rate increases, though still being lower and of a different order of magnitude from the throughput.** Besides, under these conditions, pre-copy migration performs similarly to post-copy in terms of total migration time. Nonetheless, the increment in the page dirtying rate from configuration A to B significantly prolongs the total migration times for pre-copy and hybrid migrations. Therefore, these two techniques are strongly affected in terms of total migration time by a page dirtying rate that, as for example under configuration B, reaches the same order of magnitude of the available throughput. The reason for this is that, under these conditions, the amount of pages dirtied during the pre-copy phase is comparable to that of pages transferred in that phase, namely to the whole state. This results in a state dump of considerable size for pre-copy migration and in a significant number of faulted pages for hybrid migration (see Table 1) and therefore in a substantial protraction of the total migration times.

Finally, we remark that hybrid migration always has the longest total migration times. This is because this technique inherits both the drawbacks of pre-copy and post-copy migrations in terms of total migration time, namely: (i) the fact that a memory page may be transferred more than once, as in pre-copy migration; (ii) the fact that also the time needed to request faulted pages needs to be considered, as in post-copy migration.

#### 5.2.2. Downtimes

Figure 8 depicts the downtimes for the migration techniques under the four considered configurations. As shown, the downtime for any technique is given by the sequence of the following times: (i) dump time; (ii) dump transfer time; and (iii) resume time. Cold migration always presents the highest downtime. This even coincides with the total migration time and proves to be unacceptable for most applications, especially for the critical ones that may exist in a Fog environment.

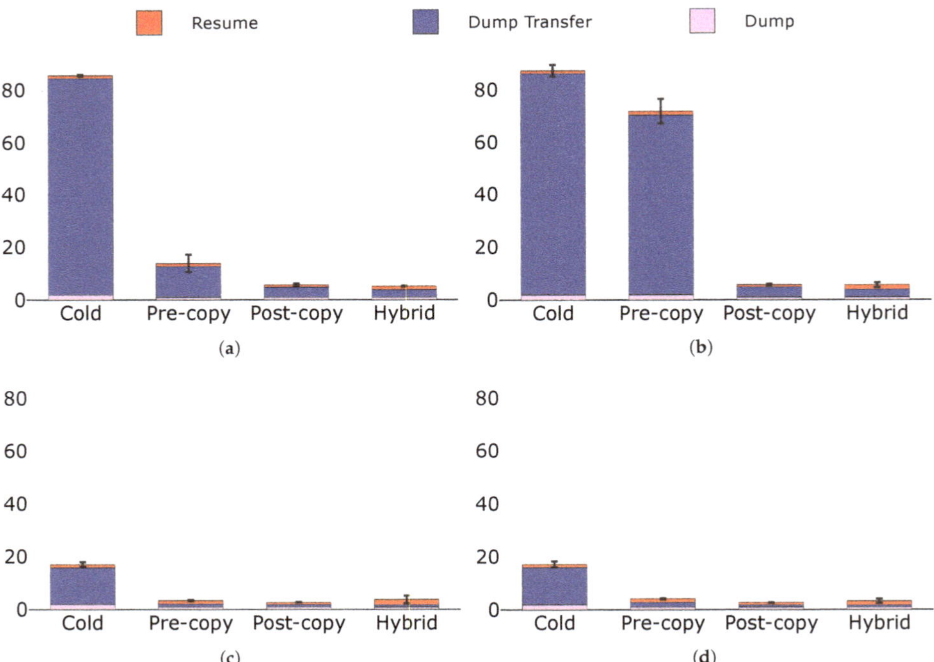


**Figure 8.** Downtimes (s) with their components in evidence under: (**a**) configuration A; (**b**) configuration B; (**c**) configuration C; and (**d**) configuration D.

Under configurations C and D, the other three migration techniques show similar performances in terms of downtime. This is because, under these conditions, few memory pages are modified during the pre-copy phase; therefore, the dump size (and hence the dump transfer time) in pre-copy migration is comparable to those in post-copy and hybrid migrations. Besides, none of the four techniques seems to be affected in terms of downtime by an increase in the page dirtying rate from configuration C to D, as already noticed and commented with regard to the total migration times in Section 5.2.1. However, under configurations A and B, pre-copy presents significantly higher downtimes than post-copy and hybrid migrations. More specifically, under A, the downtime for pre-copy migration is longer because, considering the lower throughput, the size of the runtime state prolongs the pre-dump transfer time, giving the service more time to modify pages than under C or D. Therefore, the dump size in pre-copy migration grows as it strongly depends on the number of dirty pages, and the downtime does the same. A higher page dirtying rate under configuration B further increases the number of dirty pages and thus lengthens the pre-copy migration downtime. This even tends to that of cold migration, with the total migration time that, in addition, noticeably exceeds that of cold migration (see Section 5.2.1). It is evident, instead, how downtimes for post-copy and hybrid migrations are not influenced by the conditions characterising configurations A and B. This result was expected since neither of these two techniques includes dirty pages in its dumps, as reported in Table 1.

We now analyse dump times and resume times in more depth. Both these times are not clear by looking at Figure 8; therefore, we illustrate them in Figures 9 and 10, respectively. In general, dump times only depend on the amount of data that need to be dumped, while resume times depend on the amount of data from which the container must be restored. By looking at Figure 9, it is possible to notice how average dump times are always equal or less than 1 s except from those of cold migration (under all configurations) and that of pre-copy migration under configuration B. In these situations, indeed, the amount of data to be dumped (i.e., the dump size) is significantly higher than in all the others: the dump in cold migration is the whole container state, while that in pre-copy migration is of considerable size because of the conditions characterising configuration B. The condition on the runtime state size under configuration A, instead, is not sufficient on its own to cause an increase in the dump time for pre-copy migration. We also highlight that, with the only exception of pre-copy migration from configuration A to B, dump times are not affected by an increase in the page dirtying rate. This is due to the fact that, as reported in Table 1 and discussed in the previous sections, only the state dump in pre-copy migration includes dirty pages, and the increase in the page dirtying rate from configuration C to D does not determine an increase in the dump size that is significant enough to prolong the dump time.

Post-copy migration presents the shortest resume times (see Figure 10), as it restores the container at destination by only applying a very limited dump size to the base container image. All the other techniques have longer resume times. Going into detail, the cold and pre-copy techniques show similar values except from that of pre-copy migration under configuration B, which is caused by a higher amount of data to be applied to the base service image. We would have expected similar values also for hybrid migration; however, results show that, in general, resume times for this technique are greater. A possible explanation of this outcome is that jointly applying the pre-dump and the dump to the base image in hybrid migration is computationally more intensive.

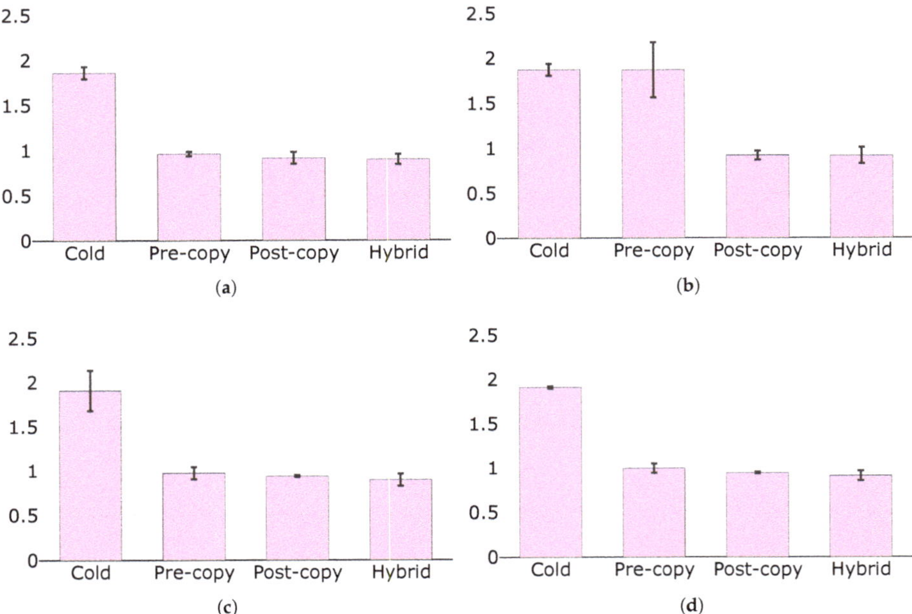


**Figure 9.** Dump times (s) under: (**a**) configuration A; (**b**) configuration B; (**c**) configuration C; and (**d**) configuration D.

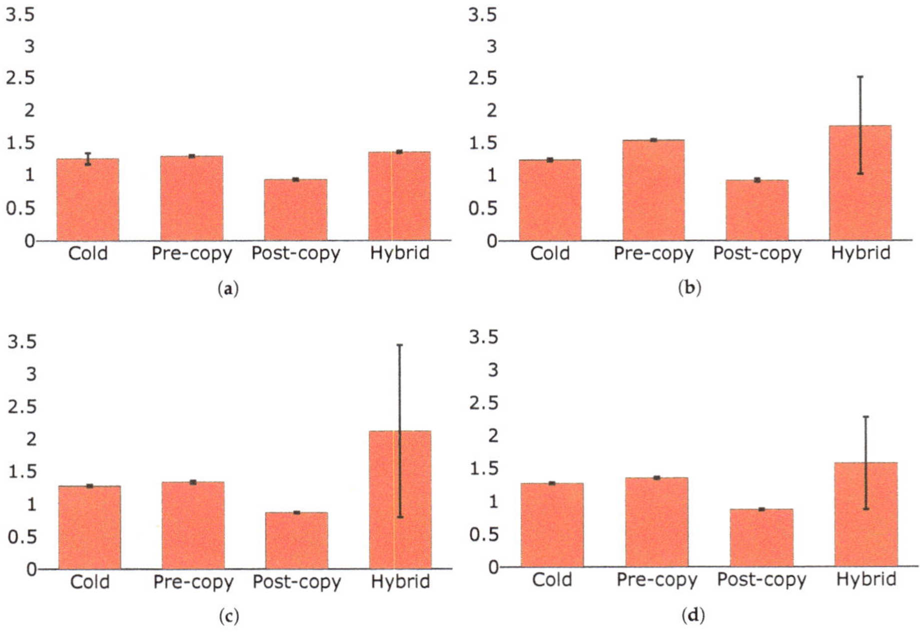


**Figure 10.** Resume times (s) under: (**a**) configuration A; (**b**) configuration B; (**c**) configuration C; and (**d**) configuration D.

#### 5.2.3. Transferred Data

In Figure 11, we illustrate the amounts of data transferred during the experiments. Firstly, it is easy to notice how most of the transferred state is: (i) the pre-dump for pre-copy and hybrid migrations; (ii) the dump for cold migration; and (iii) the faulted pages for post-copy migration. This is in line with what reported in Table 1. Secondly, the thinness of the dump layer, which is almost invisible, in the bar charts of post-copy migration shows another detail: **the execution state of a container is markedly negligible with respect to the size of memory pages.** Another consideration that can be made by looking at Figure 11 is that **a container is an environment that occupies and updates more memory than that of the application running inside it**. Indeed, as discussed in Section 5.1, the server running in the container allocates 75 MB of RAM, but more than 110 MB are transferred on average during container migration. Similarly, the dump size for pre-copy migration and the faulted pages size for hybrid migration are greater than what we were expecting, given the two considered page dirtying rates.

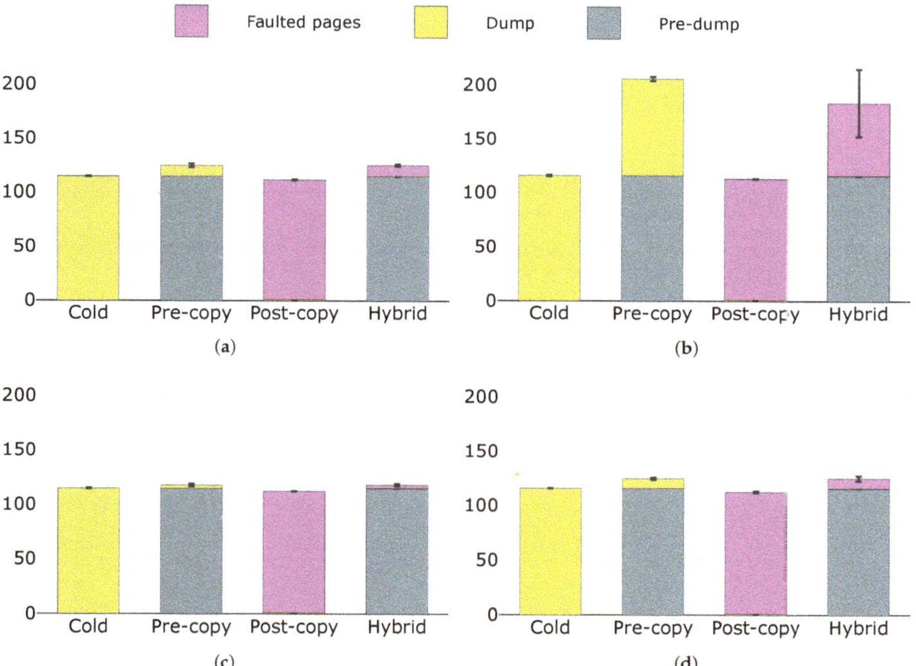


**Figure 11.** Amounts of transferred data (MB) with their components in evidence under: (**a**) configuration A; (**b**) configuration B; (**c**) configuration C; and (**d**) configuration D.

Let us now compare the migration techniques. Cold and post-copy migrations transfer the lowest amounts of data, irrespective of the specific configuration. This is due to the fact that they both transfer each memory page only once. Under configurations A, C, and D, pre-copy and hybrid migrations generate volumes of data that are comparable to those of cold and post-copy migrations, even though slightly higher. Going into detail, pre-copy and hybrid migrations perform at their best under configuration C, where there is the maximum difference between the throughput and the page dirtying rate (i.e., about three orders of magnitude) and, considering the available throughput, the runtime state size leads to a limited pre-copy phase. An increase in the page dirtying rate from configuration C to D augments the amounts of transferred data only for pre-copy and hybrid migrations, but these increases are limited because the page dirtying rate is still of a different order of magnitude from the throughput. Instead, under B, a page dirtying rate of the same order of magnitude of the throughput causes these two migration techniques to modify a quantity of memory pages that is comparable to the whole state and thus transfer significantly greater volumes of data than those of cold and post-copy migrations. In particular, the dump size is what grows in pre-copy migration and the faulted pages size is what increases in hybrid migration, as these are the parts of the state containing dirty pages (see Table 1).

#### 5.2.4. Lessons Learnt

For convenience, in Table 4 we summarise the most salient results relative to the total migration time, the downtime, and the amount of transferred data. As a closing remark, this work shows that **no migration technique is the very best** under all network and service conditions. However, based on the discussed results, we can conclude by stating that in general:

**Table 4.** Summary of the evaluated migration techniques.

| Technique | Total Migration Time | Downtime | Transferred Data |
|---|---|---|---|
| Cold | Always the lowest. | Always the highest. Coincides with the total migration time. | Always the least. Comparable to those of post-copy. |
| Pre-copy | Comparable or lower than that of post-copy when page dirtying rate and throughput are of different orders of magnitude, with page dirtying rate lower than throughput. Higher than that of post-copy when page dirtying rate is of the same order of magnitude of throughput. | Higher than those of post-copy and hybrid when page dirtying rate is of the same order of magnitude of throughput and/or pre-copy phase has a prolonged duration. | Much more than cold or post-copy when page dirtying rate is of the same order of magnitude of throughput. |
| Post-copy | Higher than that of cold, especially with a very high RTT between nodes. Comparable or higher than that of pre-copy when page dirtying rate and throughput are of different orders of magnitude, with page dirtying rate lower than throughput. Lower than that of pre-copy when page dirtying rate is of the same order of magnitude of throughput. | Always low and comparable to that of hybrid. | Always the least. Comparable to those of cold. |
| Hybrid | Always the highest. | Always low and comparable to that of post-copy. | Much more than cold or post-copy when page dirtying rate is of the same order of magnitude of throughput. |

- **Cold** migration is to be avoided under all conditions because it always causes downtimes that are considerably higher than those of the other techniques;
- In situations where the throughput between nodes and the page dirtying rate are of different orders of magnitude, with the former higher than the latter, and the pre-copy phase does not have a prolonged duration (e.g., under configurations C and D), **pre-copy** migration may be the best option. Indeed, it has similar performances to those of post-copy and hybrid migrations, but it is not afflicted by the issues characterising these other two techniques (see Section 4.2.2);
- In situations where the page dirtying rate is of the same order of magnitude of the throughput and/or the pre-copy phase has a prolonged duration (e.g., under configurations A and B), pre-copy is to be avoided mainly because of rather long downtimes. **Post-copy** could be the best alternative, considering that it provides downtimes comparable to those of hybrid migration but performs

better in terms of total migration time and amount of transferred data. It is worth noting, though, that post-copy presents a couple of non-negligible issues, which are explained in Section 4.2.2.

## 6. Conclusions

Containerisation and container migration are fundamental aspects of fog computing, a paradigm that provides resources and services of computing, storage, and networking near to where sensors and actuators are deployed. In this paper, we critically analysed the existing container migration techniques, with a specific focus on their suitability for fog computing environments. We carried out a comprehensive performance evaluation of these techniques over a real fog computing testbed, observing their behaviour under different conditions of throughput and page dirtying rate. After the assessment of total migration time, downtime, and overall amount of transferred data, we can conclude that no container migration technique is the very best under all network and service conditions. Therefore, we identified categories of conditions and pointed out which technique could be the most suitable under each of those categories. The results show that, in general, cold migration is mostly afflicted by long downtimes, while the main shortcoming of hybrid migration is a prolonged total migration time. Pre-copy and post-copy migrations might therefore represent the best options, under different conditions.

As a future work, we plan to evaluate and compare container migration techniques also within cloud computing data centres, where conditions are different than those in fog environments (e.g., throughputs are in general much higher, there is usually the need to also migrate the persistent state). In addition, we would like to propose a novel migration approach that better adapts to the diverse network and service conditions.

**Author Contributions:** Conceptualization, C.P., C.V., E.M., G.M., F.L. and A.P.; methodology, Carlo Puliafito, C.V., E.M., G.M., F.L. and A.P.; software, C.P. and G.M.; investigation, C.P., G.M. and F.L.; data curation, C.P.; writing—original draft preparation, C.P., G.M. and F.L.; writing—review and editing, C.P., C.V., E.M. and A.P.; supervision, E.M.; funding acquisition, E.M.

**Funding:** This research received no external funding.

**Conflicts of Interest:** The authors declare no conflict of interest.

## References

1. Atzori, L.; Iera, A.; Morabito, G. The Internet of Things: A Survey. *Comput. Netw.* **2010**, *54*, 2787–2805,. [CrossRef]
2. Al-Fuqaha, A.; Guizani, M.; Mohammadi, M.; Aledhari, M.; Ayyash, M. Internet of Things: A Survey on Enabling Technologies, Protocols, and Applications. *IEEE Commun. Surv. Tutor.* **2015**, *17*, 2347–2376. [CrossRef]
3. Manyika, J.; Chui, M.; Bisson, P.; Woetzel, J.; Dobbs, R.; Bughin, J.; Aharon, D. *The Internet of Things: Mapping the Value Beyond the Hype*; Technical Report; McKinsey Global Institute: New York, NY, USA, 2015. Available online: http://www.mckinsey.com/business-functions/digital-mckinsey/our-insights/the-internet-of-things-the-value-of-digitizing-the-physical-world (accessed on 20 March 2019).
4. Delicato, F.C.; Pires, P.F.; Batista, T. The Resource Management Challenge in IoT. In *Resource Management for Internet of Things*; Springer International Publishing: Cham, Switzerland, 2017; pp. 7–18. [CrossRef]
5. Chen, M.; Mao, S.; Liu, Y. Big Data: A Survey. *Mob. Netw. Appl.* **2014**, *19*, 171–209. [CrossRef]
6. Bonomi, F.; Milito, R.; Zhu, J.; Addepalli, S. Fog Computing and its Role in the Internet of Things. In Proceedings of the 1st Workshop on Mobile Cloud Computing (MCC), Helsinki, Finland, 17 August 2012; pp. 13–16. [CrossRef]
7. Puliafito, C.; Mingozzi, E.; Longo, F.; Puliafito, A.; Rana, O. Fog Computing for the Internet of Things: A Survey. *ACM Trans. Internet Tech.* **2019**, *19*, 2. [CrossRef]
8. Satyanarayanan, M.; Lewis, G.; Morris, E.; Simanta, S.; Boleng, J.; Ha, K. The Role of Cloudlets in Hostile Environments. *IEEE Pervasive Comput.* **2013**, *12*, 40–49. [CrossRef]
9. Satyanarayanan, M. The Emergence of Edge Computing. *Computer* **2017**, *50*, 30–39. [CrossRef]

10. Shi, W.; Dustdar, S. The Promise of Edge Computing. *Computer* **2016**, *49*, 78–81. [CrossRef]
11. Cisco. Fog Computing and the Internet of Things: Extend the Cloud to Where the Things Are; Technical Report. 2015. Available online: https://www.cisco.com/c/dam/en_us/solutions/trends/iot/docs/computing-overview.pdf (accessed on 14 January 2019).
12. Zhang, F.; Liu, G.; Fu, X.; Yahyapour, R. A Survey on Virtual Machine Migration: Challenges, Techniques, and Open Issues. *IEEE Commun. Surv. Tutor.* **2018**, *20*, 1206–1243. [CrossRef]
13. Choudhary, A.; Govil, M.C.; Singh, G.; Awasthi, L.K.; Pilli, E.S.; Kapil, D. A Critical Survey of Live Virtual Machine Migration Techniques. *J. Cloud Comput.* **2017**, *6*, 1–41. [CrossRef]
14. Ha, K.; Abe, Y.; Eiszler, T.; Chen, Z.; Hu, W.; Amos, B.; Upadhyaya, R.; Pillai, P.; Satyanarayanan, M. You Can Teach Elephants to Dance: Agile VM Handoff for Edge Computing. In Proceedings of the ACM/IEEE 2nd Symposium on Edge Computing (SEC), San Jose, CA, USA, 28 July 2017; pp. 1–14. [CrossRef]
15. Morabito, R.; Cozzolino, V.; Ding, A.Y.; Beijar, N.; Ott, J. Consolidate IoT Edge Computing with Lightweight Virtualization. *IEEE Netw.* **2018**, *32*, 102–111. [CrossRef]
16. Wang, N.; Varghese, B.; Matthaiou, M.; Nikolopoulos, D.S. ENORM: A Framework For Edge NOde Resource Management. *IEEE Trans. Serv. Comput.* **2018**. [CrossRef]
17. Tang, Z.; Zhou, X.; Zhang, F.; Jia, W.; Zhao, W. Migration Modeling and Learning Algorithms for Containers in Fog Computing. *IEEE Trans. Serv. Comput.* **2018**. [CrossRef]
18. Ma, L.; Yi, S.; Carter, N.; Li, Q. Efficient Live Migration of Edge Services Leveraging Container Layered Storage. *IEEE Trans. Mob. Comput.* **2018**. [CrossRef]
19. Puliafito, C.; Mingozzi, E.; Vallati, C.; Longo, F.; Merlino, G. Companion Fog Computing: Supporting Things Mobility Through Container Migration at the Edge. In Proceedings of the IEEE 4th International Conference on Smart Computing (SMARTCOMP), Taormina, Italy, 18–20 June 2018; pp. 97–105. [CrossRef]
20. Nadgowda, S.; Suneja, S.; Bila, N.; Isci, C. Voyager: Complete Container State Migration. In Proceedings of the IEEE 37th International Conference on Distributed Computing Systems (ICDCS), Atlanta, GA, USA, 5–8 June 2017; pp. 2137–2142. [CrossRef]
21. Puliafito, C.; Mingozzi, E.; Vallati, C.; Longo, F.; Merlino, G. Virtualization and Migration at the Network Edge: An Overview. In Proceedings of the IEEE 4th International Conference on Smart Computing (SMARTCOMP), Taormina, Italy, 18–20 June 2018; pp. 368–374. [CrossRef]
22. Kakakhel, S.R.U.; Mukkala, L.; Westerlund, T.; Plosila, J. Virtualization at the Network Edge: A Technology Perspective. In Proceedings of the IEEE 3rd International Conference on Fog and Mobile Edge Computing (FMEC), Barcelona, Spain, 23–26 April 2018; pp. 87–92. [CrossRef]
23. Puliafito, C.; Mingozzi, E.; Anastasi, G. Fog Computing for the Internet of Mobile Things: Issues and Challenges. In Proceedings of the IEEE 3rd International Conference on Smart Computing (SMARTCOMP), Hong Kong, China, 29–31 May 2017; pp. 1–6. [CrossRef]
24. Jiang, Y.; Huang, Z.; Tsang, D.H.K. Challenges and Solutions in Fog Computing Orchestration. *IEEE Netw.* **2018**, *32*, 122–129. [CrossRef]
25. Zhu, C.; Tao, J.; Pastor, G.; Xiao, Y.; Ji, Y.; Zhou, Q.; Li, Y.; Ylä-Jääski, A. Folo: Latency and Quality Optimized Task Allocation in Vehicular Fog Computing. *IEEE Internet Things J.* **2018**. [CrossRef]
26. Fernández-Caramés, T.M.; Fraga-Lamas, P.; Suárez-Albela, M.; Vilar-Montesinos, M. A Fog Computing and Cloudlet Based Augmented Reality System for the Industry 4.0 Shipyard. *Sensors* **2018**, *18*, 1798. [CrossRef]
27. Du, J.; Zhao, L.; Feng, J.; Chu, X. Computation Offloading and Resource Allocation in Mixed Fog/Cloud Computing Systems With Min-Max Fairness Guarantee. *IEEE Trans. Commun.* **2018**, *66*, 1594–1608. [CrossRef]
28. Puthal, D.; Obaidat, M.S.; Nanda, P.; Prasad, M.; Mohanty, S.P.; Zomaya, A.Y. Secure and Sustainable Load Balancing of Edge Data Centers in Fog Computing. *IEEE Commun. Mag.* **2018**, *56*, 60–65. [CrossRef]
29. Nan, Y.; Li, W.; Bao, W.; Delicato, F.C.; Pires, P.F.; Dou, Y.; Zomaya, A.Y. Adaptive Energy-Aware Computation Offloading for Cloud of Things Systems. *IEEE Access* **2017**, *5*, 23947–23957. [CrossRef]
30. Dastjerdi, A.V.; Gupta, H.; Calheiros, R.N.; Ghosh, S.K.; Buyya, R. Fog Computing: Principles, Architectures, and Applications. In *Internet of Things*; Elsevier: Amsterdam, The Netherlands, 2016; pp. 61–75.
31. Habib, I. Virtualization with KVM. *Linux J.* **2008**, *2008*, 8. Available online: http://dl.acm.org/citation.cfm?id=1344209.1344217 (accessed on 15 January 2019).
32. Asvija, B.; Eswari, R.; Bijoy, M.B. Security in Hardware Assisted Virtualization for Cloud Computing—State of the Art Issues and Challenges. *Comput. Netw.* **2019**, *151*, 68–92. [CrossRef]

33. Desai, A.; Oza, R.; Sharma, P.; Patel, B. Hypervisor: A Survey on Concepts and Taxonomy. *Int. J. Innov. Technol. Explor. Eng.* **2013**, *2*, 222–225.
34. Soltesz, S.; Pötzl, H.; Fiuczynski, M.E.; Bavier, A.; Peterson, L. Container-based Operating System Virtualization: A Scalable, High-performance Alternative to Hypervisors. *ACM SIGOPS Oper. Syst. Rev.* **2007**, *41*, 275–287. [CrossRef]
35. Morabito, R. Virtualization on Internet of Things Edge Devices with Container Technologies: A Performance Evaluation. *IEEE Access* **2017**, *5*, 8835–8850. [CrossRef]
36. Bellavista, P.; Zanni, A. Feasibility of Fog Computing Deployment Based on Docker Containerization over RaspberryPi. In Proceedings of the 18th International Conference on Distributed Computing and Networking (ICDCN), Hyderabad, India, 5–7 January 2017; pp. 1–10. [CrossRef]
37. Ismail, B.I.; Goortani, E.M.; Ab Karim, M.B.; Tat, W.M.; Setapa, S.; Luke, J.Y.; Hoe, O.H. Evaluation of Docker as Edge Computing Platform. In Proceedings of the IEEE Conference on Open Systems (ICOS), Melaka, Malaysia, 24–26 August 2015; pp. 130–135. [CrossRef]
38. Kozhirbayev, Z.; Sinnott, R.O. A Performance Comparison of Container-based Technologies for the Cloud. *Future Gen. Comput. Syst.* **2017**, *68*, 175–182. [CrossRef]
39. Biederman, E.W.; Networx, L. Multiple Instances of the Global Linux Namespaces. In Proceedings of the Linux Symposium, Ottawa, ON, Canada, 19–22 July 2006; Volume 1, pp. 101–112.
40. Ali Babar, M.; Ramsey, B. Understanding Container Isolation Mechanisms for Building Security-Sensitive Private Cloud; Technical Report. 2017. Available online: https://www.researchgate.net/publication/316602321_Understanding_Container_Isolation_Mechanisms_for_Building_Security-Sensitive_Private_Cloud (accessed on 25 March 2019).

Article

# A Fog Computing Solution for Context-Based Privacy Leakage Detection for Android Healthcare Devices

**Jingjing Gu [1,*], Ruicong Huang [1], Li Jiang [1], Gongzhe Qiao [1], Xiaojiang Du [2] and Mohsen Guizani [3]**

1 MIIT Key Laboratory of Pattern Analysis and Machine Intelligence, College of Computer Science and Technology, Nanjing University of Aeronautics and Astronautics, Nanjing 211106, China; huangruicong@nuaa.edu.cn (R.H.); nuaa_jiangli@nuaa.edu.cn (L.J.); 1002351818@alumni.sjtu.edu.cn (G.Q.)
2 Department of Computer and Information Sciences, Temple University, Philadelphia, PA 19122, USA; dux@temple.edu
3 College of Engineering, Qatar University, Doha 2713, Qatar; mguizani@ieee.org
* Correspondence: gujingjing@nuaa.edu.cn

Received: 16 January 2019; Accepted: 4 March 2019; Published: 8 March 2019

**Abstract:** Intelligent medical service system integrates wireless internet of things (WIoT), including medical sensors, wireless communications, and middleware techniques, so as to collect and analyze patients' data to examine their physical conditions by many personal health devices (PHDs) in real time. However, large amount of malicious codes on the Android system can compromise consumers' privacy, and further threat the hospital management or even the patients' health. Furthermore, this sensor-rich system keeps generating large amounts of data and saturates the middleware system. To address these challenges, we propose a fog computing security and privacy protection solution. Specifically, first, we design the security and privacy protection framework based on the fog computing to improve tele-health and tele-medicine infrastructure. Then, we propose a context-based privacy leakage detection method based on the combination of dynamic and static information. Experimental results show that the proposed method can achieve higher detection accuracy and lower energy consumption compared with other state-of-art methods.

**Keywords:** privacy leakage detection; intelligent medical service; fog computing; Android; context information

## 1. Introduction

Intelligent medical service systems integrate the wireless internet of things (WIoT), such as medical sensors, wireless communications, and middleware techniques to monitor and analyze the patient's physical health in the form of portable, wearable or body-embedding micro-intelligent personal health devices (PHDs). It can also collect and analyze a large amount of patients' data by various PHDs to perform the disease diagnosis and prevention both inside and outside hospitals with a flexible doctor-patient communication way. In various PHDs with diverse uses, there are a great quantity of devices with Android installed. With the open-source flexibility, strong content delivery system, and numerous Android's consumers, Android-based PHDs present significant advantages for both designers and consumers.

However, most intelligent medical devices are vulnerable to external attacks, especially when connected to the network or to different types of custom cloud servers, where malicious attackers are ubiquitous. According to the report of Android malicious apps by the 360 Cyber Security Center in 2016 [1], a cumulative of 14.033 million new samples of malicious programs on Android platforms were intercepted. Things are even worse in the medical field and caused great security concerns. There have been many security accidents caused by hacking of medical equipment or related mobile devices [2,3].

For example, a blackmail software attacked some hospitals both in the USA and Germany [4,5] to invade patient monitors and drug distribution systems. As reported by Kaspersky Lab's global research and analysis team [6], hackers can easily find wireless devices in hospitals and control the network, or even some PHDs for obtaining the patients' information. What's more, a large number of mobile health applications have actively collected users' sensitive information and sent it to their vendors or other third-party domains over HTTP using plaintext [7], which greatly increases the risk of consumers' privacy being leaked.

Therefore, how to build a secure intelligent medical service system and protect patient's privacy still remains a very challenging research issue. There have been some works on privacy leakage detection and privacy protection in the wireless sensor network [8–11], which are generally considered from three aspects: static analysis, dynamic analysis, and integrated analysis of static and dynamic. (1) Static analysis uses static data flow to analyze the direction of the sensitive data flow in the program with the Android package (APK) file [10,12–14]. It could detect efficiently with high code coverage, but is not applicable to the analysis of apps with multi-thread methods. (2) Dynamic analysis, on the contrary, could avoid the shortcomings of static analysis when monitoring the running state of software [11,15–18]. It compensates for static analysis in detection accuracy, but costs much more code coverage, and often lags behind leakage events during the detection. (3) Integrated analysis combines static and dynamic analysis [19], which consists of software piling, automated testing, and protective systems. By the integrated analysis, monitoring codes are inserted through static code piling to obtain data flow information and sensitive application programming interfaces (APIs) usage data. Then the repackaged software is automatically tested and a protective layer is provided to protect devices from malicious software attacks [20].

Our proposal is motivated by such the integration of static and dynamic analysis. However, most of traditional privacy protection methods are unsuitable because PHDs based on WIoT need a strong technological foundation for their rapid development from both the hospitals and patients. Therefore, in this paper, we develop a novel context-based privacy leakage detection method, which is based on an invented fog computing solution [21] for Android PHDs and services. Specifically, first, we design a privacy protection framework for intelligent medical service systems based on fog computing. In this framework, we can monitor privacy leakage of PHDs with the Android system in real time, and process user's privacy data and the real-time operation status at the fog. Second, we propose an privacy leakage detection method based on Android application by utilizing the context information (described in Section 4). The proposed method combines the static stain analysis with the dynamic hook monitoring, which could effectively detect privacy leakage and provide protection. The experimental results show that our method can achieve higher detection accuracy and lower energy consumption compared with other state-of-art ones.

## 2. Related Work

### 2.1. Intelligent Medicine and Fog Computing

Intelligent medical service and fog computing are hot research topics recently. For example, Ref. [22] proposed an architecture of personalized medical service based on fog computing, and optimized it by the clustering method. Ref. [23] proposed a method of combining drivers' mHealth data and vehicular data for improving the vehicle safety system to solve the problem of road accidents and incidents, due to various factors, in particular the health conditions of the driver. To deal with the increasing false alarms in frequently changing activities, Ref. [24] presented a user-feedback system for use in activity recognition, which improved alarm accuracy and helped sensors to reduce the frequency of transactions and transmissions in wireless body area networks. Ref. [25] addressed some threats of mHealth networks and focused on the security provisioning of the communication path between the patient terminal and the monitoring devices. To solve the problem of response delay and resources waste in the case of increasing complexity, Ref. [26] put forward a fog-based cloud model for time-sensitive medical applications. Ref. [27] designed a medical and health system based on fog aided computing, which classified user infection categories by decision tree and generates diagnostic alerts in fog layer. To diagnose and prevent the outbreak of Chikungunya virus, Ref. [28] put forward a medical and healthcare privacy data protection scheme based on fog calculation. Ref. [29] proposed a multilevel architecture based on fog computing. Ref. [21] suggested that fog computing will be widely used in intelligent medicine in the future.

### 2.2. Security Based on Android Platforms

Security issues have always been the focus of network research [30–33]. With the popularity of Android medical equipment and the emergence of malicious software, the privacy protection of Android platform has caused widespread concern in the academic field in recent years. Generally, the research of privacy leakage detection is considered from three aspects: static, dynamic, and integrated analysis of static and dynamic. Static analysis analyzes the APK file, and uses static data flow to analyze the direction of the static sensitive data flow in the program. For instance, Ref. [12] proposed a static privacy leakage analysis system, which first created the mapping between API functions and required permissions. Refs. [10,13,14] used inter-application interactions in Android to mark the components in security. By the static analysis, privacy leakage detection of Android had high code coverage of the software. However, static analyses are incapable of analyzing apps with reflection, multi-threaded, or reference methods. Since static analysis cannot obtain the running state of the softwares, its accuracy may be unsatisfied. Dynamic analysis can avoid such a shortcomings when monitoring the running state of a software. Ref. [15] designed the TaintDroid to perform a dynamic analysis. Ref. [11] performed a dynamic stain analysis of the running mode defaulted by Google in Android 5.0 and the above systems. Similar methods, namely to detect privacy leaks by modifying system codes are DroidBox [16], Mobile-Sandbox [17,34], VetDroid [35], AppFence [36], FlaskDroid [37]. Ref. [38] proposed a privacy leakage monitoring system to repackage the software and insert the monitoring logic codes. Similar systems are AppGuard [39] and Uranine [18]. But detection results of the dynamic analysis possibly lagged behind leakage events [40]. Therefore, some works combined static and dynamic analysis [19,20]. For instance, AspectDroid [19] inserted monitoring codes through static bytecode instrumentation to automatically test the repackaged software, and added a protective layer to protect the device from malicious software attacks. AppIntent [20] combined static data flow analysis and dynamic symbolic execution to verify privacy disclosure, which reduced the search space without sacrificing code coverage.

The works above can solve the problem of privacy leakage to a certain extent, but there are still some shortcomings, as follows: (1) static analysis is unable to get the dynamic running information of the software. Many malicious apps can download executable codes to avoid from the static detection; (2) dynamic analysis usually sacrifices code coverage, and some methods require modification of the source codes of Android systems, which increases difficulty of development at the expense of some system resource; (3) analysis based on app repackaging has some impact on the original app, and some apps are resistant to these methods by using encrypted packers.

## 3. The Privacy Leakage Detection Framework Based on Fog Computing

In this section, we propose a fog computing framework for privacy leakage detection of healthcare networks to protect the intelligent medical service systems. Basically, it monitors various applications on PHDs in real time, detects malicious codes, and feeds detection results back to users. Moreover, this framework is combined with fog computing to conduct encryption, decryption, and identity authentication of the user's privacy data.

### 3.1. Intelligent Personal Health Devices

The applications of intelligent PHDs can be divided into two parts.

For data collection, PHDs collect data through diverse sensors on users and upload data to a healthcare monitoring and management center, so as to perform 24-h health monitoring. PHDs collect and send users' physiological health data, such as electrocardiography (ECG), heart rate, and blood pressure. Figure 1 is a diagram of commonly used PHDs, which fall into three categories: (1) portable small medical devices, or micro-intelligent chip devices which can be worn on or embedded into the human body; (2) indoor care devices, such as smart medicine boxes; (3) medical equipment used in hospitals, such as intelligent film extractors.

For data application, the collected data is sent to the fog for customizing different treatments to users. For example, smart medicine box reminds the user to take medicine on time. The intelligent infusion pump adjusts the infusion rate by observing the changes of blood pressure and other information. The intelligent atomizer can revise the atomization time and dose according to the body condition. The intelligent film taker reduces the queue waiting time, and outputs corresponding medical images based on patients' biological characteristics. The intelligent dispenser can precisely configure user-defined drugs.

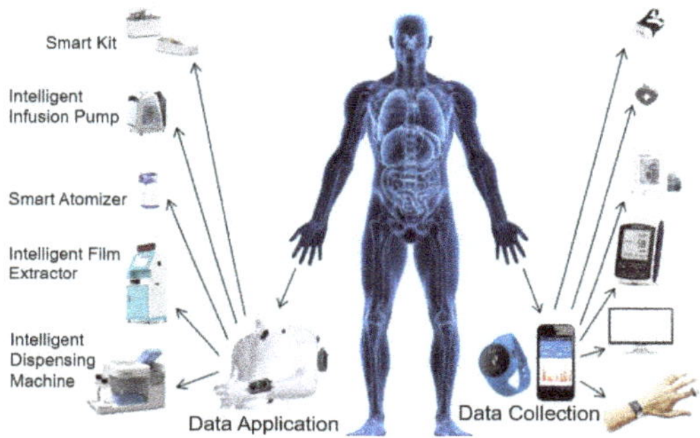


**Figure 1.** Healthcare devices.

### 3.2. System Architecture

In this paper, we propose a privacy leakage detection method based on fog computing framework, which is a higy vitalized platform that provides computing, storage, and network services. The architecture of the designed system here is shown in Figure 2, which shows the components and interrelationships, namely cloud, fog interface and intelligent medical terminal. Generally, fog nodes work between terminal devices and traditional cloud computing data centers, which means that they are physically closer to the users. Besides, fog has less requirements of network bandwidth, so it reduces the network costs and time delay. Due to these features, fog not only extends the capabilities of the cloud, but also reduces the requirements for the organization to apply it.

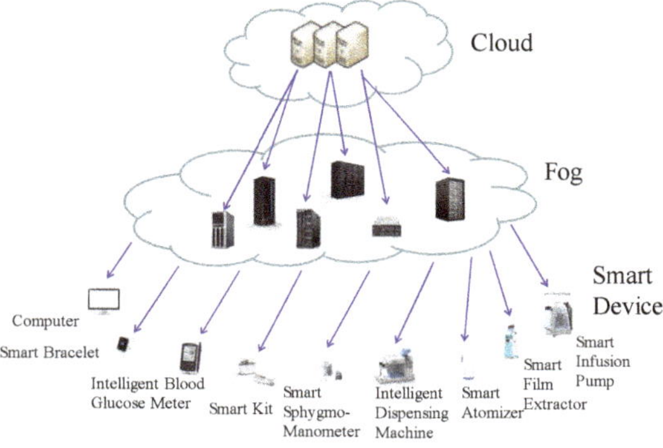


**Figure 2.** Three-layer architecture.

As illustrated in the Figure 2, the cloud manages the data storage, computing, and information processing, while the fog mainly provides computing and storage resources for the lower level, including making the information of upper and lower levels inter-operable, data analysis and management, and security protection. As the services could require excessive computing and storage resources beyond the capacity of the fog, the cloud will provide a replacement service at this situation. In the lower level, terminal devices carry out the collection of data, transmission and simple processing of information, and so on.

Figure 3 is the logical architecture of fog computing for the healthcare network. The first layer is the intelligent PHDs layer, including smart wristbands, smart blood glucose meters, smart film extractors, etc. They are mainly used for collecting health information of users, and sending the information to the fog computing layer for the further processing.

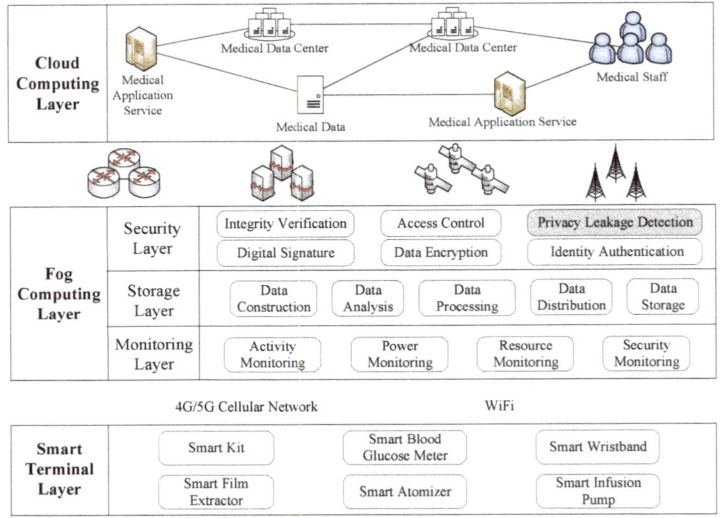


**Figure 3.** Logical architecture of fog computing for the healthcare network.

The second layer is the fog computing layer, composed of three sub-layers: monitoring layer, data storage layer, and security protection layer. The monitoring layer includes activity's monitoring, power status monitoring, resource monitoring, and service monitoring. In this sub-layer, monitoring information is sent to users and abnormal information is detected. In the data storage layer, data received from the intelligent PHDs layer are filtered and pruned for data analysis to extract the privacy information. In the security protection layer, there are integrity verification, access control, digital signature, data encryption, identity authentication and privacy leakage detection which is the main issue in this paper. Specifically, based on fog computing, we design an Android malicious code monitoring scheme to prevent intrusion by illegal users and dynamic malware monitoring of various applications on devices.

The third layer is the medical cloud computing layer for processing, storing and generating user personal files.

### *3.3. Privacy Leakage Detection*

In this paper, we design a privacy leakage detection method with the combination of the fog and users due to the strong capabilities of data processing and network control of the fog. It is mainly for intelligent PHDs based on Android systems, and protects the user's private information by monitoring in real time.

Basically, we use context analysis technology to design the detection scheme, including static privacy leakage analysis and dynamic privacy disclosure monitoring, as shown in Figure 4. First, we use static analysis to analyze the permissions mapping to various API functions, system and user interface events, static taint propagation path, and function calls. Next, we perform dynamic privacy leakage monitoring, which mainly includes the following four stages: (1) the users' information and system working status are collected by PHDs at the user terminal for constructing the context information and transmitting to the fog; (2) on the fog, the privacy data is extracted for encryption, and the monitoring data collected in the user terminal is analyzed for performing the privacy leakage detection by the access control technology (Section 4.2); (3) if a privacy leakage is detected, the next data-transmission will be blocked. Meanwhile, the fog will intercept the behavior of the privacy leakage and notify the user for protecting the user's information; (4) the fog uploads the user information and the system status to the cloud periodically.

Note that, for the convenience, in this paper we use MyPrivacy to present the collection mechanism of the privacy and system information on the user terminal, and FogPrivacy to present the privacy protection mechanism (such as privacy leakage detection) on the fog.

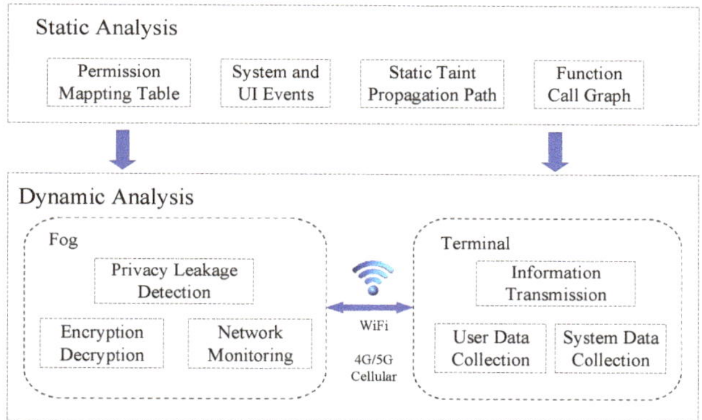


**Figure 4.** Framework of privacy leakage detection.

## 4. Context-Based Privacy Leakage Detection

Our context-based privacy leakage detection method includes two parts: static privacy leakage analysis and dynamic privacy leakage monitoring, which are carried out by the combination of the fog and the user terminal. The static analysis constructs the context of the privacy-related API function to predict trigger events and the possible privacy leakage of the API call. Dynamic monitoring intercepts privacy API by using hook technology to predict the privacy leaks which may be caused by API calls. If there is a privacy leakage, it will be automatically blocked.

### *4.1. Static Privacy Leakage Analysis*

The static analysis is used to construct the context of the software privacy-related API functions, which is based on the FlowDroid [41]. From the static analysis, the path between sources and sinks can be found, and the sequence of sensitive function calls could be extracted. The framework of the static privacy leakage analysis mechanism is shown in Figure 5, which contains five parts:

(1) Static taint propagation path: it inputs the original APK application installation package, and configure the sources and sinks function files. Then the system performs the static stain analysis through the FlowDroid platform and finds out the possible privacy leakage path. Here,

we use the data of the Susi project [42] to mark the sources and sinks functions for increasing the coverage of the privacy functions.

(2) Function call graph: it extracts the Java codes from the decompiler APK, then constructs the function call graph using Soot [43], a Java language optimization framework.

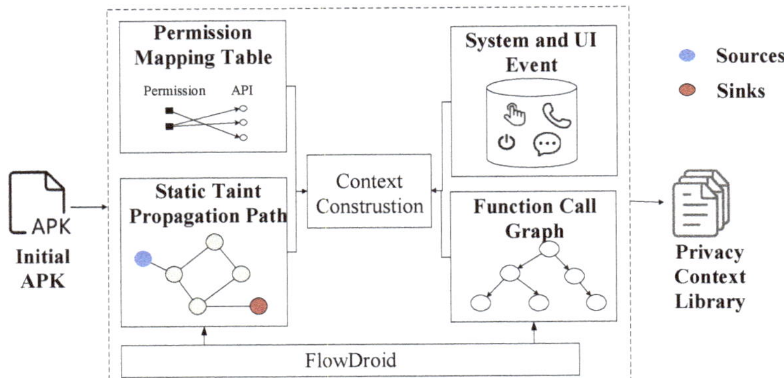


**Figure 5.** Frame of static privacy leakage analysis mechanism.

(3) Permission mapping table: it generates the permission-API mapping table based on the PScout project [44], which describes the relationship between an API and its corresponding permissions by scanning the Android code and has a more accurate result than the Google's official API.

(4) System and UI event: it analyzes privacy entry functions and triggering conditions from a large amount of system and UI event information. Here, the system events include various callback functions of Android system (such as receiving text messages and changing network state), as well as the lifecycle functions of Android components (*onCreate*, *onRestart*, etc.). The UI events contain the user's interaction with the software interface (such as clicking a button, pressing the volume key).

(5) Context construction: it builds the corresponding context information based on Algorithm 1.

Here, we present the definitions and description of Algorithm 1 as follows.

**Definition 1.** *A function call graph is a directed graph $CG = (N, E)$, where N represents the set of functions in the software and E is the set of edges. For example, $e(a, b) \in E$ represents that the function a calls function b.*

**Definition 2.** *In a path $p_{s2s}$ from source to sink, $p_{s2s} = n_{source}n_1n_2...n_{sink}$, where $n_i \in N(i = source, 1, 2, ..., sink)$ are called a privacy leakage path.*

**Definition 3.** *In a function call graph $CG = (N, E)$, if there is a path $p = n_e n_1 n_2 ... n_{source}$, and there are no edges that go into $n_e$ ($\forall n \in N, e(n, n_e) \notin E$), then $n_e$ is called a privacy entry point, which means that no functions in N call $n_e$. Since Android is an event-driven operating system and its components used for development have their own lifecycles, the privacy entry point functions generally consist of various message response functions and lifecycle functions.*

**Definition 4.** *A privacy API function context information PrivacyContext is a triple shown in Equation (1),*

$$PrivacyContext = (api, permission, context), \tag{1}$$

where 'api' represents the name of the privacy-related function. The set 'permissions' is the set of permissions that the privacy-related function requires. *context* is the set of $< p_{s2s}, n_e >$ pairs, where each pair contains a privacy leakage path $p_{s2s}$ and its corresponding entry point function $n_e$.

**Algorithm 1** Context construction algorithm.

**Input**
Function Call Graph $CG$
Privacy Disclosure Path $Paths$
System and UI Events $Events$
Permission Mapping Table $Table$
**Output**
Context $PrivacyContext$
**Begin**
$PrivacyContext = null$
$context = null$
**for all** $e \in E$ **do**
**if** $e.target \in PrivacyAPI$ **then**
$n_e = getEntryPointFromCG(e.origin)$ //Retrieval entry function from CG
$permission = getPermissionFromTable(e.target)$ //Retrieval API permission
**for all** $path \in Paths$ **do**
**if** $path.source == e.target$ **then**
$context.add(< path, n_e >)$ //Add context
**end if**
**end for**
$PrivacyContext.add(e.target, permission, context)$
$context.clear()$
**end if**
**end for**
**return** $PrivacyContext$

The main idea of the Algorithm 1 includes: (1) traverse each edge from the function call graph $CG$, and locate the privacy-related function and its corresponding permission using the permission mapping table; (2) get the privacy entry point functions for the API call from the function call graph. Privacy entry point functions are defined in Definition 2; (3) for a chosen edge, find out all subsequent edges in static taint propagation path as possible privacy leakage path for the API call, and generate the context information for the privacy-related API function. After context construction, the context information PrivacyContext is loaded into the privacy context library, which will be deployed on the fog.

Here, we list the general categories of the privacy data in the system of PHDs in Table 1, including device resource, system information, login data and user data. Here, device resource represents the information of devices, which depends on the specific input from external devices (e.g., GPS). Its privacy-related API functions are getLatitude() and getLongitude(). System information describes attributes and labels of the device systems (e.g., international mobile equipment identity (IMEI)). Its privacy-related API function is getDeviceId(). Login data is the login data entered by the user, which mainly includes the account password. Its privacy-related API function is getPasswd(). User data is user-related information, such as step count, heart beat and sleep status, with the privacy-related API functions stepListener, heartListener and sleepListener respectively.

**Table 1.** Some sensitive privacy-related API functions.

| Classification | Example | API |
|---|---|---|
| Device Resource | GPS | getLatitude, getLongitude |
| System Infomation | IMEI | getDeviceId |
| Login Data | Password | getPasswd |
| User Data | StepCount | stepListener |
| User Data | SleepStatus | sleepListener |
| User Data | Heartbeat | heartListener |

## 4.2. Dynamic Privacy Leakage Monitoring

Static analysis cannot reflect the real state of the app. Besides, some malicious apps can download malicious third party libraries and executable programs and execute them dynamically to steal privacy information. Static analysis can not detect this kind of attack efficiently. Therefore, we proposed a dynamic monitoring scheme for privacy leakage based on fog computing, which is realized by the combination of the fog and user terminal. On the user terminal, dynamic behaviors of the app are monitored by the key privacy-related API function of dynamic hook technology. Then the real state of the app is obtained and the relevant information is sent to the fog. At the fog, similarity between the static analysis results and the dynamic behavior information is calculated to find out the possible privacy leakage risks of the dynamic API calling behavior. Finally, the result of similarity comparison is sent back to the user terminal and the software behavior which poses a risk of privacy leakage would be blocked and informed back to the user. The framework of the proposed dynamic monitoring mechanism is shown in Figure 6.

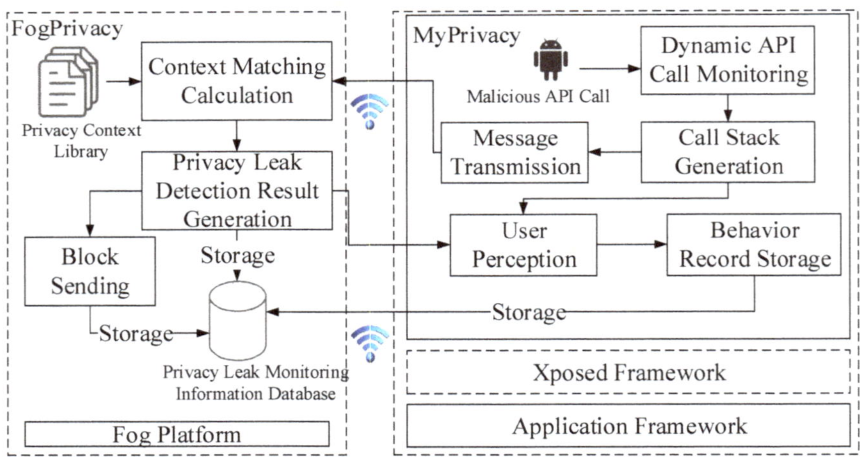


**Figure 6.** Dynamic privacy leakage monitoring.

The main modules in the Figure 6 have the following functions:

(1) Dynamic API call monitoring: dynamic API call monitoring module uses Xposed-based hook technology to write privacy-related API monitoring code. By collecting API function call stack information, the dynamic API execution context information is constructed and sent to the fog.
(2) Context matching calculation: on the fog, the context information database of privacy API functions derived from the static analysis is matched with the context information of API dynamic execution. The matching calculation algorithm is shown in detail below.
(3) User perception: when the detection result of the fog indicates that the suspicious call may cause privacy leakage, the fog will send the detection results to the user terminal. Through the user

perception module, the event information that triggers the API and the risk of possible privacy leakages will be prompted to the user. The system intercepts the invocation of the related API and blocks the privacy leakage.

(4) Behavioral log module: the behavioral log module makes quick judgments about similar situations during follow-up monitoring. The module would format and store the information of each API call and the user's choice. Then it feeds the information back to the fog, and the fog will store these information in the privacy leakage monitoring information database.

(5) Privacy leakage monitoring information database: the privacy leakage monitoring information database keeps information on privacy leakage monitoring of all PHDs on the fog and uploads it regularly to the cloud for permanent preservation.

The context matching algorithm (Algorithm 2) and its related definitions are introduced below.

**Algorithm 2** Dynamic context matching algorithm.

```
Input
  Context DynamicContext, PrivacyContext
Output
  The closest pc of the API call
Begin
  similarity = 0
  result = null
  for all pc ∈ PrivacyContext do
    if pc.api == DynamicContext.api then
      simTemp = Similarity(DynamicContext.stack, pc.context) //Calculate similarity
      if similarity < simTemp then
        similarity = simTemp
        result = pc// Update result
      else if similarity == simTemp then
        result.add(pc)// Add result
      end if
    end if
  end for
  return result
```

**Definition 5.** *The execution context information of a dynamic API (DynamicContext) represents the api and call stack information, as shown in expression (2),*

$$DynamicContext = (api, stack < funcs >) \quad (2)$$

where $api$ represents the system function api calls. $stack < funcs >$ represents the call stack information of the function.

**Definition 6.** *Given an API dynamic execution context information* $DynamicContext = (api, < f_1, f_2, ..., f_n >)$ *and a call path* $p = n_1 n_2 ... n_m$ *in the static function call graph CG, the similarity between them is calculated according to Equations (3) and (4).*

$$Similarity = \frac{\sum_{i=1}^{n} \sum_{j=1}^{m} F(f_i, n_j)}{n} \quad (3)$$

$$F(f_i, n_j) = \begin{cases} 1, & f_i = n_j \\ 0, & f_i \neq n_j \end{cases} (1 \leq i \leq n, 1 \leq j \leq m), \tag{4}$$

where $n$ is the length of the function call stack in DynamicContext, that is, the number of functions in the function call stack. $m$ is the length of the call path $p$. Function $F$ is used to identify whether two functions for similarity calculation are equal.

From the Equations (3) and (4), the similarity is calculated. Then, we can get the context information which is closest to the API function call from Algorithm 2. With the information, it is possible to predict the privacy disclosure that may occur when the API is called. We extract Android system events from the dynamic execution context information of the API, which directly lead to the API call. When a privacy leakage occurs, calls are blocked and the leakage is prompted for privacy protection.

## 5. Experimental Verification and Results Analysis

In order to verify the effectiveness of the proposed method, we conducted two kinds of experiments, i.e., the static analysis experiment and the dynamic monitoring experiment. We used a testing PC machine (CPU: Intel Core i5-6500, RAM: 8GB, OS: 64-bit Ubuntu 16.04) as a fog node and a *Samsung GT-I9500, Samsung Electronics Co., Ltd., Korea* (https://www.samsung.com) (Android 4.4.4 system with Xposed) to conduct the experiments. The dataset consisted of 397 malicious samples and 300 benign samples. The malicious samples were selected from the DREBIN dataset [45,46], and the benign ones included 10 types of apps, all of which are downloaded from Google Store and China Mall. We use this dataset to verify the algorithm and the reliability of the proposed scheme.

### 5.1. Privacy Leakage Monitoring Experiments

First, we performed the static privacy leakage detection experiment. We constructed the context information of the privacy-related API function of the software. In our experiments, we considered on following types of privacy: international mobile equipment identity (IMEI), international mobile subscriber identification number (IMSI), integrated circuit card identity (ICCID), short messaging service (SMS), contacts, phone number, and location. For leakage events, we focus on the network transmissions, logs and SMS messages. From the context information, we can obtain kinds and proportions of the privacy data, as shown in Figure 7.

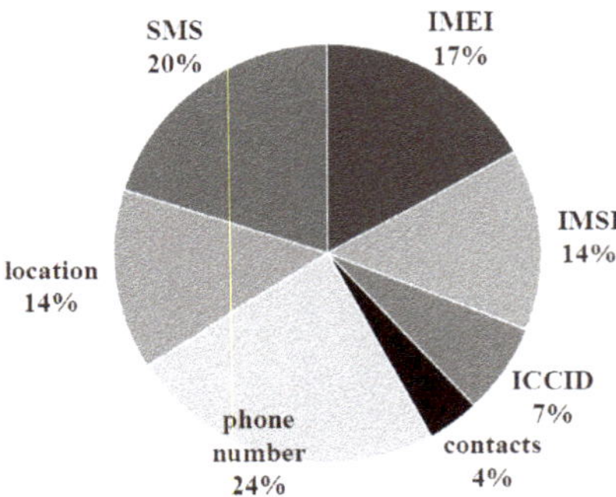


**Figure 7.** Proportion of privacy leakage types.

From the Figure 7, we can find that leakage of phone number is the most common, reaching about one quarter (24%), followed by short messages (20%), IMEI (17%), IMSI (14%) and location (14%).

From the privacy-related API context information, we also can get the following data, as shown in Table 2 by counting the data of each privacy entry.

**Table 2.** Entry point statistics for privacy leakages.

| Entrypoint | Lifecycle Method | System Event | UI Event |
|---|---|---|---|
| Proportion | 84.5% | 9% | 6.5% |

Table 2 shows that the privacy entry point functions dominated by the lifecycle functions (described in the Section 4.1). For several apps (package name: com.gp.lights and com.keji.danti607), the static analysis of them finds that, in order to disguise themselves, malware actions often occur when the status of Android components change. Privacy leakages that happened on the first run of the app were rarely seen, for reducing the probability of being discovered by users.

Furthermore, we installed the test application on the real machine, and built the Xposed framework with the coded MyPrivacy and the FogPrivacy program, which includes:

(1) Detection platform: in order to compare the accuracy of our method, we used the test results of DroidBox platform as a baseline, which modifies Android systems based on TaintDroid [15] and has extra functions of stain analysis and call monitoring. With its output (the log), we can analyze the detection results.
(2) Behavior triggers: Generally, malware actions (such as privacy theft) were set to be triggered under certain conditions for hiding their sensitive behavior. This type of malware makes function calls by tapping of system events, which is declared in the AndroidManifest.xml file of all tapping events in Android. Thus we decompiled the APK file, and extracted the tapping events to be stored in the the database.
(3) Result analysis: We used manual de-compilation to analyze the results by using JEB2 [47] to find out the reason that caused the differences.

We compared results of two platforms, as shown in Table 3. From the table, we can observe that MyPrivacy detects more privacy leakages than DroidBox. MyPrivacy had 2876 pop-up windows and DroidBox had 2431 MyPrivacy leaks in its log records. There are 1780 same leakage events in the same operation, which means that both detection platforms successfully detected the same 1780 leakages. After analyzing the results and manual de-compiling of the software, we found that DroidBox run the tests with an emulator, which could be detected by some malwares through IMEI number, telephone number and other information. As a consequence, some malicious codes could not successfully triggered. MyPrivacy, however, is installed in a real mobile phone, making this type of privacy to be detected.

**Table 3.** Accuracy comparison

| Platform | Total Leakage Count | Same Results |
|---|---|---|
| MyPrivacy(FogPrivacy) | 2876 | 1780 |
| DroidBox | 2431 | |

### 5.2. Comparison of Experimental Results

As *hook* technology used in the proposed method, we tested the system performance before and after the installation of MyPrivacy on the Antutu Benchmark, Quadrant Benchmark respectively, on the Android phones (*Samsung GT-I9500, Samsung Electronics Co., Ltd., Korea*, Google Nexus (www.google.com/nexus/), *Xiaomi M6, Xiaomi Technology Co., Ltd., China* (https://www.mi.com/global/mi6/) and *Huawei STF-AL10, Huawei Technologies Co., Ltd., China* (https://www.huawei.com/cn/)). The results are shown in Table 4.

**Table 4.** System performance.

| Platform | Before | After | Phone Type | Extra Energy Consumption |
|---|---|---|---|---|
| AnTuTu | 56,333 | 53,629 | SamSung SM-N900 | 4.8% |
| | 40,479 | 39,186 | Google Nexus 5 | 3.19% |
| | 55,186 | 52,923 | Xiaomi MI6 | 4.10% |
| | 68,357 | 65,212 | Huawei STF-AL10 | 4.60% |
| Quadrant | 49430 | 47670 | Samsung SM-N900 | 3.56% |
| | 36,320 | 34,849 | Google Nexus 5 | 4.05% |
| | 39,035 | 37,559 | Xiaomi MI6 | 3.78% |
| | 39,882 | 38,729 | Huawei STF-AL10 | 2.89% |

In Table 4, "*before*" and "*after*" mean the benchmark results of evaluating the performance of each hardware before and after MyPrivacy is installed on a device. According to the evaluation results, MyPrivacy (FogPrivacy) caused a little extra energy consumption (no more than 5%) on all phone systems and platforms, which was within the allowable range.

Table 5 shows the comparison results between our method and some other privacy leakage detection methods on the Android system: LeakMiner [15], FlowDroid [11], TaintDroid [35] and Aurasium [41]. Since most of the comparative methods do not provide test data and system source codes, we conducted the comparison from the perspective of the analytic methods and features, i.e., whether customized system was needed, whether modification of the application itself was needed and whether this method was able to prevent the leakage.

**Table 5.** Comparison with other systems.

| System | Method Type | Feature | Customized System Is Needed | Modification of Application Is Needed | Able to Prevent Leakages |
|---|---|---|---|---|---|
| LeakMiner | Static Analysis | Function Call Graph | No | No | No |
| FlowDroid | Static Analysis | Static Taint Analysis, etc. | No | No | No |
| TaintDroid | Dynamic Analysis | Dynamic Analysis, etc. | Yes | No | No |
| Aurasium | Dynamic Analysis | App Repackaging | No | Yes | Yes |
| Our Method | Static and Dynamic Analysis | Function Call Graph, Dynamic Analysis, etc. | No | No | Yes |

In the systems shown in Table 5, the methods of privacy leakage detection based on different analysis and detection strategies (i.e., features) were selected to perform the detection. LeakMiner used the static function call graph as the basic analysis data, by calling the reachable relationship of the marked function in the graph to determine whether there is a privacy leakage. The method was simple to practice, and with high code coverage. Similarly, FlowDroid used static taint analysis, taking Android lifecycle functions into consideration. These two systems did not need to modify the app or Android system. However, they were both unable to prevent leakages when the app was running due to the shortcomings of static analysis, i.e., offline analysis. TaintDroid and Aurasium were two privacy leakage detection schemes based on dynamic analysis. TaintDroid modified the system and inserted taint analysis code, and Aurasium repackaged the software itself to add privacy disclosure decision logic. Both of them can carried out real-time privacy data usage monitoring. However, TaintDroid can just conduct privacy leakage reports in the form of system notifications, and it modifies the system codes, making it less adaptive. Aurasium allows users to intercept the leakage, but repackaging may affect the app, or may fail if the app uses some reinforcement methods. In this experiment, our method consistently performed the best in all conditions, due to the combination of the static and dynamic analysis, which ensures the code coverage and the real-time performance. As *hook* technology is non-intrusive to Android system, our method could guarantee detection without sacrificing adaptability.

### 5.3. Security Experiment of Smart Wristband App

To further verify the validity of our method, we conducted a vulnerability analysis of a smart wristband application to find possible privacy leakage problems. The wristband and its app are shown in Figure 8. The experiment was carried out as follows:

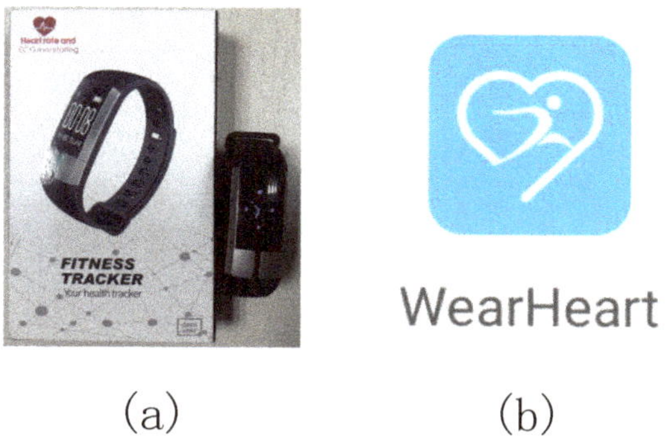


**Figure 8.** *Example* of a smart wristband. (**a**) The wristband. (**b**) App of clients.

(1) Vulnerability analysis: we used FlowDroid to perform static analysis of the original app to verify its vulnerability. We did find out some vulnerabilities that might be exploited. For instance, the app uses hard-coded URL, data transmission by HyperText transfer (HTTP) protocol without encryption, the "send data" function follows immediately after a Java built-in AES cipher function.

(2) Malicious code insertion: based on the above analysis in (1), we found that the app would send user data to a specified URL. If the specified URL was modified maliciously to be the server address of a hacker, the user's privacy data would be obtained constantly. To simulate such an attack, we used Apktool, an apk analysis tool to unpack the app, and some malicious codes were inserted in the unpacked smali file. Specifically, first, we searched for functions that use the hard-coded URL as a parameter, because those functions may call network transmission APIs and send the information to this address. Then, we modified the parameter to our server address and call those functions again. The malicious code copies the user data and sends it to the server we experimented with. Figure 9 shows the malicious codes in detail.

```
public static void a(Context paramContext, String paramString1, String paramString2, JSONObject paramJSONObject, a parama)
{
  try
  {
    c = com.zjw.wearheart.j.a.a(paramJSONObject.toString(), "wo.szzhkjyxgs.20");
    d = new JSONObject();
    d.put("body", c.replace("\n", ""));
    b = new s(1, paramString1, d, parama.a(), parama.b());
    b.a(paramString2);
    b.a(new f(15000, 0, 1.0F));
    BaseApplication.a().a(b);
    BaseApplication.a().a();
    b = new s(1, "172.18.128.116:8089/zh", paramJSONObject, parama.a(), parama.b());
    b.a(paramString2);
    b.a(new f(15000, 0, 1.0F));
    BaseApplication.a().a(b);
    BaseApplication.a().a();
    return;
  }
  catch (Exception paramContext)
  {
    for (;;)
    {
      paramContext.printStackTrace();
    }
  }
}
```

obfuscated cipher function
information sending function
original codes
our server's address
malicious codes we add

**Figure 9.** Malicious codes in Java form.

(3) Privacy leakage detection and display: the malicious code inserted in (2) would divulge user's privacy, including a user's login name and password, and health data. Leaked data is illustrated in Figure 10. Some sensitive data are marked in red. As can be seen from the figure, login password and the phone number was disclosed in plaintext.

```
ubuntu@ubuntu-VirtualBox:~/projects/go/src/wh-server$ ./wh-server
data: map[c:ctl000001 m:upd t:1546948485265 data:map[c_app_version:1.0.43 c_ip:5
8.213.91.6 c_eq_os:5.1.1 c_imei:355799054711989 c_internet_type:WIFI c_sim_type:
 c_market_sources:Umeng c_phone_type:samsung c_eq_type:samsung  GT-I9500]]
data: map[data:map[c_password:t      4 c_mail: c_app_version:1.0.43 c_offer:Andr
oid c_mobile:156      69 c_eq_id:355799054711989 c_eq_type:samsung  GT-I9500 c_e
q_os:5.1.1 c_imei:355799054711989] c:ctl000001 m:gL t:1546948525710]
data: map[data:map[c_phone_type:samsung  GT-I9500 c_app_version:1.0.43 c_app_nam
e:WearHeart c_uid:823655 c_offer:Android c_eq_os:5.1.1] c:ctl000001 m:upds]
data: map[c:ctl000016 m:getHomepageInfo data:map[c_date:2019-01-08 c_uid:823655]
]
data: map[c:ctl000001 m:giosv data:map[model:1]]
data: map[c:ctl000016 m:getMyInfo data:map[c_uid:823655]]
```

**Figure 10.** Decoded information received from the malicious application.

(4) Detection and interception test: MyPrivacy focused on accessing the dynamic environment and running information of the user-interface. FogPrivacy was responsible for privacy leakage detection and interception. When a running application attempts to send information over the network, the user will be informed of the location to which the data would be sent to determine target trustworthiness. If the request of an app was rejected by most users, FogPrivacy and MyPrivacy would mark it as a malicious behaviour and send the rejection record to the user-interface. From Figure 11, we can see that MyPrivacy poped up a window, because FogPrivacy had successfully intercepted the suspicious network transmission. The context information is also shown in the pop-up window, which is in the red rectangle.

In this case, the complete context is:

*API*
HttpURLConnection(<constructor>)
*Permissions*
android.permission.INTERNET
*CallStack*
com.zjw.wearheart.g.d.a(<Xposed>)
com.zjw.wearheart.home.exercise.RecordDayFragment.a
com.zjw.wearheart.home.exercise.RecordDayFragment.b
com.zjw.wearheart.base.BaseFragment.onCreateView

(5) Discussion: this attack is a kind of app piggybacking (repackaging) attack, and should have been the meal of research [48,49]. However, the original app (benign one) does not appear in most popular app markets, so most detection systems have little knowledge of this app at the beginning (especially signature information). When they analyze this pair of apps, they could hardly tell which app is benign or malicious. We once uploaded the original app and the maliciously modified one to VirusTotal, a malware analysis website, for detection, and found that both applications passed all security tests. In this situation, wristband users could be easily cheated to install a compromised app, if it is uploaded to app markets.

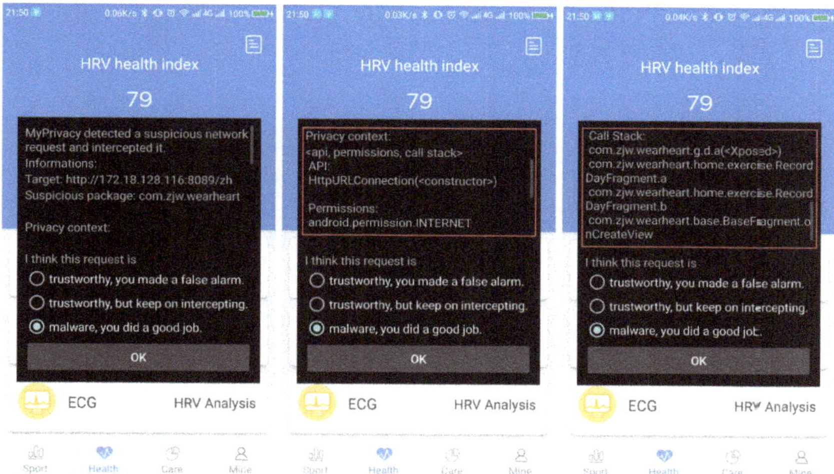


**Figure 11.** Interception to the malicious app.

## 6. Conclusions

In this paper, we studied the privacy protection method of PHDs based on fog computing. We proposed a framework for security and privacy protection based on fog computing in IoT healthcare networks. We analyzed the internal mechanism of software accessing private data, presented the method of constructing the context information base of privacy-related API functions and proposed a new method of privacy leakage detection method. Experiment results showed that our method had a efficient detection of privacy leakage and outperformed state-of-the-art methods.

**Author Contributions:** Conceptualization, J.G.; Data curation, G.Q.; Formal analysis, J.G., X.D. and M.G.; Supervision, J.G.; Validation, R.H.; Visualization, L.J.; Writing—original draft, J.G.

**Funding:** This work was supported by the National Natural Science Foundation of China (General Program) under Grant No.61572253, the 13th Five-Year Plan Equipment Pre-Research Projects Fund under Grant No.61402420101HK02001, and the Aviation Science Fund under Grant No. 2016ZC52030.

**Conflicts of Interest:** The authors declare no conflict of interest.

## References


1. Cybersecurity Center. Special Report of Android Malicious App (2016). Technical Report, 360 Cybersecurity Center. 2017. Available online: http://zt.360.cn/1101061855.php?dtid=1101061451&did=490301065 (accessed on 27 February 2017).
2. Xiao, Y.; Rayi, V.K.; Sun, B.; Du, X.; Hu, F.; Galloway, M. A survey of key management schemes in wireless sensor networks. *Comput. Commun.* **2007**, *30*, 2314–2341. [CrossRef]
3. Zhou, Z.; Zhang, H.; Du, X.; Li, P.; Yu, X. Prometheus: Privacy-aware data retrieval on hybrid cloud. In Proceedings of the 2013 Proceedings IEEE INFOCOM, Turin, Italy, 14–19 April 2013; pp. 2643–2651.
4. Mohney, G. Hospital Hack Spotlights How Medical Devices and Systems Are at Risk. *ABC News*, 19 February 2016.
5. Millman, R. Ransomware holds data hostage in two German hospitals. *SC Media*, 29 February 2016.
6. Brook, C. Sergey Lozhkin on How He Hacked His Hospital. Technical Report, Kaspersky Lab, 2016. Available online: https://threatpost.com/sergey-lozhkin-on-how-he-hacked-his-hospital/116314/ (accessed on 18 February 2016).
7. Papageorgiou, A.; Strigkos, M.; Politou, E.; Alepis, E.; Solanas, A.; Patsakis, C. Security and privacy analysis of mobile health applications: The alarming state of practice. *IEEE Access* **2018**, *6*, 9390–9403. [CrossRef]
8. Du, X.; Chen, H.H. Security in wireless sensor networks. *IEEE Wirel. Commun.* **2008**, *15*, 60–66.

9. Sicari, S.; Rizzardi, A.; Grieco, L.A.; Coen-Porisini, A. Security, privacy and trust in Internet of Things: The road ahead. *Comput. Netw.* **2015**, *76*, 146–164. [CrossRef]
10. Chin, E.; Felt, A.P.; Greenwood, K.; Wagner, D. Analyzing inter-application communication in Android. In Proceedings of the 9th international conference on Mobile Systems, Applications, and Services, Bethesda, MD, USA, 28 June–1 July 2011; pp. 239–252.
11. Sun, M.; Wei, T.; Lui, J. Taintart: A practical multi-level information-flow tracking system for android runtime. In Proceedings of the 2016 ACM SIGSAC Conference on Computer and Communications Security, Vienna, Austria, 24–28 October 2016; pp. 331–342.
12. Gibler, C.; Crussell, J.; Erickson, J.; Chen, H. AndroidLeaks: Automatically detecting potential privacy leaks in android applications on a large scale. In *International Conference on Trust and Trustworthy Computing*; Springer: Berlin, Germany, 2012; pp. 291–307.
13. Lu, L.; Li, Z.; Wu, Z.; Lee, W.; Jiang, G. Chex: Statically vetting android apps for component hijacking vulnerabilities. In Proceedings of the 2012 ACM Conference on Computer and Communications Security, Raleigh, NC, USA, 16–18 October 2012; pp. 229–240.
14. Yang, Z.; Yang, M. Leakminer: Detect information leakage on android with static taint analysis. In Proceedings of the 2012 Third World Congress on Software Engineering (WCSE), Wuhan, China, 6–8 November 2012; pp. 101–104.
15. Enck, W.; Gilbert, P.; Han, S.; Tendulkar, V.; Chun, B.G.; Cox, L.P.; Jung, J.; McDaniel, P.; Sheth, A.N. TaintDroid: An information-flow tracking system for realtime privacy monitoring on smartphones. *ACM Trans. Comput. Syst.* **2014**, *32*, 5. [CrossRef]
16. Lantz, P.; Desnos, A.; Yang, K. DroidBox: An Android Application Sandbox for Dynamic Analysis. Available online: https://github.com/pjlantz/droidbox (accessed on 25 August 2014).
17. Spreitzenbarth, M.; Schreck, T.; Echtler, F.; Arp, D.; Hoffmann, J. Mobile-Sandbox: Combining static and dynamic analysis with machine-learning techniques. *Int. J. Inf. Secur.* **2015**, *14*, 141–153. [CrossRef]
18. Rastogi, V.; Qu, Z.; McClurg, J.; Cao, Y.; Chen, Y. Uranine: Real-time privacy leakage monitoring without system modification for android. In *International Conference on Security and Privacy in Communication Systems*; Springer: Berlin, Germany, 2015; pp. 256–276.
19. Ali-Gombe, A.; Ahmed, I.; Richard, G.G., III; Roussev, V. Aspectdroid: Android app analysis system. In Proceedings of the Sixth ACM Conference on Data and Application Security and Privacy, New Orleans, LA, USA, 9–11 March 2016; pp. 145–147.
20. Yang, Z.; Yang, M.; Zhang, Y.; Gu, G.; Ning, P.; Wang, X.S. Appintent: Analyzing sensitive data transmission in android for privacy leakage detection. In Proceedings of the 2013 ACM SIGSAC conference on Computer & Communications Security, Berlin, Germany, 4–8 November 2013; pp. 1043–1054.
21. Yi, S.; Hao, Z.; Qin, Z.; Li, Q. Fog computing: Platform and applications. In Proceedings of the 2015 Third IEEE Workshop on Hot Topics in Web Systems and Technologies (HotWeb), Washington, DC, USA, 12–13 November 2015; pp. 73–78.
22. He, S.; Cheng, B.; Wang, H.; Huang, Y.; Chen, J. Proactive personalized services through fog-cloud computing in large-scale IoT-based healthcare application. *China Commun.* **2017**, *14*, 1–16. [CrossRef]
23. Kang, J.J.; Venkatraman, S. An Integrated mHealth and Vehicular Sensor Based Alarm System Emergency Alarm Notification System for Long Distance Drivers using Smart Devices and Cloud Networks. In Proceedings of the 2018 28th International Telecommunication Networks and Applications Conference (ITNAC), Sydney, Australia, 21–23 November 2018; pp. 1–6. [CrossRef]
24. Kang, J.; Larkin, H. Application of an Emergency Alarm System for Physiological Sensors Utilizing Smart Devices. *Technologies* **2017**, 5. [CrossRef]
25. Kang, J.; Adibi, S. A Review of Security Protocols in mHealth Wireless Body Area Networks (WBAN). In *Future Network Systems and Security*; Doss, R., Piramuthu, S., Zhou, W., Eds.; Springer International Publishing: Cham, Switzerland, 2015; pp. 61–83.
26. Chakraborty, S.; Bhowmick, S.; Talaga, P.; Agrawal, D.P. Fog networks in healthcare application. In Proceedings of the 2016 IEEE 13th International Conference on Mobile Ad Hoc and Sensor Systems (MASS), Brasilia, Brazil, 10–13 October 2016; pp. 386–387.
27. Sood, S.K.; Mahajan, I. A Fog-Based Healthcare Framework for Chikungunya. *IEEE Internet Things J.* **2018**, *5*, 794–801. [CrossRef]

28. Al Hamid, H.A.; Rahman, S.M.M.; Hossain, M.S.; Almogren, A.; Alamri, A. A security model for preserving the privacy of medical big data in a healthcare cloud using a fog computing facility with pairing-based cryptography. *IEEE Access* **2017**, *5*, 22313–22328. [CrossRef]
29. Cerina, L.; Notargiacomo, S.; Paccanit, M.G.; Santambrogio, M.D. A fog-computing architecture for preventive healthcare and assisted living in smart ambients. In Proceedings of the 2017 IEEE 3rd International Forum on Research and Technologies for Society and Industry (RTSI), Modena, Italy, 11–13 September 2017; pp. 1–6.
30. Du, X.; Xiao, Y.; Guizani, M.; Chen, H.H. An effective key management scheme for heterogeneous sensor networks. *Ad Hoc Netw.* **2007**, *5*, 24–34. [CrossRef]
31. Suo, H.; Wan, J.; Zou, C.; Liu, J. Security in the internet of things: A review. In Proceedings of the 2012 International Conference on Computer Science and Electronics Engineering (ICCSEE), Hangzhou, China, 23–25 March 2012; Volume 3, pp. 648–651.
32. Du, X.; Guizani, M.; Xiao, Y.; Chen, H.H. Transactions papers a routing-driven Elliptic Curve Cryptography based key management scheme for Heterogeneous Sensor Networks. *IEEE Trans. Wirel. Commun.* **2009**, *8*, 1223–1229. [CrossRef]
33. Xiao, Y.; Du, X.; Zhang, J.;Hu, F.;Guizani, S. Internet protocol television (IPTV): the killer application for the next-generation internet. *IEEE Commun. Mag.* **2007**, *45*, 126–134. [CrossRef]
34. Spreitzenbarth, M.; Freiling, F.; Echtler, F.; Schreck, T.; Hoffmann, J. Mobile-sandbox: Having a deeper look into android applications. In Proceedings of the 28th Annual ACM Symposium on Applied Computing, Coimbra, Portugal, 18–22 March 2013; pp. 1808–1815.
35. Pravin, M.N.P. Vetdroid: Analysis using permission for vetting undesirable behaviours in android applications. *Int. J. Innov. Emerg. Res. Eng.* **2015**, *2*, 131–136.
36. Hornyack, P.; Han, S.; Jung, J.; Schechter, S.; Wetherall, D. These aren't the droids you're looking for: Retrofitting android to protect data from imperious applications. In Proceedings of the 18th ACM conference on Computer and Communications Security, Chicago, IL, USA, 17–21 October 2011; pp. 639–652.
37. Bugiel, S.; Heuser, S.; Sadeghi, A.R. Flexible and Fine-grained Mandatory Access Control on Android for Diverse Security and Privacy Policies. In Proceedings of the USENIX Security Symposium, Washington, DC, USA, 14–16 August 2013; pp. 131–146.
38. Xu, R.; Saïdi, H.; Anderson, R.J. Aurasium: Practical Policy Enforcement for Android Applications. In Proceedings of the USENIX Security Symposium, Bellevue, WA, USA, 8–10 August 2012; Volume 2012.
39. Backes, M.; Gerling, S.; Hammer, C.; Maffei, M.; von Styp-Rekowsky, P. Appguard–enforcing user requirements on android apps. In *International Conference on TOOLS and Algorithms for the Construction and Analysis of Systems*; Springer: Berlin, Germany, 2013; pp. 543–548.
40. Zhang, L.; Zhu, D.; Yang, Z.; Sun, L.; Yang, M. A survey of privacy protection techniques for mobile devices. *J. Commun. Inf. Netw.* **2016**, *1*, 86–92. [CrossRef]
41. Arzt, S.; Rasthofer, S.; Fritz, C.; Bodden, E.; Bartel, A.; Klein, J.; Le Traon, Y.; Octeau, D.; McDaniel, P. Flowdroid: Precise context, flow, field, object-sensitive and lifecycle-aware taint analysis for android apps. *ACM Sigplan Not.* **2014**, *49*, 259–269. [CrossRef]
42. Arzt, S.; Rasthofer, S.; Bodden, E. *Susi: A Tool for the Fully Automated Classification and Categorization of Android Sources and Sinks*; Tech. Rep. TUDCS-2013-0114; University of Darmstadt: Darmstadt, Germany, 2013.
43. Sable Research Group. Soot: A Framework for Analyzing and Transforming Java and Android Applications. 2016. Available online: https://sable.github.io/soot/ (accessed on 1 March 2019).
44. Au, K.W.Y.; Zhou, Y.F.; Huang, Z.; Lie, D. Pscout: Analyzing the android permission specification. In Proceedings of the 2012 ACM Conference on Computer and Communications Security, Raleigh, NC, USA, 16–18 October 2012; pp. 217–228.
45. Arp, D.; Spreitzenbarth, M.; Hubner, M.; Gascon, H.; Rieck, K.; Siemens, C. DREBIN: Effective and Explainable Detection of Android Malware in Your Pocket. In Proceedings of the NDSS'14, San Diego, CA, USA, 23–26 February 2014; Volume 14, pp. 23–26.
46. Michael, S.; Florian, E.; Thomas, S.; Felix, C.F.; Hoffmann, J. Mobilesandbox: Looking deeper into android applications. In Proceedings of the 28th International ACM Symposium on Applied Computing (SAC), Coimbra, Portugal, 18–22 March 2013.
47. PNF Software. JEB Decompiler by PNF Software. 2017. Available online: https://www.pnfsoftware.com/ (accessed on 26 December 2017).

48. Zhang, F.; Huang, H.; Zhu, S.; Wu, D.; Liu, P. ViewDroid: Towards Obfuscation-resilient Mobile Application Repackaging Detection. In Proceedings of the 2014 ACM Conference on Security and Privacy in Wireless & Mobile Networks (WiSec '14), Oxford, UK, 23–25 July 2014; ACM: New York, NY, USA, 2014; pp. 25–36. [CrossRef]
49. Zhou, W.; Zhang, X.; Jiang, X. AppInk: Watermarking Android Apps for Repackaging Deterrence. In Proceedings of the 8th ACM SIGSAC Symposium on Information, Computer and Communications Security (ASIA CCS '13), Hangzhou, China, 8–10 May 2013; ACM: New York, NY, USA, 2013; pp. 1–12. [CrossRef]

*Review*

# Review and Evaluation of MAC Protocols for Satellite IoT Systems Using Nanosatellites

**Tomás Ferrer [1], Sandra Céspedes [1,2,*] and Alex Becerra [3]**

1 Department of Electrical Engineering, Universidad de Chile, Av. Tupper 2007, Santiago 8370451, Chile; tomas.ferrer@ing.uchile.cl
2 NIC Chile Research Labs, Universidad de Chile, Santiago 8370403, Chile
3 Aurora Space, Santiago 7750053, Chile; abecerra@auroraspace.cl
* Correspondence: scespedes@ing.uchile.cl

Received: 16 January 2019; Accepted: 14 February 2019; Published: 25 April 2019

**Abstract:** Extending the internet of things (IoT) networks to remote areas under extreme conditions or for serving sometimes unpredictable mobile applications has increased the need for satellite technology to provide effective connectivity. However, existent medium access control (MAC) protocols deployed in commercial satellite networks were not designed to offer scalable solutions for the increasing number of devices predicted for IoT in the near future, nor do they consider other specific IoT characteristics. In particular, CubeSats—a low-cost solution for space technology—have the potential to become a wireless access network for the IoT, if additional requirements, including simplicity and low demands in processing, storage, and energy consumption are incorporated into MAC protocol design for satellite IoT systems. Here we review MAC protocols employed or proposed for satellite systems and evaluate their performance considering the IoT scenario along with the trend of using CubeSats for IoT connectivity. Criteria include channel load, throughput, energy efficiency, and complexity. We have found that Aloha-based protocols and interference cancellation-based protocols stand out on some of the performance metrics. However, the tradeoffs among communications performance, energy consumption, and complexity require improvements in future designs, for which we identify specific challenges and open research areas for MAC protocols deployed with next low-cost nanosatellite IoT systems.

**Keywords:** CubeSats; internet of things; medium access control; nanosatellites; sensor networks; wireless access networks

---

## 1. Introduction

From the beginnings of space exploration, satellites were large objects that took years to construct and cost billions of dollars for a single unit. With more advanced and smaller technologies, cheaper spacecraft (stand alone satellites and constellations of satellites) have evolved for diverse applications, telecommunication applications being most prominent. Commercial satellite companies like Iridium, Intelsat, O3b, and others offer a portfolio of products, including voice services, broadband, and sensor data collection, with extensive coverage of the Earth's surface. For example, Figure 1 shows the approximate coverage of just one geostationary satellite located at a longitude of 91° W.

**Figure 1.** Approximate coverage of a geostationary satellite located at 91° W.

With the internet of things (IoT), the paradigm that promises to revolutionize our world with the collection of enormous quantities of data, the connectivity demands are being increased around the globe. It is estimated that the IoT communications market will have an impact in the economy close to three to 11 trillion dollars per year in 2025 [1]. Nonetheless, terrestrial technologies do not fully cover the Earth's surface yet. It is in such a scenario that satellite technology seems to offer the critical solution to the problem of global connectivity. However, traditional satellites are expensive—Iridium's NEXT constellation of 75 satellites costs three billion dollars [2]—and thus novel, cheaper satellite solutions have become the focus of growing interest.

With the need for more coverage of the IoT networks and the search for cheaper solutions, nanosatellites may be the best answer for the global connectivity that the IoT demands. The nanosatellite standard, the CubeSat with a volume of less than one liter and a weight of less than one kilogram, also offers access to space and satellite development for countries that previously had no experience in space sciences. Nevertheless, the performance of such a solution will largely depend on the low-level protocols selected for the network architecture.

At the core of network architecture are the medium access control (MAC) protocols. Given the broadcast nature of channels in satellite communications, a MAC protocol ensures the proper coordination of frame transmissions, together with the logic for retransmissions and the recovery of data in case of collisions. In the past, there have been comprehensive reviews related to MAC protocols for satellite technology and also in the context of IoT. Peyravi [3] compiled a thorough revision and evaluation of MAC protocols for satellite communications. Although the study includes an evaluation with objective metrics such as throughput, buffer occupancy, scalability, stability, and reconfigurability, these metrics have been defined in the context of a constellation of geostationary satellites, which highly differ from the network conditions provided by smaller satellites deployed in lower orbits. Other similar surveys that focus on resource allocation and MAC protocol comparisons in conventional satellite systems are presented in Gaudenzi et al. [4] and Herrero et al. [5].

A more recent survey by De Sanctis et al. [6] makes the case of applicability of satellite communications for IoT and machine to machine (M2M) systems, also mentioning the potential of employing CubeSats within this context. Although the authors do provide a review of MAC protocols, no quantitative or comparative evaluation is provided for the MAC protocols reviewed in the work. Other works discussing the applicability of (small/nano) satellites in broadband internet access, IoT, and M2M communications can be found in [6–10]. The mentioned works, however, do not cover specific evaluations related to MAC protocols. Reviews devoted to the IoT, the enabling technologies, and services are also found in [11–14]. The focus of such works is more general except for the revision of MAC protocols for IoT presented by Oliveira et al. [14]; nevertheless, besides the fact that the mentioned works do not include a quantitative performance evaluation and comparison, a good part only discuss terrestrial IoT wireless technologies; hence, the discussion is oriented to different channel and network conditions from the ones addressed in this survey.

The contributions of this paper are threefold: (1) we review MAC protocols employed or proposed for satellite systems from a novel viewpoint that considers the restricted characteristics of CubeSat technology for wireless communications together with the particular requirements of IoT services and applications; (2) we provide a comparative quantitative and qualitative evaluation of the current protocols with metrics including communications performance (i.e., throughput, channel load, packet loss), dependency of network topology, implementation complexity, and energy consumption; and (3) we discuss the open research and implementation challenges to address by the next generation of nanosatellite networks for IoT environments.

The remainder of this paper is organized into seven sections. In Section 2 we present the fundamental aspects of satellites, space communication systems, and nanosatellites technology. It also includes a discussion about the IoT requirements. In Section 3 we introduce the specifics about the proposed IoT scenario served by a constellation of CubeSats. In Section 4 we provide the backgrounds on MAC layer protocols and introduce the metrics for evaluation. Section 5 presents a detailed review of MAC protocols designed and developed for satellite systems and other IoT-related technologies. It also includes a performance evaluation with objective metrics relevant to the IoT study scenario. In Section 6 we discuss the advantages and shortcomings of the existent protocols and identify the open challenges. Section 7 presents the final remarks.

## 2. Overview of Space and Communications Systems

### 2.1. Satellites Evolution

In the 1950s, the Soviet Union launched Sputnik I, the first artificial satellite that orbited the Earth. This milestone marked the beginning of a competition between two powerful countries that had, as one of its consequences, an accelerated technological development in aerospace sciences. Satellites created in the decades after the beginning of this competition were designed for very specific missions, and each development had its own subsystems—energy, command and data handling, attitude control, etc.—which allowed the particular requirements of a given project to be met. Such a design methodology involved an extremely expensive process due to the constant iterations necessary to create a new device, and the difficulties in reusing previous versions and designs.

One early shift occurred when, due to the use of modular systems, the main bus was designed to be flexible and reconfigurable according to the goal of each mission. As a result of reducing the costs in developing one unit, constellations of these spacecraft began to be feasibly designed and used by those countries and companies that could afford the still enormous cost of development. Depending on the configuration, these formations could increase the instantaneous global coverage and reducing revisiting times, among others benefits.

A large number of satellites now dot the skies for diverse applications such as navigation, imaging, meteorology, and communications. Some of the most significant applications are the following:

- Positioning systems: constellation of satellites located in medium height orbits (approximately 20,000 km) that make it possible to determine the position of an object on the Earth's surface in a given coordinate system. There are several systems belonging to different countries, namely: GPS (USA—24 satellites), global orbiting navigation satellite system (GLONASS) (Russia—24 operational satellites), GALILEO (European Union—24 satellites), BEIDOU (China—17 operational satellites).
- Earth observation: Several satellites with a wide variety of cameras in different spectral bands have been sent to space. Defense and security, cartography, and meteorology are some of the disciplines that have benefited from these types of missions.
- Communications: Satellite systems that provide voice services, satellite television service, and narrowband/broadband connectivity through standalone satellites or constellations.

The Union of Concerned Scientists (UCS) maintains a count of operational satellites orbiting the Earth. In Figure 2 we illustrate the distribution, according to the country of origin and the type of orbit, for the 1957 active satellites reported up to 30th November 2018 [15].

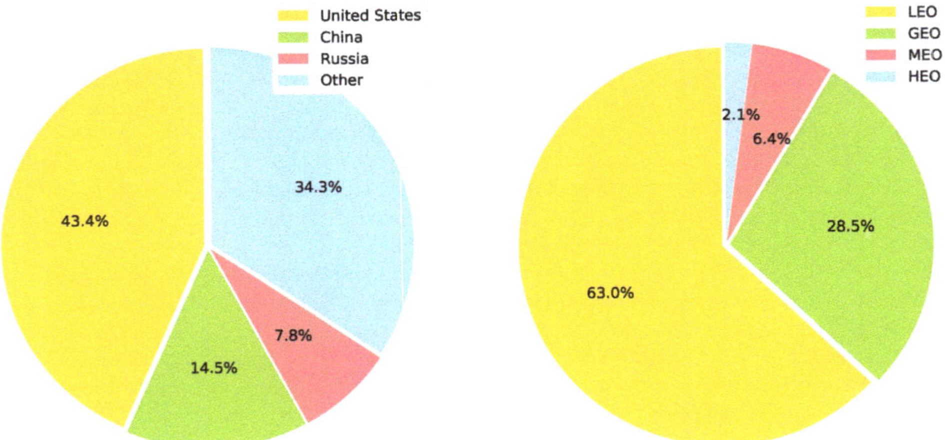


**Figure 2.** Active Earth orbiting satellites, separated by country of origin and type of orbit, from a total of 1957 active satellites reported until 30th November 2018. Data published by the Union of Concerned Scientists (UCS) in its annual report [15].

### 2.2. Communication Satellite Systems

One of the areas in which satellites have been relevant is in communication networks. Due to the innate capacity of these spacecraft to cover the whole terrestrial surface, satellite systems are able to provide connectivity to remote or isolated areas that by other means are almost impossible to connect.

There are three main types of architecture used in satellite communication systems: store and forward, bent-pipe, and crosslink [8]. In the first, the satellite retrieves data from one point, stores it for some time, and then downloads it to the first ground station it establishes a connection with. In the second case, the satellite acts as a relay, collecting data and retransmitting it to another point on Earth. In the crosslink architecture, the data is transmitted immediately through a satellite network via inter-satellite links.

Satellite communication systems can be deployed in different orbits, offering a different set of services according to the channel/network conditions derived from the characteristics of the orbit of deployment. The types of orbits are the following:

- Geosynchronous equatorial orbit (GEO): This corresponds to an orbit whose rotation period is the same as the Earth's. Consequently, the satellite seems to "stand still" to an observer at one point on the planet. To achieve this effect, the satellites are placed at a distance of approximately 35,786 km from Earth. Given such a long distance, the communication delays are considerable, in the order of 120 ms, in the satellite-ground direction or vice versa, for the best scenario; also, the transmission power required to establish effective links is high. Nevertheless, these systems have an excellent and broad coverage, reaching a 30% of the Earth's surface. The placement process of a satellite into this orbit is an expensive task, and in order to remain at that position, the crew on the ground must perform orbital maneuvers from time to time.
- Low Earth orbit (LEO): Most of the satellites in space today are placed in this type of orbit. Its height ranges from 300–2000 km and, therefore, the delay in communications is low, in the order of tens of milliseconds for the worst case. The transmission power required to establish the links from this orbit are as low as hundreds of milliwatts [16]. Satellites in this orbit have low

temporal and spatial coverage. Because of the speed—about 7.5 km per second for a satellite in a 500 km orbit—the Doppler effect has to be considered in these systems.

- Medium Earth orbit (MEO): Heights are between the low and geostationary orbits—2000–35,786 km. One example system, the O3b network, is placed at a height of 8000 km and has a theoretical minimum delay of 26 ms satellite-ground, or vice versa. All global navigation satellite systems (GNSS) constellations are placed in this orbit.
- Highly elliptical orbit (HEO): Orbits with a large apogee and a small perigee. The most famous of this kind is the Molniya orbit, which offers large coverage for high latitudes. Another example is the Tundra orbit. In Molniya, the apogee is greater than a geostationary—about 40,000 km. Satellites in this particular orbit have an approximate period of 12 h. The Soviet Union was the first country to use it to provide communication services throughout its territory and also to obtain meteorological images.

The classical services provided by satellite communication systems are the following:

- Broadband communications: The commercial satellite networks providing this service offer connectivity with broadband data rates. For example, the new Iridium's NEXT constellation offers connectivity at 1.5 Mbps [17], whereas the Inmarsat's BGAN HDR offers connections at 800 kbps [18]. Generally, stations on the ground require a large antenna along with a high transmission power to establish effective links. Satellites serving broadband communications usually operate in the Ka, Ku, L, and C bands.
- Voice services: Using small devices such as satellite telephones, these satellite systems offer voice connectivity on almost any part of the planet.
- Signaling services: In this area, some of the highlight services are the reception of automatic identification system (AIS) and automatic dependent surveillance broadcast (ADS-B) signals, which can track the path of vessels and aircraft, respectively.
- Sensor data collection: These satellite systems offer services at low data transfer rates, which allow data to be retrieved from small sensors placed on the ground.

Table 1 provides a list of some of the commercial constellations providing communication services in different orbits.

**Table 1.** A set of commercial constellations providing communication services as of 2018.

| Company | Number of Satellites | Orbit | Services |
|---|---|---|---|
| Inmarsat | 13 | GEO | Broadband |
| Viasat | 4 | GEO | Broadband |
| Intelsat | 52 | GEO | Broadband |
| O3b | 16 | MEO | Broadband |
| Iridium | 66 | LEO | Voice, broadband |
| Globalstar | 24 | LEO | Voice, broadband |

### 2.3. CubeSats

Traditionally, most of the projects for designing and building satellites have been excessively expensive. They involved complex designs and, consequently, long development time spans. However, starting in the 1980s, a new paradigm was established that significantly reduced the size of some satellites, leading to the appearance of microsatellites and, in the 2000s, the creation of nanosatellites or CubeSats: aircraft whose weight is equal to or less than one kilogram.

The CubeSat standard was created in 1999 at the California Polytechnic State University in conjunction with the Stanford University's Space Systems Development Lab. The development of this standard aimed at improving access to space by providing opportunities for satellite development, design, and construction to institutions that could not do so with the classical paradigm. Figure 3 shows the number of Cubesats launched and operational to date.

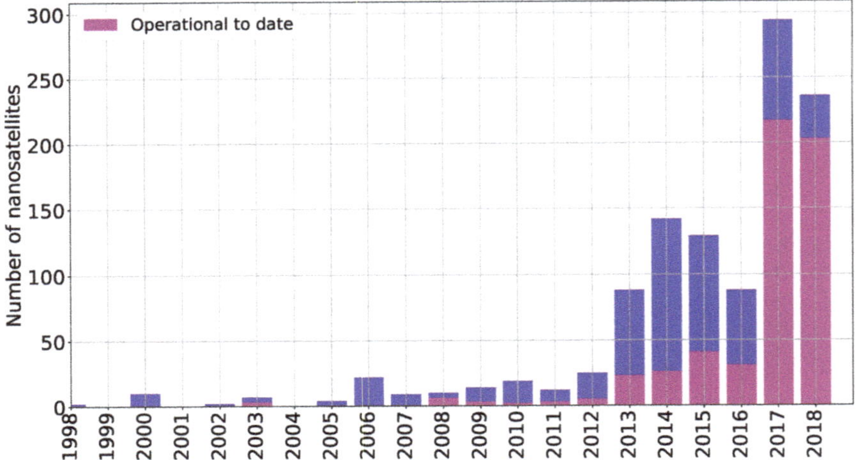


**Figure 3.** CubeSats launched since 1998. Data taken from the database at [19].

The basic design of a CubeSat consists of a 10-cm cube—called 1U—which must contain the primary subsystems for the operation: an onboard computer, batteries, transmitters and receivers for communication, and attitude determination and control system (ADCS), among others. The cubic shape and volume defined for this new standard considerably reduce launching costs, but, at the same time, incorporate restrictions regarding availability of computational resources, energy, and volume, among others.

Initially, the development of nanosatellites was intended to test components and study their behavior in the space environment. Nowadays, applications have spread widely with projects led by universities, governments, and commercial entities. Moreover, and of interest to the authors, this technology represents an excellent opportunity for developing countries to exploit space resources in addition to providing a tool to democratize the use of space [20].

In South America, for example, several countries have taken advantage of nanosatellite technology to promote educational initiatives within universities, including, the Libertad-1 in Colombia (Sergio Arboleda University) [21], PUCPSAT in Peru (Catholic University of Peru) [22], SUCHAI-1 in Chile (University of Chile) [20], to mention just a few. In the commercial field, new companies have appeared in the market for developing and selling CubeSat parts and pieces; other companies make use of CubeSats for applications such as satellite imaging collection. Government agencies, such as NASA and ESA, developed nanosatellite-related missions. One of the most notable examples is the experimental use of two CubeSats—Mars Cube One (MarCO) A and B—as communication relays for the InSight-1 probe that landed on Mars in November 2018.

### 2.4. IoT and M2M Requirements

Cisco forecasts that by the year 2020 the number of devices connected to the internet will exceed 50 billion [23], an increase that raises a connectivity challenge for these new massive networks. It is in this field that the capabilities of the new low-cost nanosatellite networks could be instrumental in achieving a global connectivity, as demanded by the fifth generation networks.

The IoT and machine-to-machine networks are characterized by their intention to meet one or more of the following requirements:

- Efficient performance against explosive traffics
- Low data rates in terminals
- Energy efficiency

- Low cost terminals
- Mobility and scalability
- Minimization in the use of spectrum
- Minimum signaling
- Data security
- Data integrity
- Reliability
- Robustness
- Flexibility

In the case of terrestrial wireless access networks, various solutions have been developed to meet the above requirements. For low consumption sensor arrays deployed across extensive areas, technologies such as LoRaWAN [24] and Sigfox [25] are available; for sensors and actuators networks deployed in urban environments, there are Wi-Sun [26] and NB-IoT [27], to mention some of the available technologies. There are also autonomous sensor networks interconnected to provide solutions to specific applications involving short-range technologies, such as IEEE 802.15.4 [28]. Despite the advances with the introduction of these new technologies, there are scenarios for which existing networks do not offer feasible solutions. Remote places without connectivity still exist; also, some monitoring applications in isolated places require devices with high and unpredictable mobility to collect information on-the-move (e.g., monitoring of wild animals in areas of difficult access). Considering the scenarios mentioned above, microsatellites and CubeSats appear as viable alternatives to cover the gap in providing fully connected global communications networks for the IoT [6,10]. An example scenario of a CubeSat providing IoT connectivity is illustrated in Figure 4.

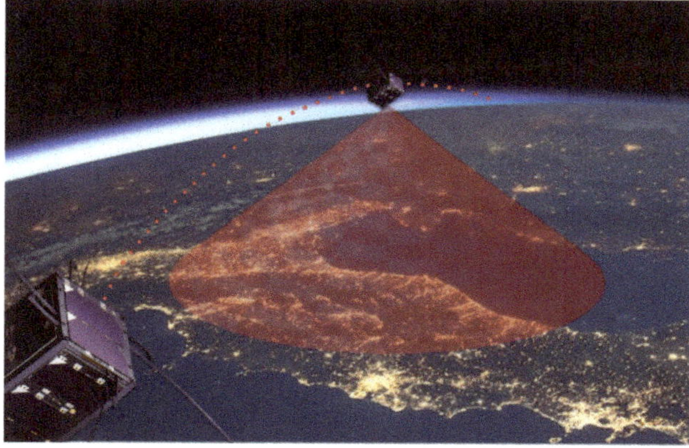

**Figure 4.** Illustrative scenario in which a CubeSat (or a constellation of them) provides connectivity for internet of things (IoT) applications.

One of the challenges of these new massive networks is to enable the many terminals to share a physical resource—the broadcast communications channel—in an efficient and orderly manner. Such a challenge would necessarily make use of the medium access control layer, which corresponds to a sub-layer of the link layer of the open system interconnection (OSI) model and is responsible for coordinating frame transmissions in broadcast links. The specific MAC protocol used for IoT applications will need to fulfil a number of requirements including increased average throughput, to meet a minimum level of fairness as well as to comply with the resources, requirements, and limitations of the access technology in use. Another critical aspect to consider in the choice of a MAC protocol is the network topology and how much knowledge the nodes have or need about that topology.

To examine the fulfillment of the IoT and M2M networks requirements, from a MAC layer perspective in the case of this study, together with the restrictions imposed by the capabilities of the CubeSats, will shed light about the viability to provide IoT connectivity using nanosatellites. The reviews and discussion presented in the coming sections address all of the IoT and M2M requirements listed above, except the ones related to data security and data integrity. Whilst security aspects are of paramount importance in the IoT ecosystem, we direct the interested reader to specialized works on this subject discussing security threats and mitigations for a variety of IoT technologies and architectures [29–32] and specifically for satellite communications [33–35].

## 3. IoT Scenario of Study

In order to exploit the full capabilities of the IoT, connectivity is a major issue to be solved in the task of recovering the amount of data generated by the—expected—billions of sensors forecasted to be deployed in coming years. Although some existent IoT technologies, like LoRa and SigFox, claim to have large coverage—40 and 20 km in rural environments, respectively [36]—they are not even close to the coverage that satellite systems can provide. Nevertheless, satellite connectivity is still considered very expensive and poor in terms of energy efficiency. It is in this context that researchers consider that the CubeSat standard could be a feasible solution to mitigate the above mentioned disadvantages of traditional satellite networks, lowering the costs of satellite systems and making it a viable alternative to current wireless technologies for IoT connectivity.

In this context, the scenario to be considered in this review corresponds to a CubeSat constellation, with no inter-satellite connectivity, whose main purpose is retrieving small amounts of data from sensors placed on the ground at a low data rate. The constellation will be deployed in several orbital planes belonging to the Low Earth orbit; each orbit with a height ranging between 500–600 km and with an inclination close to 97°. Each nanosatellite from the constellation will face the same problem: as it orbits around the planet, it will have to recover data from a network on Earth whose number of nodes and geographic distribution is unknown and (possibly) changing continuously. Analyzing the case for one satellite—the master—and several ground sensors—the slaves—will be representative of the problem to be faced by the complete fleet.

The satellite communication system uses the 400 MHz band, which has low propagation losses compared with the typical bands employed by commercial companies offering satellite broadband services. Such frequencies are in the range of the amateur frequency band used and proven to work by most of the CubeSat projects deployed to date [16]. The communications are half-duplex and have an expected maximum data transmission rate of 100 kbps, which is similar to the rate offered by commercial developments of transceivers for nanosatellites [37]. It is assumed that the antennas in use, as well as the transmission power and the receivers' sensitivity, are adequate to establish effective data links for most nodes under the coverage area of the nanosatellite. However, it is expected that the furthest nodes from the nanosatellite are less likely to generate a correct link due to the greater distances to be covered.

As mentioned above, the sensors are distributed randomly in any geographical area on Earth. A sensor node is not aware of the network topology, and the spacecraft does not know in advance how many devices needs to serve in an area of coverage. Each sensor generates a quantity of data independent of the others. It is also assumed there is no temporal synchronization among the sensors nodes, nor between the sensor nodes and the nanosatellite.

## 4. Background on MAC Protocols

MAC layer groups a set of protocols and mechanisms in order to distribute the resources for the nodes to make an effective (and efficient) use of the communications channel. The resources are typically distributed in terms of time assignment, frequency assignment or code assignment. In the particular case of broadcast links, a MAC protocol is in charge of coordinating the frame transmission.

Each MAC protocol is designed to cover different requirements, and its performance can be quantified with different metrics. In some cases, the priority is set to the performance concerning data transmission rate, for which the normalized offered load and the normalized throughput are measured. Other priorities may include measurements of delays in sending data or the packet loss ratio (PLR). In the particular case of IoT applications, there may be limitations regarding processing capabilities, available storage, hardware complexity, and energy consumption.

In this section, the authors provide a set of metrics that can quantify the fulfillment of the different requirements objectively. We also present the traditional categorization employed to classify the existent MAC protocols for broadcast channels.

### 4.1. Evaluation Metrics

#### 4.1.1. Normalized Offered Load (C)

The normalized offered load (C) is the quotient between all the data injected into the network and the maximum data that could be sent at the transmission rate of the link. The latter corresponds to the product of the transmission rate and the total transmission time. The normalized offered load is calculated according to the following formula:

$$C = \frac{\sum D_i}{T_x \cdot t_t}, \tag{1}$$

where $D_i$ is the data sent to the satellite by sensor $i$, $T_x$ is the link transmission rate, and $t_t$ is the total transmission time.

#### 4.1.2. Normalized Throughput (S)

The normalized throughput is the quotient between the data received by the satellite in a given time and all the data that could be sent continuously at the transmission rate of the link. It can be interpreted as how effective is the use of the channel. It is always true that $S \leq C$. The normalized throughput is calculated according to the following equation:

$$S = \frac{D_r}{T_x \cdot t}, \tag{2}$$

where $D_r$ is the amount of data received by the satellite, $T_x$ is the link transmission rate, and $t$ is an arbitrary time.

#### 4.1.3. Packet Loss Ratio (PLR)

PLR corresponds to the proportion of data lost or received with errors due to miscoordinations of frame transmissions, and that cannot be recovered over the total amount of data sent. The PLR is calculated as follows:

$$PLR = \frac{P_l}{P_s}, \tag{3}$$

where $P_l$ is the number of lost packets and $P_s$ is the number of packets sent. This ratio turns out to be important when energy efficiency is required, since a high PLR may trigger a high number of retransmissions when implementing a reliable link layer, which may mean more waste of energy. In general, the channel performance is analyzed by examining the supported channel load for a target PLR, which is commonly considered on the order of $10^{-3}$ in the literature. In some cases, the normalized load achieved with a target $PLR = 10^{-3}$ is very low, making it necessary to consider worse PLR values in the analysis, e.g., $PLR = 10^{-2}$.

The relation among the three metrics presented above is described by the following equation:

$$S = C(1 - PLR). \tag{4}$$

#### 4.1.4. Energy Consumption

From the point of view of MAC protocols, energy consumption is directly affected by the length of time in which data is being sent and received; to a lesser degree, energy consumption is also affected by the amount of processing required by the protocol. To evaluate the energy consumption, the length of time the transceiver is in transmission, reception, and idle modes should be compared. The consumption on each state depends specifically on the model of transmitter/receiver that is being used and the chosen MAC protocol. For example, in SigFox, the current consumption is 11 mA in reception mode and 125 mA in transmission mode [38]. The peak current consumption is about 32 mA and a range from 120 mA to 300 mA, in the cases of LoRa and NB-IoT, respectively [39].

In general scenarios, the main energy limitation is found in the terminal nodes, since in most cases the receiving station has a virtually infinite energy source (e.g., a base station in a cellular network, a WiFi access point, etc.). In our study scenario, the case is different since CubeSats may also have energy limitations. Nevertheless, it is expected that energy limitation in the sensor nodes will be considerably higher than in the spacecraft.

#### 4.1.5. Complexity of Implementation

In the context of CubeSats and low-cost satellite solutions, the complexity of implementation turns out to be a relevant factor. For this reason, aspects such as the need of high processing availability, the presence of very specialized hardware, and large amounts of required storage, should be considered as directly impacting the complexity of a given MAC protocol.

Usually, on-board computers (OBC) employed on CubeSats are microcontrollers such as the Microchip PIC24 or the Texas Instruments MSP430, which are very limited in terms of computational resources. Newer OBCs using the ARM Cortex family or ATMEL devices are already available in the market for nanosatellites, but they still are in the category of modest processors.

### *4.2. MAC Protocols Categories*

A brief categorization of the MAC protocols is provided as follows [3]:

- **Fixed Assignment**: Protocols in this category are characterized by assigning a limited resource equitably and fixedly between different interlocutors. The resource can be a frequency channel, a time interval or a code, deriving in the well-known mechanisms frequency division multiple access (FDMA), time division multiple access (TDMA), and code division multiple access (CDMA). These protocols are characterized by being easy to implement, as well as being efficient in link usage when they occupy all or most of their resources. However, protocols following a fixed assignment are not very flexible to changes in data rates, nor are they tolerant to variations in the number of stations since they require a coordinated allocation among all the stations involved.
- **Random Access**: These protocols are characterized by having a non-fixed number of users that, without prior coordination, make use of the same channel (i.e., contention-based protocols). Since the allocation of resources is random, more than one device may win the right to use the channel at the same time, causing frame collisions. Therefore, protocols in this category cannot guarantee the successful arrival of frames. Depending on the scenario, these protocols may waste system capacity in failed transmissions (and retransmissions). However, they have a fundamental role in networks whose previous characteristics (number of nodes, nature of traffic, etc.) are not known in advance.
- **On-demand**: These protocols are designed for scenarios in which the terminals require sending an unequal and variable amount of data; in that case, on-demand protocols can vary the allocation of resources depending on the nodes requirements. For example, a TDMA-based protocol may assign additional time-slots to nodes with higher requirements regarding data rate. To manage the variable assignment of resources, these protocols usually require extra control signaling, such as

the incorporation of the packet generation rate of each terminal as additional control information in every message.

- **Adaptive**: These correspond to protocols designed to manage variable network conditions. These protocols are intended to change the MAC logic dynamically. For example, when communication is carried out among a few terminals, the MAC employs a random access scheme; conversely, when the number of devices increases, it uses a fixed allocation scheme.
- **Reservation**: The goal of these protocols is to achieve a collision-free allocation of resources. A typical way to achieve the collision-free scheme is the use of a subchannel dedicated to the coordination of access for each station, in such a way that only one station transmits at a given time. In that subchannel, the devices may rotate a testimony (i.e., a token) that indicates who has the right to transmit on the channel. Most of these protocols make use of TDMA or variations of Aloha to assign the token.

The last three categories are, in a general way, hybrids of the first two. This is mainly because the network characteristics—number of nodes, data generation rates, network explosiveness, etc.—have a nature that is essentially either random or deterministic. In this way, the dominant categories that match the network characteristics are either random access or fixed assignment protocols.

In this work, the MAC protocols selected for review corresponded mainly to random access and its derivations. These schemes were selected because, in the study scenario, it is infeasible to predict the state of channel congestion at all times, which and how many nodes are within nanosatellite coverage, and the amount of data each node wants to transmit.

## 5. MAC Protocols for Satellite IoT and M2M

The early satellite solutions traditionally employed protocols mainly based on fixed assignment (e.g., CDMA, FDMA, and TDMA). In some cases, the protocol was combined with a random access scheme to perform the adaptive assignment according to the demand of the nodes. An example of an early protocol is the demand assignment multiple access [40]. Nonetheless, as mentioned before, considering the nature of our study scenario, protocols in the random access category are more relevant and suitable for the comparative evaluation.

The protocols selected in this section included several descendants of the well-known Aloha protocol, since such derivations are present in current satellite systems and modern IoT technologies such as LoRa and Sigfox. The selection also included other significant—and more modern—random access protocols that were considered suitable for the IoT scenario described in Section 3, all applicable to satellite environments and other IoT technologies. Such protocols make use of advanced techniques like interference cancellation, and adaptiveness, among others.

### *5.1. Aloha-Based Protocols*

The most representative random access protocol—and the inspiration for many other MAC protocols—is Aloha, developed in 1970. Although this protocol is quite old and simple, in current IoT developments Aloha plays a fundamental role. For example, leading IoT technologies that use variations of this protocol are LoRa and SigFox. Furthermore, there are several applications and modifications to Aloha reported for satellite environments in the literature. Some of them can be found in [41–44] for the interested reader.

In Pure Aloha, nodes send data when they have data to send, hoping that a collision does not occur. When the reception of a packet is successful, the receiver sends an acknowledgement (ACK); otherwise, nodes retransmit the same packet after a random time [45,46]. The performance of this protocol is quite modest. It achieves a maximum normalized throughput of $S = 0.18$ when $C = 0.5$. In terms of packet loss rate, it achieves a $PLR = 10^{-3}$ for an extremely low normalized load of $C < 10^{-3}$. The advantage of this protocol lies in its simplicity of implementation, since it does not require pre-coordination or extra access control signaling [47]. When there is low load in the channel, the energy consumption of Aloha is efficient, since it only requires sending data and the reception of an ACK, so the active consumption due to transmission and reception is proportional to the data and ACKs transmission delays. Nevertheless, for high channel loads, the packet losses due to collisions become high, which causes more retransmissions, overloading the channel with the associated wasting of energy.

#### 5.1.1. Slotted Aloha (S-Aloha)

The most similar version to Aloha is S-Aloha. It consists of discretizing the channel, where each time slot has the duration of a packet transmission time [46]. The purpose of discretizing is to avoid partial collisions among packets. S-Aloha is used, for example, in the sending of short packets and requests to initiate communications in the DVB-RCS standard [48].

S-Aloha achieves a low normalized throughput of $S = 0.368$ when the normalized load is $C = 1$. Similar to the case of pure Aloha, a channel load of $C < 10^{-3}$ is supported when the $PLR = 10^{-3}$. In the case of a higher packet loss, $PLR = 10^{-2}$, the normalized load is increased to $C \simeq 0.01$. Regarding the complexity of implementation, it can be said that S-Aloha adds additional complexity since it requires all the nodes to be synchronized, both on the ground and also on the satellite. Such a synchronization requires us to consider the time margins in order to align the time-slots among nodes that have different delays. Similar to pure Aloha, this protocol proves to be quite inefficient for high channel loads due to the need for retransmissions.

For IoT aplications, S-Aloha is a good option in scenarios where the offered load is low and the delays between nodes and base station do not have a large variation, but it becomes impractical if the delays imposed on a satellite link are considered. An application of this protocol on the recent Weightless-N IoT technology is reported in [49].

#### 5.1.2. Diversity Aloha/Slotted Aloha (DA/DSA)

This protocol is considered for systems that have large transmission delays (e.g., satellites in GEO orbits) and for which confirmations of packet reception are impractical [50]. In the diversity Aloha/slotted Aloha (DA/DSA) protocol each terminal sends two or more copies of a packet at different randomly selected times, without waiting for the reception of an ACK. The idea is to increase the probability of packet reception and to avoid retransmissions; however, the consequence is an overloaded channel. An application of this protocol in satellite environments is observed in the IP-over-Satellite system for sending short packets and registration [51,52].

The maximum performance of this protocol is reported to be lower than for slotted Aloha. In DSA, a maximum normalized throughput of $S \simeq 0.3$ is achieved for a $C \simeq 0.6$. However, when $C < 0.5$, the performance of DSA is slightly better. In the case of a $PLR = 10^{-2}$, the protocol supports a normalized load of $C \simeq 0.05$ (compared to a $C \simeq 0.01$ in S-Aloha). DSA is similar in behavior to S-Aloha in terms of implementation complexity and energy efficiency.

DA/DSA is suitable for links with large delays and offered loads less than $C = 0.5$. Nonetheless, in IoT scenarios the channel load will tend to increase progressively with time, consequently DA/DSA may not meet the scalability requirement, a crucial aspect in an IoT system like the one described in the study scenario.

#### 5.1.3. Spread Spectrum Aloha (SS-Aloha)

The protocol is proposed to provide random multiple access over an unsynchronized channel. SS-Aloha uses spread spectrum techniques to send messages; it is similar to a CDMA protocol where each terminal uses the same code to spread the signal and accesses the channel without coordination (like in Aloha) [53]. The multiple access capability is given by the large bandwidth employed instead of the assignment of different codes. In Figure 5, we illustrate the responses of a correlator detector (at the satellite's receiver) applied to signals from one terminal (see Figure 5a) and four terminals (see Figure 5b). In this example, a spreading factor $SF = 60$ is used, in consequence, 60 chips are placed between two consecutive bits from the same terminal. In order to achieve multiple access, spread spectrum Aloha (SSA) makes use of the offset in chips between two signals, so in Figure 5b, the messages from the four terminals are still decodable. Previous evaluations of the SS-Aloha protocol in a satellite environment are reported in [54].

Regarding the performance of SS-Aloha, thanks to the use of proper error correction codes, the protocol achieves a maximum normalized throughput close to $S \simeq 0.6$ for a given load of $C \simeq 0.7$. In the case of $PLR = 10^{-3}$, the system supports a load of $C \simeq 0.5$ [5]. This protocol has a reduced level of complexity because it does not require synchronization. However, the spread spectrum technique has a strong dependency on the signal to noise plus interference ratio (SNIR) threshold in the demodulator to operate correctly. SS-Aloha shows to be efficient in terms of energy consumption; similar to Aloha, SS-Aloha only requires the sending of data packets with no need to send extra control signaling or synchronization information.

The simplicity required in transmitters in addition to the good performance of SS-Aloha, compared to previous protocols, may make this protocol suitable for IoT scenarios. However, when employed in LEO satellite links, where the expected power imbalance is high, the performance of this protocol drops drastically, behaving similarly to S-Aloha.

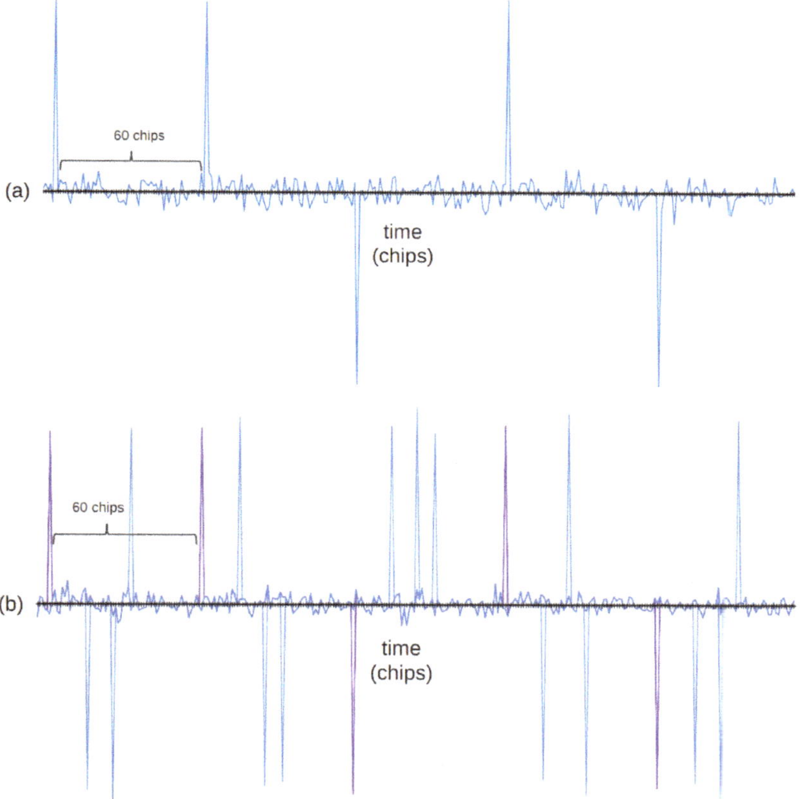


**Figure 5.** In (**a**), the signal received by the satellite from one transmitter. In (**b**), the signal received by the satellite from four transmitters; signals are still decodable because of the offset of chips between the different terminals. Figure adapted from [55].

#### 5.1.4. Enhanced Aloha (E-Aloha)

When sensor nodes of a telemetry application transmit data readings, they are usually configured to send packets in a periodical way, with a fixed time interval between messages. In this scenario, the E-Aloha protocol has been proposed as a simplified version of Aloha. At the time of sending, nodes simply initiate transmission, with no additional control to avoid collisions; a confirmation of reception is also not considered in the protocol. For some applications, nodes may end up with the same time interval between packets, in which case the nodes will collide permanently, with no effective communication. To avoid this situation, E-Aloha defines a time window—considerably longer than the packet transmission time—located around the fixed sending time. Within this window, each node selects randomly a new sending time, thus reducing the chances of permanent collisions among nodes using the same time interval.

Figure 6 illustrates the behavior of the protocol for nodes originally colliding over the same time interval (see Figure 6a), and the corrections made through the use of a time window around the fixed sending times (see Figure 6b). E-Aloha was introduced in [43] for use in satellite systems devoted to telemetry such as Argos.

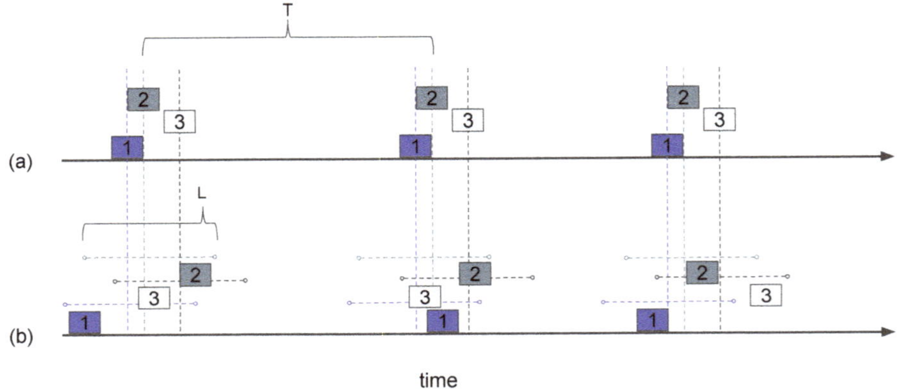


**Figure 6.** Example of E-Aloha operation where three users send packets at a rate of $\frac{1}{T}$ packets/s. In (**a**), the users wait a fixed time ($T$), in consequence, there is a permanent collision between the users 1 and 2; whereas in (**b**) there is a time window ($L$) for selecting a new random sending time in order to avoid permanent collisions. Figure adapted from [43].

The performance evaluation of this protocol considered the periodic traffic characteristic of typical telemetry systems [43]. The reported results indicate that for a $PLR = 10^{-1}$, the protocol achieves a normalized throughput of $S = 0.091$, considering a channel load $C = 0.101$. The performance is, then, very similar to the one reported for S-Aloha ($C = 0.1$ for a $PLR = 10^{-1}$), with the advantage that E-Aloha does not require time synchronization. Furthermore, there is less complexity in implementing E-Aloha than for Aloha, in particular since E-Aloha does not require a specific logic implemented at the receiver. By not requiring ACK, this protocol is energy efficient when the load on the channel is very low ($C \sim 0.1$). For higher loads, the performance of the protocol drastically decreases due to the high number of collisions.

The simplicity of this protocol is desirable for the implementation on the IoT terminal node. It actually behaves well for the reported channel loads found in the systems where the protocol has been implemented. However, E-Aloha may lack the necessary scalability to support the future IoT networks.

5.1.5. Random Frequency Time Division Multiple Access (RFTDMA)

Despite not being reported in satellite environments, this protocol has an important relevance in low power wide area (LPWA) technologies, more precisely for its use in SigFox. Considering the ultra narrowband (UNB) technology, this protocol acquires prominence when low cost transmitters that do not require expensive oscillators are required to perform a precise adjustment of the carrier frequency. Taking this into account, random frequency time division multiple access (RFTDMA) uses the time and frequency to send messages without discretization (as in pure Aloha) [38]. Figure 7 illustrates the process of a transmitter selecting a frequency according to the protocol, a representation of the channel resulting from many nodes transmitting at the same time and, finally, the receiver's architecture to retrieve the data out of the composite signal.

The performance of this protocol is calculated in [56]. Its maximum normalized throughput is lower than 0.1 for $C \simeq 0.25$, and the $PLR$ results are not reported as a function of $C$.

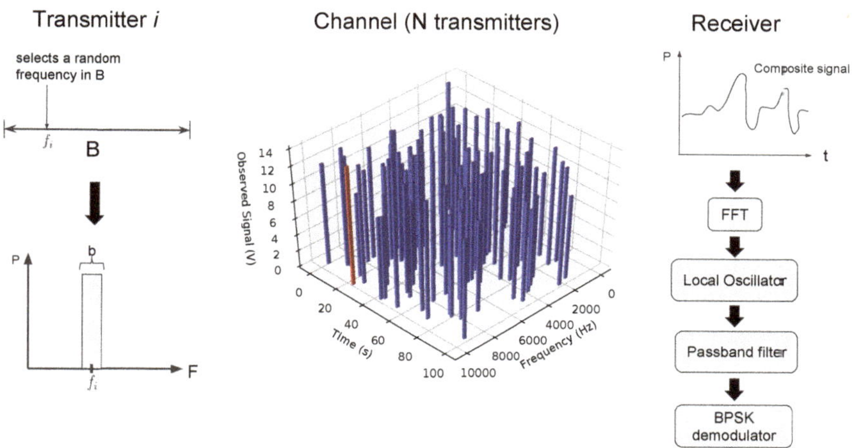


**Figure 7.** Random frequency time division multiple access (RFTDMA) communication process. Temporal and spectral random access from hundred of nodes.

## 5.2. Reservation and Adaptive Protocols

5.2.1. Reservation Aloha (R-Aloha)

This protocol divides the time into $m$ slots, each with the duration of a packet transmission time. The slots are grouped in frames, and the nodes randomly choose one slot per frame to send a packet. If a node successfully sends a packet, it proceeds to reserve the same slot in future frames. At the end of each frame, the receiver responds with an ACK, also indicating what the available slots for the next cycle are. An example of the operation of reservation Aloha (R-Aloha) is shown in Figure 8. The R-Aloha protocol was proposed for incorporating satellite communications in the ARPA network [57].

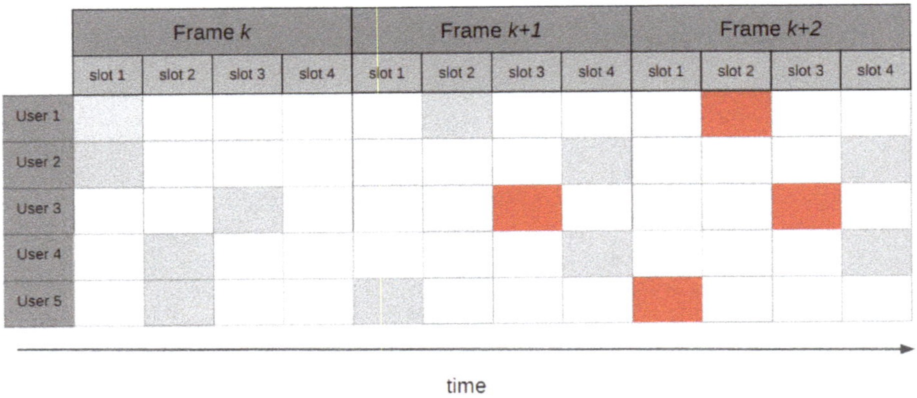

**Figure 8.** R-Aloha operation: five users contending for channel access. User three transmits successfully in slot three of frame $k$, and reserves the same slot in future frames $k+1$ and $k+2$. Users one and five achieve successful transmissions in frame $k+1$, and place their reservations in frame $k+2$ for slots two and one, respectively. Red packets correspond to active reservations.

The throughput of this protocol depends mainly on the number of packets sent by each node during a reservation. In the worst case, the performance of R-Aloha is similar to the S-Aloha. When the protocol holds reservations for a large number of frames, the normalized throughput $S$ approaches to 1 [58].

Although R-Aloha reports having a normalized throughput that tends to 1, it may not meet the requirements of IoT scenarios: the protocol has a good performance when the nodes disputing the channel resemble the number of slots in each frame, and also when the reservations made by each terminal last a large number of frames. However, when the scenario does not comply with these assumptions, R-Aloha turns out to be an unscalable protocol. Moreover, for IoT applications where nodes typically have small and/or infrequent amounts of data to send, the reservation mechanism of R-Aloha may result impractical.

#### 5.2.2. Carrier Sense Multiple Access with Collision Avoidance (CSMA/CA) with RTS/CTS

A classical protocol in wireless network, the carrier sense multiple access with collision avoidance (CSMA/CA) defines the monitoring of the channel before sending a message. If a node senses the channel is busy, it refrains from transmitting and enters an exponential backoff stage; otherwise, it issues a reservation request by means of a small broadcast message (i.e., the request-to-send (RTS) packet). When there are no collisions and the receiver decodes the RTS correctly, another broadcast message granting the reservation (i.e., the clear-to-send [CTS] packet) is sent by the receiver. Upon reception of the CTS, the sender proceeds to send the data packet [59]. Figure 9 illustrates the exchange of packets when a successful reservation is placed in CSMA/CA with RTS/CTS. The protocol has been evaluated in a LEO satellite environment considering different back-off distribution functions [60].

The maximum performance regarding throughput of this protocol varies between $S \simeq 0.5$ and $S \simeq 0.8$, since its performance depends on factors such as the packet length, number of nodes, and the number of hidden terminals, among others. As reported in [59], for an example network with 10 stations and no hidden terminals, the protocol achieves a normalized throughput of $S = 0.75$ when the load is $C = 0.8$. In the presence of hidden terminals—a common scenario for a satellite system with ground terminals distributed over a large area—if there is a 5% probability of hidden terminals for a total of 10 stations on ground, the normalized throughput is $S = 0.65$. By increasing the number of stations to 50 in the same case, the normalized throughput falls to $S = 0.57$. $PLR$ values for this protocol are not reported since packet losses are avoided when using the reservation mechanism.

In terms of complexity, CSMA/CA is still a simple protocol: it does not impose high processing demands on the nodes beyond the ability to transmit and receive. Conversely, the power consumption of CSMA/CA is high, since each node must permanently listen and monitor the channel.

CSMA/CA is widely used in wireless IoT technologies, such as ZigBee, D7AP, and other short-range wireless sensor networks [49,61]. However, for the protocol to be implemented on a satellite IoT network, it may be impractical for two reasons: first, the probability of hidden nodes in a satellite scenario is high due to a wide geographic distribution area; and second, the long delays of the different devices on ground make the channel sensing ineffective even when there are no hidden terminals.

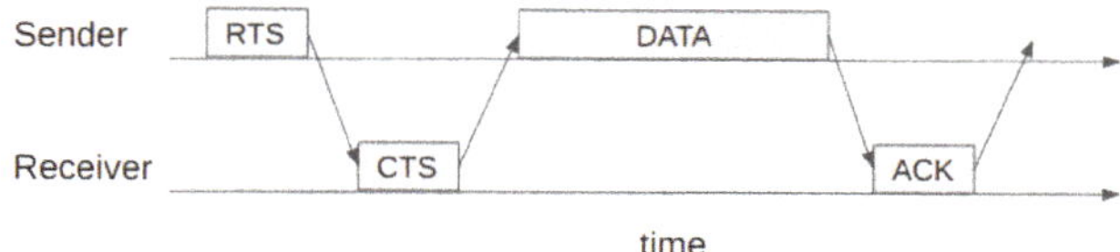


**Figure 9.** Exchange of packets for a successful channel reservation with request-to-send/clear-to-send (RTS/CTS) signaling in carrier sense multiple access with collision avoidance (CSMA/CA).

### 5.2.3. Fixed Competitive TDMA (FC-TDMA)

Similar to R-Aloha, the fixed competitive TDMA (FC-TDMA) protocol defines a set of $m$ slots grouped in frames. In each frame, the nodes select a slot for transmission in a pseudo-random manner [62]. To calculate the allocated slot for packet transmission, $n_{slot}$, a node employs its ID—a previously assigned integer number—and follows the calculation shown in (5). The receiver has to estimate the number of stations on the ground based on the colliding slots and those with successful transmissions; the frame is further divided according to that estimation. An example of the operation of FC-TDMA is illustrated in Figure 10. The authors in [62] have suggested this MAC protocol for LEO satellite systems.

$$ID\%m = n_{slot}. \tag{5}$$

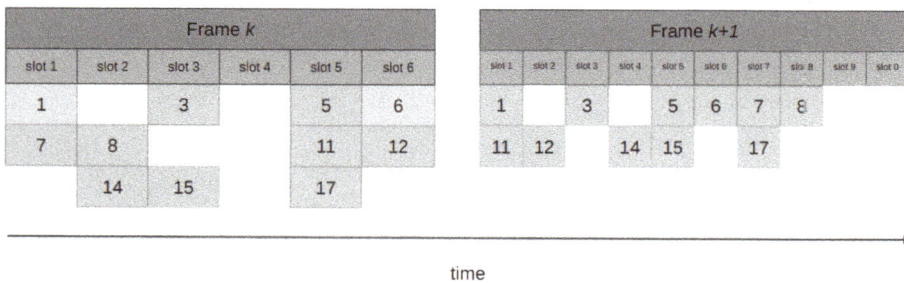

| Frame *k* | | | | | |
|---|---|---|---|---|---|
| slot 1 | slot 2 | slot 3 | slot 4 | slot 5 | slot 6 |
| 1 | | 3 | | 5 | 6 |
| 7 | 8 | | | 11 | 12 |
| | 14 | 15 | | 17 | |

| Frame *k+1* | | | | | | | | | |
|---|---|---|---|---|---|---|---|---|---|
| slot 1 | slot 2 | slot 3 | slot 4 | slot 5 | slot 6 | slot 7 | slot 8 | slot 9 | slot 0 |
| 1 | | 3 | | 5 | 6 | 7 | 8 | | |
| 11 | 12 | | 14 | 15 | | 17 | | | |

time

**Figure 10.** Fixed competitive TDMA (FC-TDMA) operation. In the first frame, packets collide in every slot. For the following frame, the number of slots is increased, resulting in five successful transmissions out of ten slots. The number in each packet represents the node's ID.

The theoretical maximum normalized throughput of the protocol is $S = 1$ with a load of $C = 1$. This performance corresponds to a scenario with a number of slots in the frame equal to the number of devices on the ground, and all the devices having an assigned ID such that there are no collisions. The complexity of this protocol is related to the TDMA functionality and the variable slot lengths. In addition, the algorithm for estimating the number of terminals on the ground may impact both the energy performance and the complexity. However, the details of this algorithm are not provided in the literature.

Similar to R-Aloha, the maximum normalized throughput of FC-TDMA is achieved under very specific conditions: (a) when the number of devices is constant; and (b) when there are no conflicting IDs among all users. However, FC-TDMA could potentially match the traffic characteristics of an IoT application as long as the estimation of the number of nodes under its coverage is accurate and fast.

### 5.3. Interference Cancellation-Based Protocols

#### 5.3.1. Contention Resolution Diversity Slotted Aloha (CRDSA)

As the name indicates, contention resolution diversity slotted Aloha (CRDSA) is based on the DSA protocol. In addition to sending two or more copies of a packet, CRDSA iteratively resolves the collisions that occur at the receiver through the use of a successive interference cancellation (SIC) mechanism [51]. For a given frame, nodes send two or three copies of the packet in different slots. The entire frame is then stored in a digital memory at the receiver. Furthermore, each packet includes control data that indicates in what slot the "twin" packet has been sent. In this way, when a packet is decoded correctly by the receiver, the latter retrieves the information regarding the arriving slot for the "twin" packet. With this information, the receiver is able to perform the interference cancellation method. The CRDSA protocol has been developed and included in the DVB-S2/RCS standard [63].

In terms of communication performance, when CRDSA uses two copies per packet, it achieves a maximum performance of $S = 0.52$ for a given load of $C = 0.65$. In the case of a $PLR = 10^{-3}$, the supported channel load is $C \simeq 0.05$, improving to $C \simeq 0.26$ when the $PLR = 10^{-2}$. Note that CRDSA achieves a high normalized throughput performance for a low PLR; consequently, the protocol is also considered energy efficient: although each node must send each packet two times, no further retransmissions are required. Nevertheless, protocols based on interference cancellation have a high dependency on a good channel estimation, which adds complexity to the implementation. In addition, CRDSA also reports high demands regarding processing and storage capabilities at the receiver, together with requiring synchronization among nodes.

To address the drawbacks reported for CRDSA, the protocol has evolved with adaptations such as multi-frequency CRDSA (MF-CRDSA) and spread spectrum CRDSA (SS-CRDSA) [64]. The former deals with the problem of requiring power peaks to send complete messages over small time slots. By dividing the available spectrum in multiple channels, the time slots in MF-CRDSA can be longer, thus avoiding very high power peaks but at the expense of the protocol's performance. The latter protocol, SS-CRDSA, addresses the problem of "loops" in the original CRDSA. A loop occurs when two different sources send replicas of their packets over the same slot, making it impossible to apply a successful interference cancellation (see for example users three and four in Figure 11). To avoid the loop, SS-CRDSA uses spread spectrum techniques and randomly associates codes to each packet.

#### 5.3.2. Irregular Repetition Slotted Aloha (IRSA)

This protocol represents an improvement over CRDSA. Although it also requires the nodes to send copies of packets in randomly chosen slots, the difference lies in that the number of duplicates for each packet varies depending on an optimized distribution probability. The protocol is intended to improve the performance over uplink satellite channels [65].

Irregular repetition slotted Aloha (IRSA) achieves a maximum performance of $S = 0.8$ for a given load of $C = 0.85$. In the case of a $PLR = 10^{-3}$, the supported normalized offered load rises to $C \simeq 0.7$. The complexity of this protocol is similar to the complexity of CRDSA, but it adds the difficulty of calculating the different number of repetitions per packet. Given its performance, IRSA is considered an energy efficient protocol for normalized offered loads near $C = 0.7$, because of its low $PLR$.

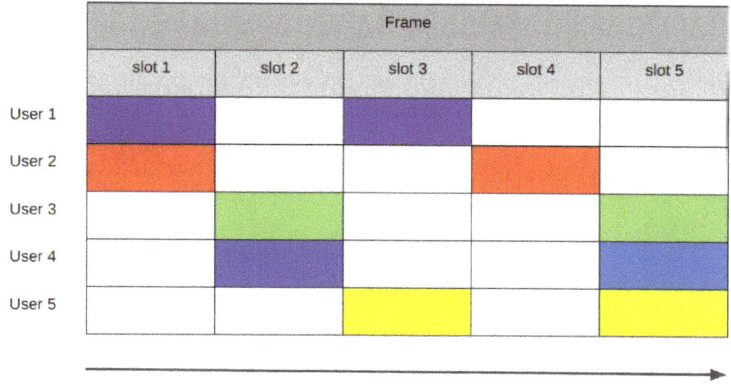


**Figure 11.** Example of the operation of contention resolution diversity slotted Aloha (CRDSA) over one frame. A collision-free packet from user two is received in the fourth slot. The packet includes the position of its "twin", located in slot one. An interference cancellation method is then applied to the packet. In this example, packets from users one, two, and five can be successfully recovered, achieving an $S = 0.6$ for a load of $C = 1$.

#### 5.3.3. Coded Slotted Aloha (CSA)

A protocol inspired by IRSA and CRDSA in which the nodes divide packets before transmission into $k$ parts of the same length. Each one of these sub-packets is then encoded with an error correction code and sent through the discrete channel [66]. Upon reception, if all the sub-packets from a sender are received with no collisions, the recovery of other packets coming from the same sender can be achieved by applying a maximum-a-posteriori (MAP) decoding scheme. In addition, the receiver also employs the interference cancellation scheme for reception from other senders. The work in [66] identifies the satellite network as a potential application for coded slotted Aloha (CSA).

The communications performance of the CSA protocol indicates a normalized throughput of $S \simeq 0.8$ for a channel offered load of $C \simeq 0.84$. The normalized load supported for a $PLR = 10^{-3}$ is not reported in the literature.

Similar to CRDSA and other interference cancellation-based protocols, CSA is also energy efficient due to its low $PLR$ for high channel loads. In terms of complexity, CSA is very similar to CRDSA, but it adds the difficulty of coding and decoding each packet with the model introduced in [66].

#### 5.3.4. Multi-Slots Coded Aloha (MuSCA)

This protocol generalizes the CRDSA protocol. By employing adequate error correction codes given a proper SNIR, multi slot coded Aloha (MuSCA) is able to decode packets even when there are collisions for all the transmissions in a frame [67]. In the example provided for CRDSA in Figure 11, MuSCA would have been able to successfully decode packets sent by users three and four, and received at slots two and five. The MuSCA protocol is designed for uplinks shared among a number of users in satellite systems [67].

MuSCA achieves a maximum normalized throughput of $S = 1.4$ when the normalized channel load does not exceed 1.42. When the $PLR = 10^{-3}$, MuSCA supports an offered load of $C = 1.22$. Considering the very high efficiency of this protocol for normalized offered loads close to one, there are no need for retransmissions. In terms of complexity, MuSCA is very similar to CSA, since the difference between them is mainly in the coding mechanism.

5.3.5. Enhanced Spread Spectrum Aloha (E-SSA)

This protocol is similar to SSA on the transmitter side. On the receiver side, enhanced spread spectrum Aloha (E-SSA) employs an iterative soft interference cancellation (ISIC) algorithm with a sliding window that captures the messages received on an unsynchronized channel. The main difference with the previous protocols is that E-SSA does not require sending multiple copies of packets, nor synchronization; thus, achieving a greater efficiency with a reduced complexity. The E-SSA protocol has been designed for integrated satellite/terrestrial mobile systems [5,68]. Figure 12 shows the operation of this protocol.

In terms of performance, the maximum normalized throughput reported for this protocol is $S = 1.2$ for a channel load of $C = 1.25$. When the $PLR = 10^{-3}$, E-SSA is able to operate with a normalized load of $C = 1.12$, assuming that the power imbalance of the transmitting nodes is equivalent. As opposed to SSA, in E-SSA, if a power imbalance of $\sigma = 3$ dB is assumed among nodes, the performance improves considerably, achieving a normalized load of $C = 1.9$ under a $PLR = 10^{-3}$.

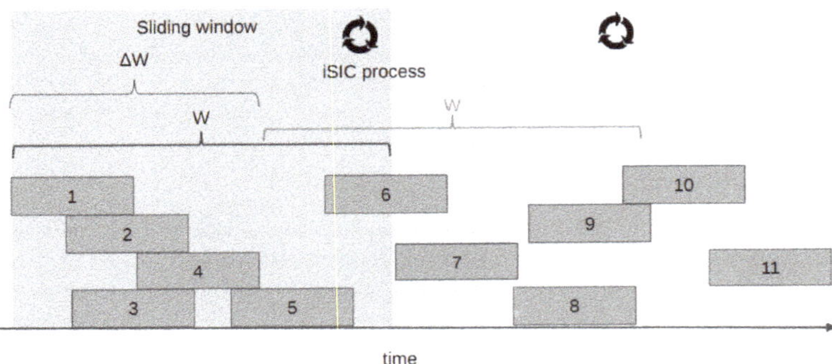


**Figure 12.** Example of operation of enhanced spread spectrum Aloha (E-SSA) protocol. The sliding window of length $W$ is shifted in $\Delta W$ after performing the iterative process of interference cancellation (ISIC) in the current window. Figure adapted from [47].

The interference-cancellation protocols presented in this section are designed mostly to solve the multiple access for the uplink of satellite systems. However, most of the control information required by the protocols is actually acquired/exchanged on the downlink. For example, a correct channel estimation is a key element for the proper operation of E-SSA. Such an estimation takes place over the downlink. Similarly, the code sharing process of CSA also takes place on the downlink. In half-duplex systems, the interference-cancellation based protocol's feasibility requires careful examination.

Regarding the suitability of this category for IoT services, it can be mentioned that despite the great performance reported for these protocols, the complexity of their correct implementation makes them hard to adapt given the capabilities of CubeSats in a LEO orbit. Among the reviewed protocols, E-SSA stands out in performance compared to the others, and significantly reduces the complexity at the ground terminal; however, it is still highly demanding of resources on the nanosatellite side.

*5.4. Hybrid Protocols*

This category corresponds to protocols whose MAC mechanisms are based on a mix of protocols belonging to the other categories previously discussed in this section. The ones already mentioned and that also match this category are SS-CRDSA, MF-CRDSA, FC-TDMA, and R-Aloha; all of them employ a mix of fixed allocation techniques in conjunction with random access mechanisms.

Aloha-LoRa

Despite not being reported in satellite environments, another protocol that is worth mentioning due to its wide use in LPWA technologies is the one employed in LoRa. In this technology, the bandwidth is divided into several channels, where the number is dependent on the regulation (e.g., 13 channels for the industrial, scientific and medical (ISM) 902–928 MHz band under federal communications commission (FCC) regulation). In these channels, nodes transmit modulating the signal with the chirp spread spectrum (CSS) and make use of different spreading factors (SF). The channels are further divided into 6 subchannels (from SF = 7 to SF = 12). The MAC behavior depends on the LoRa class of device:

- Class A: The lowest power consumption type of device. Transmits data when is necessary using Pure Aloha. To receive messages from the gateway, a listening window is open after each transmission.
- Class B: For these devices there is a schedule to transmit, which is defined through beacons.
- Class C: These nodes are always in reception mode, except during transmission.

In the case of Class A nodes, the MAC protocol performs a fixed allocation of resources (in bandwidth and code) together with a random access scheme [14].

To summarize our review, Figure 13 presents a taxonomy elaborated with the MAC protocols under study. Furthermore, the main characteristics and performance metrics when the MAC protocols are evaluated in a satellite environment are presented in Table 2.

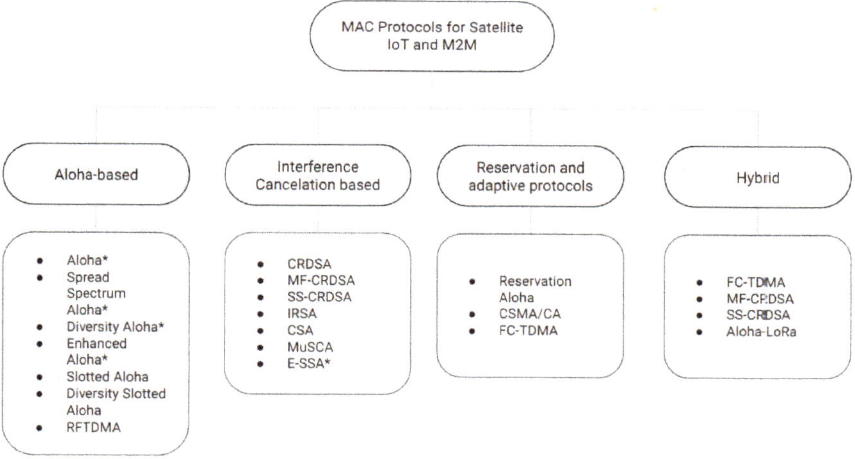


**Figure 13.** Taxonomy of random access protocols evaluated for satellite systems based on nanosatellites for IoT connectivity. Protocols with * indicate no time synchronization is needed.

**Table 2.** Comparison of medium access control (MAC) protocols considering communications performance, complexity of implementation, energy efficiency, and the topology impact.

| Protocols | $S_{max}$ | $C_{S_{max}}$ | $C_{PLR=10^{-3}}$ | Complexity | Energy Efficiency | Topology Impact * |
|---|---|---|---|---|---|---|
| MuSCA | 1.4 | 1.42 | 1.22 | high | high | different delays (−) |
| E-SSA | 1.2 | 1.25 | 1.12 | high | high | power imbalance (+) |
| FC-TDMA | 1 | 1 | - | medium | medium | variability in number of nodes (−), different delays (−) |
| R-Aloha | 1 | 1 | - | medium | medium | different delays (−) |
| CSA | 0.8 | 0.84 | - | high | high | different delays (−) |
| IRSA | 0.8 | 0.85 | 0.7 | high | high | different delays (−) |
| CSMA/CA | 0.75 | 0.8 | - | low | low | hidden nodes (−), different delays (−) |
| SS-Aloha | 0.6 | 0.7 | 0.5 | medium | medium | power imbalance (−) |
| CRDSA | 0.52 | 0.65 | 0.05 | high | medium | power imbalance (+), different delays (−) |
| S-Aloha | 0.368 | 1 | $<10^{-3}$ | low | medium | power imbalance (+), different delays (−) |
| DSA | 0.3 | 0.6 | $<10^{-3}$ | low | medium | power imbalance (+), different delays (−) |
| E-Aloha | 0.09 | 0.1 | $<10^{-3}$ | low | medium | power imbalance (+) |

* (+) means the topology characteristics help improve the communications performance of the protocol, whereas (−) indicates a negative impact on performance.

## 6. Discussion and Open Research Challenges

In this section, we discuss the performance evaluation provided in Section 5 together with additional characteristics of the IoT scenario described in Section 3. Of the protocols reviewed, although several options offer a high throughput in terms of communications performance, such a good result is associated with a medium to high cost in terms of implementation complexity and energy consumption (see the first four protocols in Table 2). Our comparative analysis of protocols suggests that the Aloha-based are good candidates for the MAC layer in nanosatellites devoted to IoT connectivity. This is mainly due to their simplicity of implementation and their minimum hardware requirements. These protocols also report having a low sensitivity to delay.

In terms of network topology, the Aloha-based protocols can benefit from the power imbalance among nodes because of the so-called capture effect. In such a case, the receiver can correctly receive a packet with a high signal strength despite the existence of interference with other transmissions with lower power levels. Among the protocols in this category, the E-Aloha is the current solution for commercial telemetry satellite systems, has been extensively tested, and is operative. The performance of E-Aloha is similar to that of S-Aloha, with no need of synchronization. However, the main problem with E-Aloha is it may lack scalability for massive applications, due to its poor performance even for moderate traffic loads, as shown in Table 2. SS-Aloha, on the other hand, has much better performance than other Aloha-based protocols. However, this mechanism does not benefit from the capture effect; on the contrary, its performance falls to values similar to S-Aloha in a situation of power imbalance.

When examining the protocols based on interference cancellation (e.g., CRDSA and E-SSA in Table 2), the limitations in processing capacity in CubeSats, together with the adverse conditions to perform correct channel estimation in LEO orbits, make employing such protocols in the IoT study scenario infeasible. Other protocols in this category, such as MuSCA, CSA, and E-SSA, require the exchange of coordination or channel estimation information delivered in advance or through a separated channel, making it more demanding regarding channel resources.

As for protocols that require the carrier sensing mechanism, i.e., CSMA/CA, they have been shown to be highly inefficient given a topology with a moderate to a high number of hidden nodes, which will be a reasonably common scenario given the random distributions of ground sensors and devices serving a variety of IoT applications' requirements. The performance of such protocols also decreases when the delays in transmissions among the nodes are highly uneven. In fact, the uneven delays are also critical for the operation of TDMA-inspired protocols such as R-Aloha and FC-TDMA. The reason is that each slot, to synchronize the channel, must incorporate a guard time of the order of the inequalities among the delays, which may result in a considerable waste of channel resources when the variability among delays is large. Moreover, despite the good performance reported for the throughput in FC-TDMA, it requires a specialized algorithm that varies dynamically the number of slots in each frame, which has not been determined in the definition of the protocol provided in [62]. In the case of R-Aloha, although it seems to have excellent performance, that only holds when reservations duration are such that the channel is always occupied; nevertheless, in the case of a large number of nodes, and short reservation times, the scalability of the protocol falls rapidly.

As can be seen through comparative analysis, MAC protocol performance varies widely when examined in the context of CubeSats together with the characteristics of the IoT networks (number of nodes, nature of traffic, geographical distribution, etc.). A visual evaluation of the suitability of each protocol derived from our comparative analysis is shown in Figure 14. In the figure, the protocols are placed according to: (1) their fulfillment of IoT-related requirements such as the scalability (the x-axis), which relates to the communications performance when serving networks composed of a large number of nodes, the topology dependency (a larger-sized geometric figure enclosing each protocol indicates a larger dependency on terminal nodes locations and knowledge of network topology); and the energy efficiency (darker colors correspond to higher power consumption in the execution of the MAC protocol); and (2) their adaptation to the constraints of nanosatellite technology, in which case we evaluate their implementation complexity (the y-axis). A level of high complexity signifies the need of costly resources such as dynamic channeling, advanced channel estimation mechanisms, synchronization, etc.; and (once again) energy efficiency, since it is important to maintain the energy consumption on the spacecraft side within the nanosatellite capabilities.

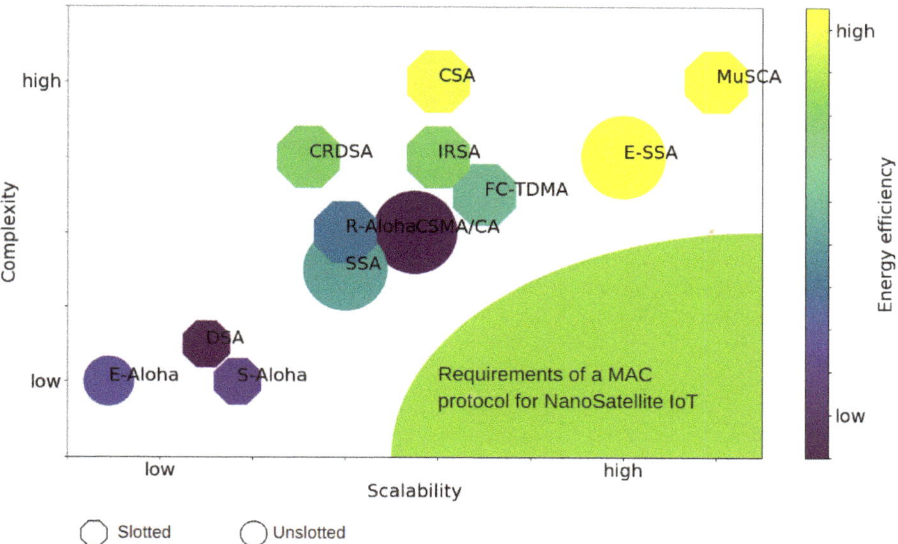


**Figure 14.** A visual comparison of the reviewed medium access control (MAC) protocols in the context of nanosatellite IoT scenarios.

To address the specific challenges derived from the utilization of nanosatellites for providing effective and cost-efficient IoT connectivity, we envision open research and implementation aspects from three perspectives: from the network protocols perspective, from the integration capabilities of CubeSats and from the evolution of the nanosatellites industry.

### 6.1. From the Network Protocols Perspective

When existing network protocols are evaluated in the context of nanosatellites technology for IoT connectivity, it is common to encounter difficulties in finding one protocol that meets all the requirements. In the case of MAC protocols, Figure 14 shows how an ideal performance zone, derived from the IoT scenario together with the CubeSat restrictions, is not yet met by any of the reviewed protocols, even though many had good performance in more traditional satellite scenarios. Additional research for MAC protocol design is needed to integrate aspects which can operate with low processing capacity demands and adapt to a variable and dynamic number of ground sensors and devices. Moreover, mitigation mechanisms should be considered for managing the high power imbalance conditions over the network links, the unequal delays derived from uneven link lengths, and the inability to provide high-quality channel estimations.

### 6.2. From the Integration Capabilities of CubeSat Connectivity with Other Wireless Technologies

The IoT ecosystem will benefit from a more integrated communications platform. In many cases, a global nanosatellite network integrated to an LPWA technology would boost the possibilities for improving connectivity at reduced costs. In recent years, there have been proposals for such a hierarchical architecture: the connected devices send data to an LPWA gateway, which in turn forwards data via the satellite network [9,69]. However, such an integration has not been explored with constellations of nanosatellites instead of traditional satellite networks. Additional research is needed to explore the requirements of compatibility in terms of MAC protocols, network architecture and united service patterns [10]. Another innovative line studied the behavior of an LPWA link to enable connectivity from the nanosatellite to a gateway on Earth [70,71]. Further research and experimentation will help understand and design an integrated platform that takes advantage of the different wireless technologies involved in these hybrid solutions.

### 6.3. From the Evolution of the Nanosatellites Industry

The enormous growth foreseen for the IoT market is highly related to the rapid evolution of low-cost wireless access technologies. In particular, with the introduction of LPWA technologies such as LoRa, Sigfox, and NB-IoT, to mention some, the massification of connected devices seems more plausible in the near future. Although nanosatellite connectivity is being identified as part of the IoT technologies ecosystem [6–9], it is still not considered as low-cost as to become part of the LPWA category [72,73]. The industry of nanosatellite construction and launching needs to keep evolving, and to evolve fast, to reduce costs even more and become another key player in the LPWA market.

## 7. Conclusions

The evolution of satellite systems and the introduction of CubeSats as low-cost satellite technology has made it possible to provide massive communications services for IoT applications, opening the opportunity for countries with no experience in space science and small corporations to participate competitively in the growing satellite communications market. However, the existent protocols in satellite technology, in particular for medium access control, were not designed with the IoT scenario, nor the low-cost technology constraints in mind.

This paper has presented a thorough review of MAC protocols designed for satellite environments, considering the specific characteristics of the IoT networks and applications together with the conditions of a wireless network served by CubeSats deployed in a low earth orbit. The study has shown that many of the reviewed protocols are not suitable for deployment in the scenario of interest, although they have been successfully implemented and deployed in other satellite systems. From the comparative evaluation, the protocols employing interference cancellation techniques are shown to have the best communications performance, but they behave poorly with regards to the demands of processing/channel resources and energy consumption. Furthermore, the Aloha-based protocols are good candidates for the MAC layer in nanosatellites devoted to IoT connectivity due to their simplicity of implementation and their minimum hardware requirements. However, these protocols report having poor communications performance when the traffic load—related to the growing number of expected nodes in the IoT—increases, and also when the delays vary greatly due to variable link lengths.

From our analysis, a better balance among performance, complexity, energy consumption, and sensitivity to topology should drive the design of future MAC protocols for nanosatellite IoT solutions. Aspects related to the network protocol design, the integration capabilities of CubeSat connectivity with other wireless technologies, and the evolution of the nanosatellites industry are some of the open challenges identified and discussed in this review.

**Author Contributions:** Conceptualization, T.F., S.C., and A.B.; Methodology, S.C. and A.B.; Investigation, T.F. and A.B.; Data curation, T.F. and A.B.; Formal analysis, T.F.; Visualization, T.F., S.C., and A.B.; Supervision, S.C. and A.B.; Funding acquisition, S.C. and A.B.; Writing—original draft, T.F. and A.B.; Writing—review & editing, S.C. and A.B.

**Funding:** This work has been funded in part by the Complex Engineering Systems Institute, ISCI (CONICYT: FB0816) and in part by the NIC Labs Chile.

**Acknowledgments:** The authors would like to thank the technical support from the Space and Planetary Exploration Laboratory (SPEL) at Universidad de Chile.

**Conflicts of Interest:** The authors declare no conflict of interest.

## Abbreviations

The following abbreviations are used in this manuscript:

| | |
|---|---|
| ACK | Acknowledgement |
| ADCS | Attitude determination and control system |
| ADS-B | Automatic dependent surveillance broadcast |
| AIS | Automatic identification system |
| CDMA | Code division multiple access |
| CRDSA | Contention resolution diversity slotted Aloha |
| CSA | Coded slotted Aloha |
| CSMA | Carrier sense multiple access |
| CSMA/CA | CSMA with collision avoidance |
| CSS | Chirp spread spectrum |
| CTS | Clear-to-send |
| DA | Diversity Aloha |
| DSA | Diversity slotted Aloha |
| E-SSA | Enhanced spread spectrum Aloha |
| ESA | European Space Agency |
| FCC | Federal communications commission |
| FC-TDMA | Fixed competitive TDMA |
| FDMA | Frequency division multiple access |
| GEO | Geosynchronous equatorial orbit |
| GLONASS | Global orbiting navigation satellite system |
| GNSS | Global navigation satellite system |
| GPS | Global positioning system |
| HEO | Highly elliptical orbit |
| ISIC | Iterative soft interference cancellation |
| IRSA | Irregular repetition slotted Aloha |
| IoT | Internet of things |
| ISM | Industrial, scientific and medical |

| | |
|---|---|
| LEO | Low Earth orbit |
| LPWA | Low power wide area |
| M2M | Machine to machine |
| MarCO | Mars Cube One |
| MAC | Medium access control |
| MEO | Medium Earth orbit |
| MF-CRDSA | Multi-frequency CRDSA |
| MuSCA | Multi slot coded Aloha |
| NASA | National Agency for Space Administration |
| OBC | On-board computer |
| OSI | Open system interconnection |
| PLR | Packet loss ratio |
| RFTDMA | Random frequency time division multiple access |
| RTS | Request-to-send |
| SNIR | Signal to noise plus interference ratio |
| SSA | Spread spectrum Aloha |
| SS-CRDSA | Spread spectrum CRDSA |
| SIC | Successive interference cancellation |
| TDMA | Time division multiple access |
| UCS | Union of Concerned Scientists |
| UNB | Ultra narrowband |

## References


1. McKinsey Global Institute. The Internet of Things: Mapping the Value Beyond the Hype. 2015. Available online: https://www.mckinsey.com/~/media/McKinsey/Business%20Functions/McKinsey%20Digital/Our%20Insights/The%20Internet%20of%20Things%20The%20value%20of%20digitizing%20the%20physical%20world/Unlocking_the_potential_of_the_Internet_of_Things_Executive_summary.ashx (accessed on 14 January 2019).
2. SpaceX Completes Iridium Next Constellation—SpaceNews.com. Available online: https://spacenews.com/spacex-completes-iridium-next-constellation/ (accessed on 14 January 2019).
3. Peyravi, H. Medium access control protocols performance in satellite communications. *IEEE Commun. Mag.* **1999**, *37*, 62–71. [CrossRef]
4. Gaudenzi, R.D.; del Rio Herrero, O. Advances in Random Access protocols for satellite networks. In Proceedings of the 2009 International Workshop on Satellite and Space Communications, Tuscany, Italy, 9–11 September 2009; pp. 331–336. [CrossRef]
5. Herrero, O.D.R.; De Gaudenzi, R. High efficiency satellite multiple access scheme for machine-to-machine communications. *IEEE Trans. Aerosp. Electron. Syst.* **2012**, *48*, 2961–2989. [CrossRef]
6. De Sanctis, M.; Cianca, E.; Araniti, G.; Bisio, I.; Prasad, R. Satellite communications supporting internet of remote things. *IEEE Internet Things J.* **2016**, *3*, 113–123. [CrossRef]
7. Kramer, H.J.; Cracknell, A.P. An overview of small satellites in remote sensing. *Int. J. Remote Sens.* **2008**, *29*, 4285–4337. [CrossRef]
8. Alvarez, J.; Walls, B. Constellations, clusters, and communication technology: Expanding small satellite access to space. In Proceedings of the IEEE Aerospace Conference, Big Sky, MT, USA, 5–12 March 2016. [CrossRef]
9. Iot UK. Satellite Technologies for IoT Applications. 2017. Available online: https://iotuk.org.uk/wp-content/uploads/2017/04/Satellite-Applications.pdf (accessed on 29 January 2019).
10. Qu, Z.; Zhang, G.; Cao, H.; Xie, J. LEO satellite constellation for Internet of Things. *IEEE Access* **2017**, *5*, 18391–18401. [CrossRef]
11. Atzori, L.; Iera, A.; Morabito, G. The Internet of Things: A survey. *Comput. Netw.* **2010**, *54*, 2787–2805. [CrossRef]
12. Gubbi, J.; Buyya, R.; Marusic, S.; Palaniswami, M. Internet of Things (IoT): A vision, architectural elements, and future directions. *Future Gener. Comput. Syst.* **2013**, *29*, 1645–1660. [CrossRef]

13. Ahmed, E.; Yaqoob, I.; Gani, A.; Imran, M.; Guizani, M. Internet-of-things-based smart environments: State of the art, taxonomy, and open research challenges. *IEEE Wirel. Commun.* **2016**, *23*, 10–16. [CrossRef]
14. Oliveira, L.; Rodrigues, J.J.; Kozlov, S.A.; Rabêlo, R.A.; Albuquerque, V.H.C.d. MAC Layer Protocols for Internet of Things: A Survey. *Future Internet* **2019**, *11*, 16. [CrossRef]
15. UCS Satellite Database | Union of Concerned Scientist. Available online: https://www.ucsusa.org/nuclear-weapons/space-weapons/satellite-database (accessed on 10 January 2019).
16. Cubesat Communications System Table. Available online: https://www.klofas.com/comm-table/ (accessed on 14 January 2019).
17. Iridium NEXT | Iridium Satellite Communications. Available online: https://www.iridium.com/file/24033/ (accessed on 12 January 2019).
18. BGAN HDR | High Data Rate Broadcasting | Inmarsat. Available online: https://www.inmarsat.com/service/bgan-hdr/ (accessed on 12 January 2019).
19. Kulu, E. Nanosatellite & Cubesat Database | Missions, Constellations, Companies, Technologies and More. Available online: https://www.nanosats.eu/index.html#database (accessed on 10 January 2019).
20. Diaz, M.; Zagal, J.; Falcon, C.; Stepanova, M.; Valdivia, J.; Martinez-Ledesma, M.; Diaz-Pena, J.; Jaramillo, F.; Romanova, N.; Pacheco, E.; et al. New opportunities offered by Cubesats for space research in Latin America: The SUCHAI project case. *Adv. Space Res.* **2016**, *58*, 2134–2147. [CrossRef]
21. Portilla, J.G. La órbita del satélite Libertad 1. *Rev. Acad. Colomb. Cienc. Exactas Fís. Nat.* **2012**, *36*, 491–500.
22. Roman-Gonzalez, A.; Vargas-Cuentas, N.I. Aerospace technology in Peru. In Proceedings of the 66th International Astronautical Congress-IAC 2015, Jerusalem, Israel, 12–16 October 2015; p. 6.
23. Evans, D. The internet of things: How the next evolution of the internet is changing everything. *CISCO White Pap.* **2011**, *1*, 1–11.
24. De Carvalho Silva, J.; Rodrigues, J.J.; Alberti, A.M.; Solic, P.; Aquino, A.L. LoRaWAN—A low power WAN protocol for Internet of Things: A review and opportunities. In Proceedings of the 2017 2nd International Multidisciplinary Conference on Computer and Energy Science (SpliTech), Split, Croatia, 12–14 July 2017; pp. 1–6.
25. Raza, U.; Kulkarni, P.; Sooriyabandara, M. Low power wide area networks: An overview. *IEEE Commun. Surv. Tutor.* **2017**, *19*, 855–873. [CrossRef]
26. Okumura, R.; Mizutani, K.; Harada, H. A Broadcast Protocol for IEEE 802.15.4e RIT Based Wi-SUN Systems. In Proceedings of the 2017 IEEE 85th Vehicular Technology Conference (VTC Spring), Sydney, Australia, 4–7 June 2017; pp. 1–5. [CrossRef]
27. Ratasuk, R.; Vejlgaard, B.; Mangalvedhe, N.; Ghosh, A. NB-IoT system for M2M communication. In Proceedings of the 2016 IEEE Wireless Communications and Networking Conference, Doha, Qatar, 3–6 April 2016; pp. 1–5. [CrossRef]
28. IEEE Standard for Local and metropolitan area networks—Part 15.4: Low-Rate Wireless Personal Area Networks (LR-WPANs). In *IEEE Std 802.15.4-2011 (Revision of IEEE Std 802.15.4-2006)*; IEEE: Piscataway, NJ, USA, 2011; pp. 1–314. [CrossRef]
29. Granjal, J.; Monteiro, E.; Silva, J.S. Security for the internet of things: A survey of existing protocols and open research issues. *IEEE Commun. Surv. Tutor.* **2015**, *17*, 1294–1312. [CrossRef]
30. Zou, Y.; Zhu, J.; Wang, X.; Hanzo, L. A Survey on Wireless Security: Technical Challenges, Recent Advances, and Future Trends. *Proc. IEEE* **2016**, *104*, 1727–1765. [CrossRef]
31. Soni, A.; Upadhyay, R.; Jain, A. Internet of Things and Wireless Physical Layer Security: A Survey. In *Computer Communication, Networking and Internet Security*; Satapathy, S.C., Bhateja, V., Raju, K.S., Janakiramaiah, B., Eds.; Springer: Singapore, 2017; pp. 115–123.
32. Frustaci, M.; Pace, P.; Aloi, G.; Fortino, G. Evaluating critical security issues of the IoT world: Present and future challenges. *IEEE Internet Things J.* **2018**, *5*, 2483–2495. [CrossRef]
33. Roy-Chowdhury, A.; Baras, J.S.; Hadjitheodosiou, M.; Papademetriou, S. Security issues in hybrid networks with a satellite component. *IEEE Wirel. Commun.* **2005**, *12*, 50–61. [CrossRef]
34. Wang, R.; Taleb, T.; Jamalipour, A.; Sun, B. Protocols for reliable data transport in space internet. *IEEE Commun. Surv. Tutor.* **2009**, *11*, 21–32. [CrossRef]
35. Jiang, C.; Wang, X.; Wang, J.; Chen, H.; Ren, Y. Security in space information networks. *IEEE Commun. Mag.* **2015**, *53*, 82–88. [CrossRef]

36. Mekki, K.; Bajic, E.; Chaxel, F.; Meyer, F. A comparative study of LPWAN technologies for large-scale IoT deployment. *ICT Express* **2018**. [CrossRef]
37. GOMspace | NanoCom AX100. Available online: https://gomspace.com/shop/subsystems/communication-(1)/nanocom-ax100.aspx (accessed on 13 January 2019).
38. Goursaud, C.; Gorce, J.M. Dedicated networks for IoT: PHY/MAC state of the art and challenges. *EAI Endorsed Trans. Internet Things* **2015**, *1*, 150597. [CrossRef]
39. Sinha, R.S.; Wei, Y.; Hwang, S.H. A survey on LPWA technology: LoRa and NB-IoT. *ICT Express* **2017**, *3*, 14–21. [CrossRef]
40. Maral, G.; Bousquet, M. *Satellite Communications Systems: Systems, Techniques And Technology*; John Wiley & Sons: Chichester, UK, 2011.
41. Abramson, N. The Throughput of Packet Broadcasting Channels. *IEEE Trans. Commun.* **1977**, *25*, 117–128. [CrossRef]
42. Ren, W.; Liu, E.; Ward, J.; Hodgart, S.; Sweeting, M. A control-centralised framed-ALOHA with capture for LEO satellite communications. In Proceedings of the Second IFIP International Conference on Wireless and Optical Communications Networks (WOCN 2005), Dubai, UAE, 6–8 March 2005; pp. 496–500. [CrossRef]
43. Ma, H.; Cai, L. Performance analysis of randomized MAC for satellite telemetry systems. In Proceedings of the 2010 5th International ICST Conference on Communications and Networking in China, Beijing, China, 25–27 August 2010; pp. 1–5.
44. Lipke, D.; Swearingen, D.; Parker, J.; Steinbrecher, E.; Calvit, T.; Dodel, H. MARISAT - A maritime satellite communications system. *COMSAT Tech. Rev.* **1977**, *7*, 351–391.
45. Abramson, N. THE ALOHA SYSTEM: Another alternative for computer communications. In Proceedings of the Fall Joint Computer Conference, Houston, TX, USA, 17–19 November 1970; pp. 281–285. [CrossRef]
46. Rom, R.; Sidi, M. *Multiple Access Protocols: Performance and Analysis*; Springer Science & Business Media: New York, NY, USA, 2012.
47. De Gaudenzi, R.; Del Rio Herrero, O.; Gallinaro, G.; Cioni, S.; Arapoglou, P.D. Random access schemes for satellite networks, from VSAT to M2M: A survey. *Int. J. Satell. Commun. Netw.* **2018**, *36*, 66–107. [CrossRef]
48. Rinaldo, R.; Ginesi, A.; De Gaudenzi, R.; Del Rio, O.; Flo, T.; Rislow, B.; Lexow, H. Advanced Physical and MAC Layer Techniques for DVB-based Interactive Satellite Terminals. In Proceedings of the IET Seminar on Digital Video Broadcasting over Satellite: Present and Future, London, UK, 28 November 2006; pp. 1–13. [CrossRef]
49. Montori, F.; Bedogni, L.; Di Felice, M.; Bononi, L. Machine-to-machine wireless communication technologies for the Internet of Things: Taxonomy, comparison and open issues. *Pervasive Mob. Comput.* **2018**, *50*, 56–81. [CrossRef]
50. Choudhury, G.; Rappaport, S. Diversity ALOHA -A random access scheme for satellite communications. *IEEE Trans. Commun.* **1983**, *31*, 450–457. [CrossRef]
51. Casini, E.; De Gaudenzi, R.; Herrero, O.D.R. Contention resolution diversity slotted ALOHA (CRDSA): An enhanced random access schemefor satellite access packet networks. *IEEE Trans. Wirel. Commun.* **2007**, *6*, 1408–1419. [CrossRef]
52. Coe, S. Internet Protocol over Satellite: Insights for telecommunications systems managers. In Proceedings of the 2009 International Conference on Wireless Communications Signal, Nanjing, China, 13–15 November 2009; pp. 1–5. [CrossRef]
53. Abramson, N. Multiple access in wireless digital networks. *Proc. IEEE* **1994**, *82*, 1360–1370. [CrossRef]
54. Herrero, O.R.; Foti, G.; Gallinaro, G. Spread-spectrum techniques for the provision of packet access on the reverse link of next-generation broadband multimedia satellite systems. *IEEE J. Sel. Areas Commun.* **2004**, *22*, 574–583. [CrossRef]
55. Abramson, N. Spread Aloha CDMA Data Communications. US Patent 5,537,397, 16 July 1996.
56. Goursaud, C.; Mo, Y. Random unslotted time-frequency aloha: Theory and application to iot unb networks. In Proceedings of the 2016 23rd International Conference on Telecommunications (ICT), Thessaloniki, Greece, 16–18 May 2016; pp. 1–5. [CrossRef]
57. Crowther, W.; Rettberg, R.; Walden, D.; Ornstein, S.; Heart, F. A system for broadcast communication: Reservation-ALOHA. In Proceedings of the International Conference on Systems Sciences, Honolulu, HI, USA, 9–11 January 1973; pp. 371–374.

58. Lam, S.S. Packet Broadcast Networks? A Performance Analysis of the R-ALOHA Protocol. *IEEE Trans. Comput.* **1980**, *C-29*, 596–603. [CrossRef]
59. Bianchi, G.; Fratta, L.; Oliveri, M. Performance evaluation and enhancement of the CSMA/CA MAC protocol for 802.11 wireless LANs. In Proceedings of the PIMRC '96—7th International Symposium on Personal, Indoor, and Mobile Communications, Taipei, Taiwan, 18 October 1996; Volume 2, pp. 392–396. [CrossRef]
60. Cawood, A.D.; Wolhuter, R. Comparison and optimisation of CSMA/CA back-off distribution functions for a low earth orbit satellite link. In Proceedings of the AFRICON 2007, Windhoek, South Africa, 26–28 September 2007; pp. 1–8. [CrossRef]
61. Demirkol, I.; Ersoy, C.; Alagoz, F. MAC protocols for wireless sensor networks: A survey. *IEEE Commun. Mag.* **2006**, *44*, 115–121. [CrossRef]
62. Luan, P.; Zhu, J.; Gao, K. An improved TDMA access protocol in LEO satellite communication system. In Proceedings of the 2016 IEEE International Conference on Signal Processing, Communications and Computing (ICSPCC), Hong Kong, China, 5–8 August 2016; pp. 1–4. [CrossRef]
63. Morlet, C.; Alamanac, A.B.; Gallinaro, G.; Erup, L.; Takats, P.; Ginesi, A. Introduction of mobility aspects for DVB-S2/RCS broadband systems. *Space Commun.* **2007**, *21*, 5–17.
64. Mengali, A.; De Gaudenzi, R.; Arapoglou, P.D. Enhancing the physical layer of contention resolution diversity slotted ALOHA. *IEEE Trans. Commun.* **2017**, *65*, 4295–4308. [CrossRef]
65. Liva, G. Graph-based analysis and optimization of contention resolution diversity slotted ALOHA. *IEEE Trans. Commun.* **2011**, *59*, 477–487. [CrossRef]
66. Paolini, E.; Liva, G.; Chiani, M. Coded slotted ALOHA: A graph-based method for uncoordinated multiple access. *IEEE Trans. Inf. Theory* **2015**, *61*, 6815–6832. [CrossRef]
67. Bui, H.C.; Lacan, J.; Boucheret, M.L. An enhanced multiple random access scheme for satellite communications. In Proceedings of the Wireless Telecommunications Symposium 2012, London, UK, 18–20 April 2012; pp. 1–6. [CrossRef]
68. De Gaudenzi, R.; del Rio Herrero, O.; Gallinaro, G. Enhanced spread Aloha physical layer design and performance. *Int. J. Satell. Commun. Netw.* **2014**, *32*, 457–473. [CrossRef]
69. Jamali-Rad, H.; Campman, X. Internet of Things-based wireless networking for seismic applications. *Geophys. Prospect.* **2018**, *66*, 833–853. [CrossRef]
70. Verspieren, Q.; Obata, T.; Nakasuka, S. Innovative approach to data gathering in remote areas using constellations of store and forward communication cubesats. In Proceedings of the 31st International Symposium on Space Technology and Science, Matsuyama, Japan, 3–6 June 2017.
71. Doroshkin, A.; Zadorozhny, A.; Kus, O.; Prokopyev, V.; Prokopyev, Y. Laboratory testing of LoRa modulation for CubeSat radio communications. *MATEC Web Conf.* **2018**, *158*, 01008. [CrossRef]
72. Sanchez-Iborra, R.; Cano, M.D. State of the art in LP-WAN solutions for industrial IoT services. *Sensors* **2016**, *16*, 708. [CrossRef]
73. Akpakwu, G.A.; Silva, B.J.; Hancke, G.P.; Abu-Mahfouz, A.M. A survey on 5G networks for the Internet of Things: Communication technologies and challenges. *IEEE Access* **2018**, *6*, 3619–3647. [CrossRef]

MDPI
St. Alban-Anlage 66
4052 Basel
Switzerland
Tel. +41 61 683 77 34
Fax +41 61 302 89 18
www.mdpi.com

*Sensors* Editorial Office
E-mail: sensors@mdpi.com
www.mdpi.com/journal/sensors

www.ingramcontent.com/pod-product-compliance
Lightning Source LLC
Chambersburg PA
CBHW051725210326
41597CB00032B/5608

*9783039210367*